MARKOV CHAINS

North-Holland Mathematical Library

VOLUME 11

NORTH-HOLLAND
AMSTERDAM · NEW YORK · OXFORD

Markov Chains

D. REVUZ

Université de Paris VII
Paris, France

Revised Edition

1984

NORTH-HOLLAND
AMSTERDAM · NEW YORK · OXFORD

ISBN: 0444 86400 8

First edition: 1975
Second (revised) edition: 1984

Published by:
Elsevier Science Publishers B.V.
P.O. Box 1991
1000 BZ Amsterdam
The Netherlands

Sole distributors for the U.S.A. and Canada:
Elsevier Science Publishing Company Inc.
52, Vanderbilt Avenue
New York, NY 10017
U.S.A.

Library of Congress Cataloging in Publication Data

Revuz, D.
Markov chains.
(North-Holland mathematical library; v. 11)
Includes bibliographical references and index.
1. Markov processes. I. Title. II. Series.
QA274.7.P.45 1983 519.2'33 83-11586
ISBN 0-444-86400-8

Transferred to digital printing 2005

PREFACE TO THE FIRST EDITION

This book grew out of a lecture course given at the "3me cycle" level of the universities of Paris during recent years. The first part is thus intended for students at that level; the later chapters are designed for more advanced readers.

Our purpose is to provide an introduction to homogeneous Markov chains with general (measurable) state space, or perhaps more accurately to the study of "sub-Markovian kernels" or "transition probabilities". This contains as special cases the study of pointwise transformations and ergodic theory, potential theory, random walks on groups and homogeneous spaces, etc. It would be preposterous to attempt a comprehensive treatment of all these topics within this book, in as much as some of them are far from being complete and offer still challenging problems.

Roughly speaking, the first four chapters – at least parts of them – are foundations that must be learnt by every beginner in the field. It will be no surprise that they include a short chapter on potential theory, since its fruitful relationship with Markov phenomena has been long and widely known. They include also the basic classification of chains according to their asymptotic behaviour and the celebrated Chacon–Ornstein Theorem. After these first chapters we proceed to some portions of the theory such as renewal theory for abelian groups, an introduction to Martin Boundary Theory and the study – the most thorough in this book – of chains recurrent in the Harris sense. The last chapter deals with the construction of chains starting from a kernel satisfying some maximum principle. These topics do not depend very strictly on one another and the reader may skip some according to his own interest. An interdependence table is given below, but this is merely a guide, and some idea of all the previous results will always come in useful.

Other important subjects such as invariant measures, pointwise transformations or random walks on non-abelian groups are barely alluded to. Relevant references will be given in the "Notes and Comments" at the end of the book.

We tried to present at the end of each section a large selection of exercises. We feel that the reader will not benefit greatly from reading the book if he does not tackle a large number of them. They are in particular designed to provide examples and counter-examples; some of them are results which might have been included in the text. With a few exceptions they can be worked out with a knowledge of the previous sections and of classical calculus.

References and credits do not appear in the text but are collected under the "Notes and Comments". We have in no way tried to draw a historical picture of the field, and apologize in advance to anyone who may feel slighted. We merely indicate the papers we have actually used; as a result insufficient tribute is paid to those who have founded the theory of chains, such as A. A. Markov, N. Kolmogoroff, W. Doeblin, J. L. Doob and K. L. Chung.

The book is divided into chapters; each chapter is divided into sections and starts with a fresh section 1. In every section, items such as theorems, examples, exercises, paragraphs, remarks, are numbered. The references are given in the following way: the nth chapter is quoted as ch. n; the pth section of ch. n is quoted as ch. n, §p. Finally, the qth numbered item in ch. n, §p is quoted as ch. n, Theorem $p.q$, etc.; however, for quotations within the same chapter we shall drop "ch. n" and write simply Theorem $p.q$, etc.

It is a pleasure to acknowledge a few of my debts. Like many French probabilists, I owe everything to J. Neveu and P. A. Meyer; but for their teaching and advice I would never have been really introduced to Markov theory. I must especially thank M. Duflo, who first aroused my interest in discrete time, and A. Brunel for the work we have done together, part of which finds its way into the present book. Many students and colleagues were kind enough to comment on preliminary drafts of the first chapters, in particular M. Brancovan, P. Jaffard and P. Priouret, and I wish to thank them all. My warm thanks go to M. Sharpe, who read through the entire manuscript and removed the most important inaccuracies in the use of English language (a very difficult language indeed); he is not responsible for any awkward points remaining. We hope they will not hamper understanding and that the native speaker will make allowances for the "exotic flavour".

Finally I must thank Martine Mirey, who typed the entire manuscript, and E. Fredriksson of the staff of North-Holland Publishing company for his understanding help.

D. REVUZ

PREFACE TO THE SECOND EDITION

This second edition differs from the first by the rearrangement of some of the material already covered and by the addition of a few new topics. Among the latter are the "zero-two" laws which were one of the most glaring omissions of the first edition and the introduction of a new section on sub-additive ergodic theory which may seem out of place in a book on Markov chains but which we feel warranted by its manifold applications to Markov chains themselves. We also tried to work more consistently with general state spaces and avoid to resort to countable spaces. Finally the number of exercises has been increased and some of them have been rewritten so as to provide more hints.

Since the first edition, some of the subjects touched upon in this book have known an important development such as ergodic theory or random walks on groups and homogeneous spaces. Attempting to include some of these advances would have increased forbiddingly the size of the book and in any case, these subjects do or will deserve books in their own right. Another omission is the theory of large deviations for Markov chains. We have increased the number of references so as to include some possible readings on these topics; but the remarks of the preface to the first edition as to references and credits are still in force.

I finally must say how grateful I am to all the colleagues, students and reviewers who have offered constructive comments on the first edition, especially to R. Durrett and R. T. Smythe. If this second edition is any better than the first, it will be, for a great part, due to them all.

D. REVUZ

INTERDEPENDENCE GUIDE

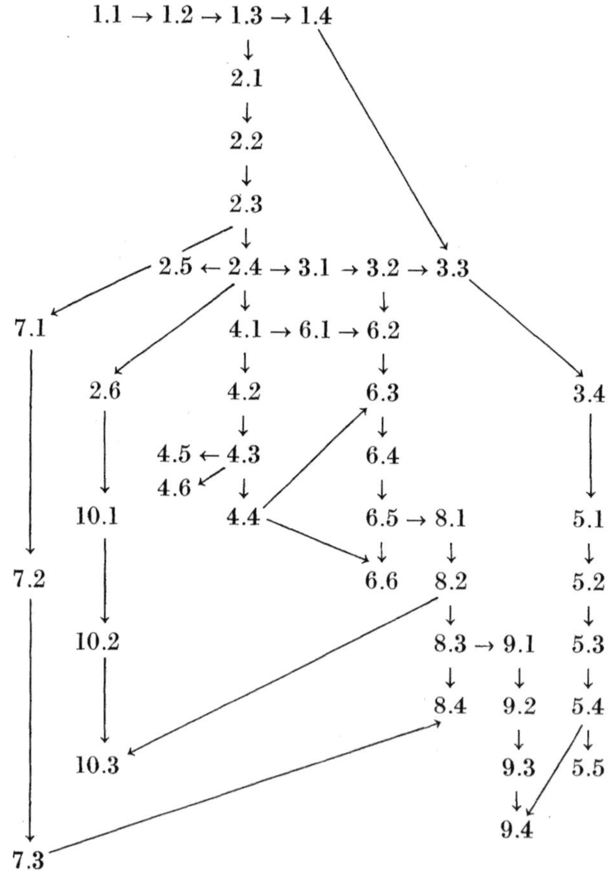

CONTENTS

CHAPTER 0

PRELIMINARIES

This chapter is devoted to a list of some basic notations and facts which will be used throughout the sequel without further explanation. We assume that the reader is familiar with measure theory and probability theory as may be found for instance in the first four chapters of the book by Neveu [2], with point set topology, and that he has some knowledge of topological groups and Fourier transforms, for which we refer to the books by Rudin [1] and by Hewitt and Ross [1].

1. Notation

1.1. Let $\mathbf{N}$ denote the set of positive integers, $\mathbf{Z}$ the set of integers and $\mathbf{R}$ the set of real numbers; $\bar{\mathbf{R}}$ will denote the extended real numbers $\{-\infty\} \cup \mathbf{R} \cup \{+\infty\}$ with the usual topology. For a and b in $\bar{\mathbf{R}}$, we write

$$a \vee b = \max(a, b), \qquad a \wedge b = \min(a, b).$$

If E is a set and f a numerical function on E, we shall write

$$\|f\| = \sup_{x \in E} |f(x)|, \qquad f^+ = f \vee 0, \qquad f^- = -(f \wedge 0).$$

1.2. If $(E, \mathscr{E})$ and $(F, \mathscr{F})$ are measurable spaces, we shall write $f \in \mathscr{E}/\mathscr{F}$ to indicate that the function $f: E \to F$ is measurable with respect to $\mathscr{E}$ and $\mathscr{F}$. In the case where F is $\mathbf{R}$ and $\mathscr{F}$ the σ-algebra of Borel subsets we write simply $f \in \mathscr{E}$. The symbol $\mathscr{E}$ thus denotes the σ-algebra as well as the set of real-valued measurable functions on E.

If A is a subset of E, we denote the indicator function of A by 1_A, so that the statements $1_A \in \mathscr{E}$ and $A \in \mathscr{E}$ have the same meaning. When $A = E$ we shall often write 1 for 1_E.

1.3. If $\mathscr{H}$ is any set of numerical functions, then $\mathscr{H}_+$ denotes the set of non-negative functions in $\mathscr{H}$ and $b\mathscr{H}$ the set of bounded functions. For instance if $(E, \mathscr{E})$ is a measurable space, $b\mathscr{E}$ is the set of bounded measurable functions and $b\mathscr{E}_+$ the set of non-negative bounded measurable functions. We recall that $b\mathscr{E}$ is a Banach space for the supremum norm $\|\cdot\|$ described above. We shall denote by $\mathscr{U}$ the unit ball of $b\mathscr{E}$. The set $\mathscr{U}_+$ is thus the set of measurable numerical functions on $(E, \mathscr{E})$ such that for every $x \in E$, $0 \leqslant f(x) \leqslant 1$.

1.4. Let Ω be a set and $\{(E_i, \mathscr{E}_i)\}_{i \in I}$ a collection of measurable sets. For each $i \in I$ let f_i be a map from Ω to E_i; the σ-algebra on Ω generated by the sets $\{f_i^{-1}(A_i): A_i \in E_i\}$ is the smallest σ-algebra on Ω relative to which all the f_i are measurable, and it will be denoted by $\sigma(f_i, i \in I)$. If $\mathscr{F}$ is any collection of subsets of E, we denote by $\sigma(\mathscr{F})$ the σ-algebra generated by $\mathscr{F}$. A measurable space $(E, \mathscr{E})$ will be said to be *separable* if $\mathscr{E}$ is generated by a countable collection of sets.

1.5. We denote by $\mathscr{M}(\mathscr{E})$ the space of σ-finite measures on $(E, \mathscr{E})$ and by $b\mathscr{M}(\mathscr{E})$ the space of finite measures. We shall call μ^+ and μ^- the positive and negative parts of the measure μ and set $|\mu| = \mu^+ + \mu^-$. The space $b\mathscr{M}(\mathscr{E})$ is a Banach space when endowed with the norm $\|\mu\| = |\mu|(E)$.

If $f \in \mathscr{E}$ and the integral of f with respect to the measure μ exists, we shall write it in any of the forms

$$\int_E f \, d\mu, \qquad \int_E f(x) \, d\mu(x), \qquad \int_E f(x) \, \mu(dx), \qquad \mu(f), \qquad \langle \mu, f \rangle, \qquad \langle f, \mu \rangle.$$

The space of integrable functions will be denoted by $\mathscr{L}^1(\mu)$, and as usual $L^1(\mu)$ will be the Banach space of equivalence classes of integrable functions endowed with the norm $\|f\|_1 = \int_E |f| \, d\mu$. There is a canonical isometry of $L^1(\mu)$ into $b\mathscr{M}(\mathscr{E})$ which maps f onto the measure $\lambda : A \to \int_A f \, d\mu$; λ is absolutely continuous with respect to μ ($\lambda \ll \mu$).

1.6. If $(\Omega, \mathscr{F}, P)$ is a probability space we shall, as is customary, use the words random variable and expectation in place of measurable function and integral, and we shall write

$$\int_\Omega X \, dP = E[X].$$

If $\mathscr{G}$ is a sub-σ-algebra of $\mathscr{E}$, we denote $E[X \mid \mathscr{G}]$ the conditional expectation of X (assumed to be quasi-integrable) given $\mathscr{G}$. When $X = 1_A$, $A \in \mathscr{E}$, we write $P(A \mid \mathscr{G})$ instead of $E[1_A \mid \mathscr{G}]$, and if $\mathscr{G}$ is the smallest σ-algebra relative to which the random variable Y is measurable, we may write $E[X \mid Y]$ and $P(A \mid Y)$. When we apply conditional expectation successively, we shall often abbreviate $E[E[X \mid \mathscr{G}_1] \mid \mathscr{G}_2]$ by $E[X \mid \mathscr{G}_1 \mid \mathscr{G}_2]$. We recall that conditional expectations are defined up to P-equivalence; in the sequel we shall often omit the qualifying phrase "almost everywhere relative to P" in the relations involving conditional expectations.

2. Martingales

Let $(\Omega, \mathscr{F}, P)$ be a probability space and $\{\mathscr{F}_n\}_{n\geqslant 0}$ an increasing sequence of sub-σ-algebras of $\mathscr{F}$. A sequence $\{X_n\}_{n\geqslant 0}$ of random variables over $(\Omega, \mathscr{F}, P)$ is said to be a *supermartingale* with respect to $\{\mathscr{F}_n\}$ if

(i) $X_n \in \mathscr{F}_n$ for every $n \geqslant 0$,

(ii) $E[X_n^-] < +\infty$ for every $n \geqslant 0$,

(iii) $E[X_n \mid \mathscr{F}_m] \leqslant X_m$ whenever $m \leqslant n$.

It is a *submartingale* if $\{-X_n\}$ is a supermartingale and a *martingale* provided it is both a supermartingale and a submartingale.

The proof of the following result, which we shall use freely in the sequel, may be found in the books by Doob [2], Loëve [1], Meyer [2] or Neveu [2].

Theorem 2.1. *Let* $\{X_n\}_{n\in\mathbf{N}}$ *be a supermartingale such that* $\sup_n E[X_n^-] < \infty$; *then* $\lim_n X_n$ *exists almost surely.*

As a result the positive supermartingales converge almost surely. We shall also need the following.

Proposition 2.2. *Let* $\mathscr{F}_n$ *be an increasing sequence of sub-σ-algebras of* $\mathscr{F}$ *such that* $\bigcup_n \mathscr{F}_n$ *generates* $\mathscr{F}$. *Let* X_n *be a sequence of random variables converging almost surely to a random variable* X, *and* Y *an integrable random variable such that* $|X_n| \leqslant Y$ *for every* n. *Then the sequence* $E[X_n \mid \mathscr{F}_n]$ *converges almost surely to* $E[X \mid \mathscr{F}]$.

Proof. Pick $\varepsilon > 0$ and set

$$U = \inf_{n\geqslant m} X_m, \qquad V = \sup_{n\geqslant m} X_n,$$

where m is chosen such that $E[V - U] < \varepsilon$. Then for $n \geqslant m$ we have

$$E[U \mid \mathscr{F}_n] \leqslant E[X_n \mid \mathscr{F}_n] \leqslant E[V \mid \mathscr{F}_n],$$

and applying Theorem 2.1 yields

$$E[U \mid \mathscr{F}] \leqslant \underline{\lim}_n E[X_n \mid \mathscr{F}_n] \leqslant \overline{\lim}_n E[X_n \mid \mathscr{F}_n] \leqslant E[V \mid \mathscr{F}]$$

as well as

$$E[U \mid \mathscr{F}] \leqslant E[X \mid \mathscr{F}] \leqslant E[V \mid \mathscr{F}].$$

It follows that

$$E[\overline{\lim}\, E[X_n \mid \mathscr{F}_n] - \underline{\lim}\, E[X_n \mid \mathscr{F}_n]] \leqslant \varepsilon,$$

which implies that $E[X_n \mid \mathscr{F}_n]$ converges almost surely and that the limit equals $E[X \mid \mathscr{F}]$.

In the sequel we write P-a.s. for "almost surely relative to P" and more simply a.s. when there is no risk of mistake.

3. The monotone class theorem

Definition 3.1. Let Ω be a set; a collection $\mathscr{S}$ of subsets of Ω will be called a *monotone class* if:

(i) $\Omega \in \mathscr{S}$;

(ii) if $A, B \in \mathscr{S}$, $A \subset B$, then $B - A \in \mathscr{S}$;

(iii) if $\{A_n\}$ is an increasing sequence of elements of $\mathscr{S}$, then $\bigcup A_n \in \mathscr{S}$.

Plainly a σ-algebra is a monotone class. It is also clear that the intersection of an arbitrary family of monotone classes is a monotone class; hence for any collection $\mathscr{F}$ of subsets of Ω there is a smallest monotone class containing $\mathscr{F}$, which we call $\mathscr{S}(\mathscr{F})$. Clearly $\mathscr{S}(\mathscr{F}) \subset \sigma(\mathscr{F})$.

Theorem 3.2. *If $\mathscr{F}$ is closed under finite intersections, then $\mathscr{S}(\mathscr{F}) = \sigma(\mathscr{F})$.*

Proof. Clearly, it suffices to show that $\mathscr{S}(\mathscr{F})$ is a σ-algebra, and for this it suffices to show that it is closed under finite intersections.

To this end we define

$$\mathscr{S}_1 = \{B \in \mathscr{S}(\mathscr{F}) \colon B \cap A \in \mathscr{S}(\mathscr{F}) \quad \text{for all } A \in \mathscr{F}\}.$$

It is easily checked that $\mathscr{S}_1$ is a monotone class, and since $\mathscr{F}$ is closed under finite intersections, $\mathscr{S}_1 \supset \mathscr{F}$, and hence $\mathscr{S}_1 \supset \mathscr{S}(\mathscr{F})$. Since by definition $\mathscr{S}_1 \subset \mathscr{S}(\mathscr{F})$, we get $\mathscr{S}_1 = \mathscr{S}(\mathscr{F})$. Now define

$$\mathscr{S}_2 = \{B \in \mathscr{S}(\mathscr{F}) \colon B \cap A \in \mathscr{S}(\mathscr{F}) \quad \text{for all } A \in \mathscr{S}(\mathscr{F})\}.$$

Again $\mathscr{S}_2$ is a monotone class, and by the preceding argument $\mathscr{S}_2 \supset \mathscr{F}$; hence $\mathscr{S}_2 \supset \mathscr{S}(\mathscr{F})$, and as before we get $\mathscr{S}_2 = \mathscr{S}(\mathscr{F})$. But this is precisely the statement that $\mathscr{S}(\mathscr{F})$ is closed under finite intersections. The proof is thus complete.

One generally has to deal with functions rather than sets and then may use the following version of Theorem 3.2.

Theorem 3.3. *Let $\mathscr{F}$ be a collection of subsets of Ω closed under finite intersections and $\mathscr{H}$ a vector space of real valued functions on Ω such that*

(i) *$1_E \in \mathscr{H}$ and $1_A \in \mathscr{H}$ for all $A \in \mathscr{F}$,*

(ii) *if h_n is an increasing sequence of elements of $\mathscr{H}_+$ such that $h = \sup h_n$ is finite (bounded), then $h \in \mathscr{H}$.*

Then $\mathscr{H}$ contains all real-valued (bounded) functions on Ω that are $\sigma(\mathscr{F})$-measurable.

Proof. Left to the reader as an exercise. [*Hint*: Set $\mathscr{D} = \{A : 1_A \in \mathscr{H}\}$ and prove that $\mathscr{D} = \mathscr{S}(\mathscr{F}) = \sigma(\mathscr{F})$.]

The above theorems will be used in the following setup: let Ω be a set and $(E_i, \mathscr{E}_i)$ a family of measurable spaces indexed by a set I. For each $i \in I$, let $\mathscr{F}_i$ be a class of subsets of E_i closed under finite intersections and such that $\mathscr{E}_i = \sigma(\mathscr{F}_i)$. Finally let f_i be a map from Ω to E_i.

Proposition 3.4. *The collection $\mathscr{F}$ of all sets of the form $\bigcap_{i \in J} f_i^{-1}(A_i)$, where $A_i \in \mathscr{F}_i$ for $i \in J$ and J ranges over all finite subsets of I, is closed under finite intersections and $\mathscr{S}(\mathscr{F}) = \sigma(f_i, i \in I)$.*

Proposition 3.5. *Let $\mathscr{H}$ be a vector space of real-valued functions on Ω such that:*

(i) *$1 \in \mathscr{H}$;*

(ii) *if h_n is an increasing sequence of elements of $\mathscr{H}_+$ such that $h = \sup h_n$ is finite (bounded), then $h \in \mathscr{H}$;*

(iii) *$\mathscr{H}$ contains all products of the form $\prod_{i \in J} 1_{A_i} \circ f_i$ where J is a finite subset of I and $A_i \in \mathscr{F}_i$.*

Then $\mathscr{H}$ contains all real-valued (bounded) functions in $\sigma(f_i, i \in I)$.

We shall also use the following corollary of the monotone class theorem.

Proposition 3.6. *Let μ and ν be two finite measures on $(E, \mathscr{E})$ and let $\mathscr{F} \subset \mathscr{E}$ be closed under finite intersections. Then if μ and ν agree on $\mathscr{F}$, they agree on $\sigma(\mathscr{F})$.*

Proof. One shows that $\mathscr{D} = \{A \in \mathscr{E}: \mu(A) = \nu(A)\}$ is a monotone class.

4. Topological spaces and groups

Except for Banach spaces of functions, the topological spaces we shall have to consider will be locally compact. If E is such a space, then $C(E) \subset C_0(E) \subset C_K(E)$ denote respectively the bounded real-valued continuous functions on E, the real-valued continuous functions vanishing at infinity, and the real-valued continuous functions with compact support. The first two of these spaces are Banach spaces under the supremum norm $\|\cdot\|$. We shall also denote by $\mathscr{E}$ the σ-algebra of Borel sets of E.

Almost always, the spaces E we shall deal with have a countable basis and we shall write for short: E is an LCCB. In that case we recall that:

(i) The σ-algebra $\mathscr{E}$ of Borel sets is generated by any open countable basis for the topology (hence $(E, \mathscr{E})$ is separable) as well as by the space $C_K(E)$.

(ii) The space E is metrizable, and one can choose a metric d compatible with the topology such that (E, d) is a separable complete metric space in which every closed and d-bounded set is compact.

(iii) The space E possesses a countable dense subset.

If E is an LCCB, a positive measure μ on $(E, \mathscr{E})$ will be called a *Radon measure* if $\mu(K) < \infty$ for every compact K. In particular, a Radon measure is σ-finite. If L is a non-negative linear functional on C_K, there is a unique Radon measure on $(E, \mathscr{E})$ such that $L(f) = \mu(f)$ for all $f \in C_K$. Every Radon measure is regular, that is for $B \in \mathscr{E}$,

$$\mu(B) = \sup\{\mu(K): K \subset B,\ K \text{ compact}\}$$
$$= \inf\{\mu(G): G \supset B,\ G \text{ open}\}.$$

The space of Radon measures on E will be endowed with the vague topology, which is the coarsest topology for which all the functions $\mu \to \mu(f)$, $f \in C_K$, are continuous. We shall need the following compactness criterion.

Proposition 4.1. *A set $\{\mu_i, i \in I\}$ of measures is vaguely relatively compact if and only if for every compact subset K of E the sets of numbers $\{\mu_i(K), i \in I\}$ is bounded.*

A locally compact metrizable group $\boldsymbol{G}$ is a group which is also an LCCB such that the transformation from $\boldsymbol{G} \times \boldsymbol{G}$ onto $\boldsymbol{G}$ which sends (x, y) into $x^{-1}y$ is continuous. (We denote by x^{-1} the inverse of the element x and by V^{-1} the set of inverses of elements x in V.)

One can choose a metric compatible with the topology of G and invariant by left translations (or right translations) in G. If e is the identity of G, the filter $\mathscr{V}$ of neighbourhoods of e has the following properties:

(i) For every $U \in \mathscr{V}$ there is a set $V \in \mathscr{V}$ such that $VV^{-1} \subset U$;

(ii) If $U \in \mathscr{V}$ and x is any element of G then there exists a set $V \in \mathscr{V}$ such that $x^{-1}Vx \subset U$.

By (i) the class of symmetric ($V = V^{-1}$) neighbourhoods is a base for $\mathscr{V}$. Finally the filter of neighbourhoods of $x \in G$ is equal to $x\mathscr{V}$.

On every LCCB group G there is a left (right) invariant Radon measure, unique up to a multiplicative constant, called the left (right) Haar measure of G. Let m be the left Haar measure on G; then for every integrable function f,

$$\int f(x) \, dm(x) = \int f(yx) \, dm(x)$$

for every y in G. The group G is *unimodular* if the left Haar measure is also a right Haar measure. Abelian and compact groups are unimodular.

Finally, the convolution $\mu * \nu$ of two measures μ and ν is defined by

$$\langle \mu * \nu, f \rangle = \iint f(xy) \, \mu(dx) \, \nu(dy),$$

provided the integral on the right is meaningful for all $f \in C_K$. The convolution of two functions f and g is defined by

$$(f * g)(x) = \int f(y) \, g(y^{-1}x) \, m(dy)$$

if m is a left Haar measure. We shall use the fact that the convolution of a function in $\mathscr{L}^\infty$ and a function in $\mathscr{L}^1$ is a continuous function.

CHAPTER 1

TRANSITION PROBABILITIES. MARKOV CHAINS

In this chapter we introduce the basic data of our study. Throughout the sequel a measurable space $(E, \mathscr{E})$ is assumed given.

1. Kernels. Transition probabilities

Definition 1.1. A *kernel* on E is a mapping N from $E \times \mathscr{E}$ into $]-\infty, +\infty]$ such that:

(i) for every x in E, the mapping $A \to N(x, A)$ is a measure on $\mathscr{E}$ which will often be denoted by $N(x, \cdot)$;

(ii) for every A in $\mathscr{E}$, the mapping $x \to N(x, A)$ is a measurable function with respect to $\mathscr{E}$ which will often be denoted by $N(\cdot, A)$.

The kernel N is said to be *positive* if its range is in $[0, +\infty]$. It is said to be *σ-finite* if all the measures $N(x, \cdot)$ are σ-finite; it is said to be *proper* if E is the union of an increasing sequence $\{E_n\}_{n \geqslant 0}$ of subsets of E such that the functions $N(\cdot, E_n)$ are bounded. The kernel N is said to be *bounded* if its range is bounded, or in other words, if there is a finite number M such that $|N(x, A)| \leqslant M < \infty$ for every x in E and A in $\mathscr{E}$. A bounded kernel is a proper kernel and a proper kernel is σ-finite, the converse statements being obviously wrong. If N is positive, then N is bounded if and only if the function $N(\cdot, E)$ is bounded.

Definition 1.2. In the sequel we shall deal mainly with positive kernels. If f is in $\mathscr{E}_+$, it is then easily seen by approximating f with simple functions that one defines a function in $\mathscr{E}_+$ denoted Nf or $N(f)$ by setting

$$Nf(x) = \int_E N(x, \mathrm{d}y)\, f(y) = \langle N(x, \cdot), f \rangle.$$

By defining $Nf = Nf^+ - Nf^-$, we may extend this to every function in $\mathscr{E}$ such that Nf^+ and Nf^- are not both infinite. We sometimes write $N(x, f)$ for $Nf(x)$; in particular $N(x, A) = N1_A(x)$.

In the same way, let m be a positive measure on $\mathscr{E}$, and define for $A \in \mathscr{E}$,

$$mN(A) = \int_E m(\mathrm{d}x)\, N(x, A) = \langle m, N(\cdot\,, A)\rangle;$$

it is easily seen that mN is a positive measure on $\mathscr{E}$, and as above this may be extended to signed measures. The two mappings thus defined are linear and, whenever the two members are meaningful, we have

$$\langle mN, f\rangle = \langle m, Nf\rangle.$$

Notice further that if we call ε_x the Dirac measure of the point x, we have $\varepsilon_x N(\cdot) = N(x, \cdot)$.

The mappings of $\mathscr{E}_+$ into itself defined by positive kernels are characterized by the following property.

Proposition 1.3. *An additive and homogeneous mapping V of $\mathscr{E}_+$ into itself is associated with a positive kernel if and only if for every increasing sequence $\{f_n\}$ of functions in $\mathscr{E}_+$ one has*

$$V\left(\lim_n f_n\right) = \lim_n V f_n.$$

Proof. Easy and left to the reader as Exercise 1.9.

We proceed to a few examples of kernels.

Examples 1.4. (i) Let λ be a positive σ-finite measure on $\mathscr{E}$ and n a positive real-valued function defined on $E \times E$ and measurable with respect to the product σ-algebra $\mathscr{E} \otimes \mathscr{E}$.

One may then define a kernel N on E by setting

$$N(x, A) = \int_A n(x, y)\, \lambda(\mathrm{d}y).$$

Such a kernel is called an *integral kernel* with basis λ. When $E = \mathbf{R}^d$, $d \geqslant 3$, λ is the Lebesgue measure, and $n(x, y) = |x - y|^{-d+2}$, we thus get the kernel of newtonian potential theory.

The kernel N is positive and for $f \in \mathscr{E}_+$,

$$Nf(x) = \int_E n(x, y)\, f(y)\, \lambda(\mathrm{d}y).$$

If $n(x, y) = a(x)\, b(y)$, where a, b are two measurable functions on E, then

$$Nf(x) = a(x) \int b(y)\, f(y)\, \lambda(\mathrm{d}y) = \langle fb, \lambda \rangle\, a(x).$$

In that case we write $N = a \otimes b\lambda$, and simply $N = a \otimes \lambda$ if $b = 1$.

A case which has been extensively studied is the case where E is countable and $\mathscr{E}$ the discrete σ-algebra on E. The measure may be taken equal to the counting measure of E ($\lambda(\{x\}) = 1$ for every x in E) and the integrals then reduce to sums, that is

$$Nf(x) = \sum_{y \in E} n(x, y)\, f(y).$$

But since in that case $N(x, \{y\}) = n(x, y)$, we shall mix the two notational devices and write more simply

$$Nf(x) = \sum_{y \in E} N(x, y)\, f(y).$$

It thus suffices to give the numbers $N(x, y)$ to define on E a kernel, which may then be viewed as an infinite square matrix indexed by the elements of E. The measures (functions) will be "row" ("column") vectors and the operators of Definition 1.2 are just the usual operators on vectors defined by square matrices.

(ii) Let $\mathbf{G}$ be a multiplicative locally compact semi-group and μ a positive Radon measure on the Borel sets of $\mathbf{G}$; one defines a positive kernel N on $\mathbf{G}$ by setting

$$N(x, A) = (\mu * \varepsilon_x)\,(A),$$

where $*$ denotes the convolution in $\mathbf{G}$. For a Borel function f

$$Nf(x) = \int_{\mathbf{G}} f(yx)\, \mu(\mathrm{d}y)$$

and for a positive measure m on $\mathbf{G}$, $mN = \mu * m$. Such a kernel is called a *convolution kernel.* The newtonian kernel of (i) is a convolution kernel.

The case where $\mathbf{G}$ is a group will of course be of special interest in the sequel. We remark that it is not true that $Nf = \mu * f$ since (whenever it makes sense)

$$(\mu * f)\,(x) = \int_{\mathbf{G}} f(y^{-1}x)\, \mu(\mathrm{d}y).$$

If we call $\hat{\mu}$ the image of μ by the mapping $x \to x^{-1}$, then $Nf = \hat{\mu} * f$.

This may be put in the following more general setting. We say that $\boldsymbol{G}$ (the elements of which are henceforth written $g, g', h, \ldots$) *operates* on the topological space $\boldsymbol{M}$, or that $\boldsymbol{M}$ is a *$\boldsymbol{G}$-space,* if there is a continuous mapping $g \times x \to gx$ from $\boldsymbol{G} \times \boldsymbol{M}$ to $\boldsymbol{M}$ such that $(g_1 g_2)\, x = g_1(g_2 x)$. With a positive measure μ on $\boldsymbol{G}$ we associate a kernel N on $\boldsymbol{M}$, by setting, for a Borel function f on $\boldsymbol{M}$,

$$Nf(x) = \int_{\boldsymbol{G}} f(gx)\, \mu(\mathrm{d}g);$$

if m is a positive measure on $\boldsymbol{M}$ then $mN = \mu * m$, where the convolution $*$ is defined by the formula

$$\langle \mu * m, f \rangle = \int_{\boldsymbol{G}\times\boldsymbol{M}} f(gx)\, \mu(\mathrm{d}g)\, m(\mathrm{d}x).$$

By letting $\boldsymbol{G}$ operate to the left on itself, we see that the former example is a special case of the latter.

(iii) A measurable mapping θ of E into itself ($\theta \in \mathscr{E}/\mathscr{E}$) is called a *point transformation* of E. With such a mapping we may associate a kernel by setting

$$N(x, A) = 1_A(\theta(x)) = 1_{\theta^{-1}(A)}(x).$$

Definition 1.5. The *composition* or *product* of two positive kernels M and N is defined by

$$MN(x, A) = \int_E M(x, \mathrm{d}y)\, N(y, A).$$

It is easily seen that MN is a kernel and we have

Proposition 1.6. *The composition of positive kernels is an associative operation.*

Proof. By approximating $f \in \mathscr{E}_+$ by simple functions, it is easily checked that

$$((LM)N)\, f = (LM)\, (\cdot\,, Nf) = L(\cdot\,, (MNf)) = (L(MN))\, f,$$

which is the desired conclusion.

From now on, unless the contrary is stated, we deal only with positive kernels. By virtue of the preceding proposition we may therefore define the powers N^n of a positive kernel N; they are the kernels defined inductively by the formula

$$N^n(x, f) = N(x, N^{n-1}f) = N^{n-1}(x, Nf).$$

For $n = 0$, we set $N^0 = I$, where $I(x, \cdot) = \varepsilon_x(\cdot)$. The convolution powers of a positive probability measure μ on a group will be denoted μ^{*n} or more simply μ^n.

Let us give yet another definition that we shall need in the sequel.

Definition 1.7. Let M and N be two kernels; we say that M is *smaller than* N and write $M \leqslant N$ if, for every $f \in \mathscr{E}_+$, we have $Mf \leqslant Nf$. We write $M < N$ if in addition there exists an $f \in \mathscr{E}_+$ such that $Mf < Nf$.

The following definitions are basic.

Definition 1.8. A kernel N such that $N(x, E) \leqslant 1$ for all x in E is called a *transition probability* or a submarkovian kernel. It is said to be *markovian* if $N(x, E) = 1$ for all x in E. In the sequel we shall often write T.P. instead of writing in full the words "transition probability."

Throughout almost all the sequel our basic datum will be a transition probability, denoted by the letter P, the properties of which we shall study from the probabilistic point of view as well as from the potential or ergodic theoretical points of view. The powers of P will be denoted P_n rather than P^n; this integer n will appear as a "time" in the sequel.

We close this section with a few useful remarks. If P is a T.P., the operators associated with P according to Definition 1.2 will also be denoted by P. They are *positive operators*, that is, they map positive functions (measures) into positive functions (measures). Moreover, they are linear and continuous operators on the Banach spaces $b\mathscr{E}$ and $b\mathscr{M}(\mathscr{E})$, and their norm is less than or equal to 1. This may be stated: P is a *positive contraction* of these Banach spaces. Finally notice that the product of two T.P's is itself a T.P.

Exercise 1.9. Prove Proposition 1.3.

[*Hint*. Define the kernel N by setting $N(x, A) = V1_A(x)$.]

Exercise 1.10. By using the convolution kernels associated with the three following measures on $\mathbf{R}$, prove that Proposition 1.6 may fail to be true if the kernels are not positive: (i) the Lebesgue measure, (ii) $\varepsilon_1 - \varepsilon_0$, (iii) the restriction of the Lebesgue measure to $\mathbf{R}_+$.

Exercise 1.11. Compute the powers of the kernels defined in Example 1.4 (ii).

Exercise 1.12. If $N = a \otimes \lambda$ compute MN and NM for an arbitrary positive kernel M.

Exercise 1.13. If N is the kernel of Example 1.4(iii), prove that $Nf = f \circ \theta$ and $mN = \theta(m)$.

Exercise 1.14. If E is countable and P is a T.P. on E written as a matrix, then P_n can be written as the nth power of this matrix.

Exercise 1.15. For any Borel function on a group G, define $T_g f$ by $T_g f(x) = f(xg)$. Prove that a kernel N on G is a convolution kernel as defined in 1.4(ii) if and only if $T_g N = N T_g$.

Exercise 1.16. Let M and N be two kernels and suppose that N is an integral kernel; prove then that MN is an integral kernel.

Exercise 1.17. Let P be a T.P. on $(E, \mathscr{E})$. A sub-σ-algebra $\mathscr{B}$ of $\mathscr{E}$ is said to be *admissible* for P if: (i) it is countably generated; (ii) for any $A \in \mathscr{B}$ the function $P(\cdot\,, A)$ is $\mathscr{B}$-measurable (in other words P is a T.P. on $(E, \mathscr{B})$).

Prove that any countable collection of sets in $\mathscr{E}$ is contained in an admissible σ-algebra.

[*Hint*: The smallest algebra $\mathscr{B}_0$ containing the given collection and all the sets $\{x: P(x, A) \leqslant r\}$ for rational r and A in $\mathscr{B}_0$ is countable. The σ-algebra $\mathscr{B}$ is generated by $\mathscr{B}_0$.]

Exercise 1.18. Let N and A be two positive kernels on $(E, \mathscr{E})$. Prove that the kernel $S = \sum_{n\geqslant 0} (NA)^n N$ satisfies the relation

$$S = N + NAS = N + SAN.$$

If B is another positive kernel, then

$$\sum_{n\geqslant 0} (SB)^n S = \sum_{n\geqslant 0} (N(A + B))^n N.$$

2. Homogeneous Markov chains

Let $(\Omega, \mathscr{F}, P_0)$ be a probability space and $X = \{X_n\}_{n\geqslant 0}$ a sequence of random variables defined on Ω with their range in E. Let $\mathscr{F}_n = \sigma(X_m, m \leqslant n)$ and $\mathscr{G}_n$ be an increasing sequence of σ-algebras such that $\mathscr{G}_n \supset \mathscr{F}_n$ for every n.

Definition 2.1. The sequence $X = \{X_n\}_{n\geq 0}$ is said to be a *Markov chain with respect to the σ-algebras* $\mathscr{G}_n$ if, for every n, the σ-algebras $\mathscr{G}_n$ and $\sigma(X_m, m \geqslant n)$ are conditionally independent with respect to X_n; in other words, if for every $A \in \mathscr{G}_n$ and $B \in \sigma(X_m, m \geqslant n)$

$$P_0[A \cap B \mid X_n] = P_0[A \mid X_n]\, P_0[B \mid X_n] \quad \text{a.s.}$$

If $\mathscr{G}_n = \mathscr{F}_n$, we say simply that X_n is a Markov chain; the "past" $\sigma(X_m, m \leqslant n)$ and the "future" $\sigma(X_m, m \geqslant n)$ then play totally symmetric roles in this definition, the intuitive meaning of which is clear: given the present X_n, the past and the future are independent.

Notice that if the above property is true with the σ-algebras $\mathscr{G}_n$, it is a fortiori true with the σ-algebras $\mathscr{F}_n$. Finally we emphasize the importance of P_0 in this definition; if we change the probability measure P_0 there is no reason why X should remain a Markov chain.

Proposition 2.2. *The sequence X is a Markov chain with respect to the σ-algebras $\mathscr{G}_n$ if and only if, for every random variable $Y \in b\sigma(X_m, m \geqslant n)$,*

$$E_0[Y \mid \mathscr{G}_n] = E_0[Y \mid X_n] \quad P_0\text{-a.s.},$$

where E_0 denotes the mathematical expectation operator with respect to P_0.

Proof. Pick A in $\mathscr{G}_n$ and B in $\sigma(X_m, m \geqslant n)$; then since $\mathscr{G}_n \supset \sigma(X_n)$

$$E_0[1_A 1_B \mid X_n] = E_0[1_A 1_B \mid \mathscr{G}_n \mid X_n] = E_0[1_A E[1_B \mid \mathscr{G}_n] \mid X_n],$$

and if the property in the statement is true, this is equal to

$$E_0[1_A E_0[1_B \mid X_n] \mid X_n] = E_0[1_A \mid X_n]\, E_0[1_B \mid X_n],$$

which proves that $\{X_n\}_{n\geq 0}$ is a Markov chain.

Conversely it suffices to show the above property when $Y = 1_B$ with B in $\sigma(X_m, m \geqslant n)$, and this amounts to showing that for every A in $\mathscr{G}_n$,

$$E_0[1_A 1_B] = E_0[1_A E_0[1_B \mid X_n]].$$

But since $\{X_n\}$ is a Markov chain, the right member is equal to

$$\begin{aligned} E_0[E_0[1_A E[1_B \mid X_n] \mid X_n]] &= E_0[E_0[1_A \mid X_n]\, E_0[1_B \mid X_n]] \\ &= E_0[E_0[1_A 1_B \mid X_n]] = E_0[1_A 1_B], \end{aligned}$$

which completes the proof.

The following definition is basic.

Definition 2.3. The sequence $X = \{X_n\}_{n\geqslant 0}$ of random variables is called a *homogeneous Markov chain with respect to the σ-algebras* $\mathcal{G}_n$ *with transition probability* P if, for any integers m, n with $m < n$ and any function $f \in b\mathcal{E}$, we have

$$E_0[f(X_m) \mid \mathcal{G}_n] = P_{n-m} f(X_m) \quad P_0\text{-a.s.}$$

The probability measure ν defined by $\nu(A) = P_0[X_0 \in A]$ is called the *starting measure.* If $\mathcal{G}_n = \mathcal{F}_n$, we say more simply that X is a homogeneous Markov chain with transition probability P.

We leave to the reader as an exercise the task of showing that a Markov chain in the sense of Definition 2.3 is a Markov chain in the sense of Definition 2.1. To this end the following proposition comes in useful.

Proposition 2.4. *The sequence* $\{X_n\}_{n\geqslant 0}$ *is a Markov chain with transition probability* P *if and only if for every finite collection of integers* $t_0 = 0 < t_1 \cdots < t_n$ *and functions* $f_0, \ldots, f_n$ *in* $b\mathcal{E}$ *one has*

$$E_0\left[\prod_{i=0}^{n} f_i(X_{t_i})\right] = \int_E \nu(dx_0)\, f_0(x_0) \int_E P_{t_1}(x_0, dx_1)\, f_1(x_1) \cdots \int_E P_{t_n - t_{n-1}}(x_{n-1}, dx_n)\, f_n(x_n).$$

Proof. If $n = 1$, the above formula follows at once from Definition 2.3. Now, applying Definition 2.3, we get

$$E_0\left[\prod_{i=0}^{n} f_i(X_{t_i})\right] = E_0\left[\prod_{i=0}^{n-1} f_i(X_{t_i})\, E_0[f_n(X_{t_n}) \mid \mathcal{G}_{t_{n-1}}]\right]$$

$$= E_0\left[\prod_{i=0}^{n-1} f_i(X_{t_i})\, P_{t_n - t_{n-1}} f_n(X_{t_{n-1}})\right].$$

Since $f_{n-1} P_{t_n - t_{n-1}} f_n$ is still a function of $b\mathcal{E}$, the "only if" part follows by induction.

To obtain the "if" part, it suffices by the monotone class theorem, to prove that for every integer $t_0 = 0 < t_1 < \cdots < t_k \leqslant m < n$ and functions $f_0, f_1, \ldots, f_k, f$ in $b\mathcal{E}$ one has

$$E_0\left[\prod_{i=0}^{k} f_i(X_{t_i})\, f(X_n)\right] = E_0\left[\prod_{i=0}^{k} f_i(X_{t_i})\, P_{n-m} f(X_m)\right];$$

but this is an immediate consequence of the above formula.

Remark 2.5. It is easily seen that the condition in Definition 2.3 is satisfied provided it is satisfied for every pair of consecutive integers.

The main goal of this section is to show that with every transition probability we may associate a homogeneous Markov chain, which will be one of the main objects of our study. We begin with a few preliminaries.

Let P be a T.P. on E. Let Δ be a point not in E and write $E_\Delta = E \cup \{\Delta\}$ and $\mathscr{E}_\Delta = \sigma(\mathscr{E}, \{\Delta\})$. We extend P to $(E_\Delta, \mathscr{E}_\Delta)$ by setting

$$P(x, \{\Delta\}) = 1 - P(x, E) \quad \text{if } x \neq \Delta, \qquad P(\Delta, \{\Delta\}) = 1.$$

If E is locally compact we choose as Δ the point at infinity in the Alexandrov's compactification of E. We introduce the following convention: any numerical function f on E will *automatically be extended to* E_Δ *by setting* $f(\Delta) = 0$.

Heuristically speaking, P is the tool which permits us to describe the random path of a "particle" in E. Starting at x at time 0, the particle hits a random point x_1 at time 1 according to the probability measure $P(x, \cdot)$, then a random point x_2 at time 2 according to $P(x_1, \cdot)$ and so forth. If $P(x, E) < 1$, that means that the particle may disappear or "die" with positive probability; by convention it then arrives at the "fictitious" point Δ, where it stays for ever after. We assume however that $P(x, E) > 0$ for every x in E; that is, the particle does not die at once with probability one.

Definition 2.6. The point Δ is called the *cemetery*. We shall rather say "point at infinity" when E is locally compact. The space $(E, \mathscr{E})$ is called the *state space*.

We are now going to give a mathematical formulation of the above description. For every integer $n \geqslant 0$, let $(E_\Delta^n, \mathscr{E}_\Delta^n)$ be a copy of $(E_\Delta, \mathscr{E}_\Delta)$; we call $(\Omega, \mathscr{F})$ their product space, namely $\Omega = \prod_{n=0}^\infty E_\Delta^n$, and $\mathscr{F}$ is the σ-algebra generated by the semi-algebra $\mathscr{S}$ of *measurable rectangles* of Ω. We recall that a measurable rectangle is a set of the form

$$\prod_{n=0}^{\infty} A_n, \qquad A_n \in \mathscr{E}_\Delta^n,$$

where it is assumed that A_n is different from E_Δ^n for only finitely many n.

Definition 2.7. The space Ω is called the *canonical probability space*. We call X_n, $n \geqslant 0$, the coordinate mappings of Ω. Let $\omega = \{x_n, n \geqslant 0\}$ be a point in Ω, then $X_n(\omega) = x_n$; we also set $X_\infty(\omega) = \Delta$ for every ω in Ω. These mappings X_n are random variables defined on Ω with range in E, and are clearly measurable with respect to the σ-algebras $\mathscr{F}_n = \sigma(X_m, m \leqslant n)$. A point ω in Ω is referred to as a *trajectory* or a *path*.

We come to our main result.

Theorem 2.8. *For every x in E there exists a unique probability measure P_x on $(\Omega, \mathscr{F})$ such that for any finite collection $n_0 = 0 < n_1 < n_2 < \cdots < n_k$ of integers and every rectangle*

$$A = \prod_{i=0}^{k} A_{n_i} \times \prod_{n \neq n_i} E_\Delta^n,$$

$$P_x[A] = 1_{A_0}(x) \int_{A_{n_1}} P_{n_1}(x, dx_1) \int_{A_{n_2}} P_{n_2 - n_1}(x_1, dx_2) \cdots$$

$$\int_{A_{n_{k-1}}} P_{n_{k-1}-n_{k-2}}(x_{k-2}, dx_{k-1})\, P_{n_k - n_{k-1}}(x_{k-1}, A_{n_k}). \tag{2.1}$$

Furthermore for every set $A \in \mathscr{F}$, the map $x \to P_x[A]$ is $\mathscr{E}_\Delta$-measurable.

Proof. Equation (2.1) defines clearly an additive set function on $\mathscr{S}$ which has a unique extension to an additive function on the Boolean algebra $\mathscr{A} = \bigcup_{n=0}^\infty \mathscr{F}_n$ and the restriction of this set function to each of the σ-algebras $\mathscr{F}_n$ is probability measure. We still call P_x the extended set function. The map $x \to P_x[A]$ is then in $\mathscr{E}$ for every $A \in \mathscr{A}$; indeed this is true for measurable rectangles and the class of sets $B \in \mathscr{F}_n$ for which it is true is, for every n, a monotone class.

To prove the first half of the theorem, we set $\Omega^n = \prod_{m=n}^\infty E_\Delta^m$, and for $A \in \mathscr{A}$ and $(x_0, x_1, \ldots, x_n)$ a finite collection of points in E, we write $A(x_0, x_1, \ldots, x_n)$ for the set of points ω^{n+1} in Ω^{n+1} such that $(x_0, x_1, \ldots, x_n, \omega^{n+1})$ is in A. The space Ω^n being isomorphic to Ω, we may define a set function on the algebra generated by the rectangles of Ω^n in the same way as we have defined P_x. This set function will be denoted by P_x^n. Again we observe that the map $x \to P_x^n[A]$ is $\mathscr{E}$-measurable for $A \in \mathscr{A}$. Moreover using the same argument as to prove Fubini's theorem, we can see that for $A \in \mathscr{A}$ we have

$$P_x[A] = \int_E P(x, dx_1)\, P_{x_1}^1[A(x)].$$

To prove that P_x can be extended to a probability measure on $\mathscr{F}$ we have to show that it is σ-additive on $\mathscr{A}$. Let $\{A^j\}$ be a sequence in $\mathscr{A}$ decreasing to $\emptyset$ and let us suppose that $\lim_j P_x[A^j] > 0$. By the last displayed formula applied to A^j we see that this implies, since $P(x, \cdot)$ is a probability measure, that there exists a point $\bar{x}_1 \in E$ such that $\lim_j P^1_{\bar{x}_1}[A^j(x)] > 0$. As a result $A^j(x)$ is non-empty for every j; moreover reasoning with $P^1_{\bar{x}_1}$ and $\{A^j(x)\}$ as we did with P_x and A^j we find that there exists a point $\bar{x}_2 \in E$ such that

$$\lim_j P^2_{\bar{x}_2}[A^j(x, \bar{x}_1)] > 0.$$

Proceeding inductively, we find that for every n there is a point $\bar{x}_n$ such that $A^j(x, \bar{x}_1, \bar{x}_2, \ldots, \bar{x}_n)$ is non-empty for every j. But since each A^j depends on only finitely many coordinates it follows that the point $\omega = (x, \bar{x}_1, \bar{x}_2, \ldots, \bar{x}_n, \ldots)$ belongs to every A^j which is a contradiction.

To prove the second half of the statement, we observe that the class of sets B such that $x \to P_x[B]$ is measurable is a monotone class which includes $\mathscr{A}$.

Definition 2.9. For every probability measure ν on $(E, \mathscr{E})$, we define a new probability measure P_ν on $(\Omega, \mathscr{F})$ by setting

$$P_\nu[A] = \int_E \nu(\mathrm{d}x)\, P_x[A].$$

Indeed, the measurability in the preceding theorem gives sense to the above formula, which clearly defines a probability measure. For $\nu = \varepsilon_x$ we have $P_\nu = P_x$, and if A is the rectangle in Theorem 2.8, then since

$$A = \{\omega\colon X_0(\omega) \in A_0, X_{n_1}(\omega) \in A_{n_1}, \ldots, X_{n_k}(\omega) \in A_{n_k}\},$$

we have

$$\begin{aligned} P_\nu[A] &= P_\nu[X_0 \in A_0, X_{n_1} \in A_1, \ldots, X_{n_k} \in A_{n_k}] \\ &= \int_{A_0} \nu(\mathrm{d}x) \int_{A_{n_1}} P_{n_1}(x, \mathrm{d}x_1) \cdots \\ &\quad \int_{A_{n_{k-1}}} P_{n_{k-1}-n_{k-2}}(x_{k-2}, \mathrm{d}x_{k-1})\, P_{n_k-n_{k-1}}(x_{k-1}, A_{n_k}). \end{aligned} \tag{2.2}$$

By comparing this relation with that of Proposition 2.4, we can state:

Proposition 2.10. *For every probability measure P_ν the sequence $X = \{X_n\}_{n\geqslant 0}$*

is a homogeneous Markov chain with transition probability P and starting measure ν. It is called the canonical Markov chain with transition probability P.

Proof. By the usual argument we may replace the sets A_{n_i} in the above relation by functions f_{n_i}, so that X satisfies the condition of Proposition 2.4.

Definition 2.11. Let $q(\omega)$ be a property of ω; then q is said to hold *almost surely* (a.s.) *on* $\Lambda \in \mathscr{F}$, if the set of ω's in Λ for which $q(\omega)$ fails to hold is contained in a set $\Lambda_0 \in \mathscr{F}$ such that for every $x \in E$, $P_x(\Lambda_0) = 0$. If $\Lambda = \Omega$ we simply say "almost surely".

We proceed with some more definitions and notation. Let Z be a positive numerical random variable on $(\Omega, \mathscr{F})$. Its mathematical expectation taken with respect to P_ν will be denoted $E_\nu[Z]$; if $\nu = \varepsilon_x$, then we write simply $E_x[Z]$. It is easily seen that the map $x \to E_x[Z]$ is in $\mathscr{E}$ and that

$$E_\nu[Z] = \int_E \nu(\mathrm{d}x)\, E_x[Z].$$

Furthermore if $f \in \mathscr{E}_+$, then $f(X_n)$ is a random variable on $(\Omega, \mathscr{F})$ and we have

$$E_x[f(X_n)] = P_n f(x) = \int_E P_n(x, \mathrm{d}y)\, f(y). \tag{2.3}$$

Indeed, for $f = 1_A$ this formula is a special case of eq. (2.1), and it may be extended to $\mathscr{E}_+$ by the usual argument.

Definition 2.12. The *shift operator* θ is the point transformation on Ω defined by

$$\theta(\{x_0, x_1, \ldots, x_n, \ldots\}) = \{x_1, x_2, \ldots, x_{n+1}, \ldots\}.$$

It is obvious that $\theta \in \mathscr{F}/\mathscr{F}$ and that $X_n(\theta(\omega)) = X_{n+1}(\omega)$. We write θ_p for the pth power of θ: $\theta_p = \theta \circ \theta \cdots \circ \theta$ p times. Clearly $X_n(\theta_p(\omega)) = X_{n+p}(\omega)$, which will also be written $X_n \circ \theta_p = X_{n+p}$. Since

$$\theta_p^{-1}\{X_n \in A\} = \{X_{n+p} \in A\},$$

it is easily seen that $\theta_p \in \sigma(X_n, n \geqslant p)/\mathscr{F}$.

The following proposition gives the handy form of the Markov property and will frequently be used in the sequel.

Proposition 2.13 (Markov property). *For every positive random variable Z on $(\Omega, \mathscr{F})$, every starting measure ν and integer n,*

$$E_\nu[Z \circ \theta_n \mid \mathscr{F}_n] = E_{X_n}[Z] \quad P_\nu\text{-}a.s.$$

on the set $\{X_n \neq \Delta\}$.

The last phrase is necessary to be consistent with the convention that functions on E, in particular the function $E.[Z]$, are extended to E by making them vanish at Δ; and there is no reason why the first member should vanish on $\{X_n = \Delta\}$. In the most frequent case however, where P is markovian, the event $\{X_n = \Delta\}$ has zero P_ν-probability for every ν and the above qualification may, and often will, be forgotten in the sequel.

Before we proceed to the proof let us also remark that the right-hand side of the above relation is indeed a random variable because it is the composite mapping of the two measurable mappings $\omega \to X_n(\omega)$ and $x \to E_x[Z]$. We also notice that if $Z = 1_{\{X_m \in A\}}$, $A \in \mathscr{E}$, the above formula becomes

$$P_\nu[X_{n+m} \in A \mid \mathscr{F}_n] = P_{X_n}[X_m \in A] \quad P_\nu\text{-a.s.},$$

which is the formula of Definition 2.3.

Proof of 2.13. We must prove that for any $B \in \mathscr{F}_n$,

$$\int_B Z \circ \theta_n 1_{\{X_n \in E\}} \, dP_\nu = \int_B E_{X_n}[Z] \, dP_\nu;$$

and by the usual extension argument it suffices to prove this relation for the case in which B is a rectangle, namely $B = \{X_{p_1} \in B_1, \ldots, X_{p_k} \in B_k\}$ with $B_i \in \mathscr{E}$, and in which $Z = 1_A$, where A is the rectangle used in Theorem 2.8. The result then follows immediately from eq. (2.2).

As much as P, our basic datum will henceforth be the canonical chain X associated with P. We shall be concerned with all probability measures P_ν, and shall use freely the Markov property of Proposition 2.13. Indeed, although the canonical chain is not the only chain with transition probability P, it has the following universal property which allows one to translate any problem on a chain Y to the analogous problem on the canonical chain X.

Proposition 2.14. *Let Y be a homogeneous Markov chain defined on the probabil-*

ity space $(W, \mathfrak{A}, Q)$ *with transition probability* P *and starting measure* $\nu(\cdot) = Q[Y_0 \in \cdot\,]$. *Then the canonical Markov chain associated with* P *is the image of* Y *by the mapping* φ *which sends* $w \in W$ *to the point*

$$\omega = \varphi(w) = (Y_0(w),\ Y_1(w), \ldots,\ Y_n(w), \ldots)$$

of $\Omega = E_\Delta^N$. *The image of* Q *by* φ *is equal to* P_ν.

Proof. The mapping φ is measurable because the composite mappings of φ with the coordinate mappings X_n are equal to the random variables Y_n. It is then easily verified that $\varphi(Q) = P_\nu$.

Exercise 2.15. Prove that a real random variable Z is $\sigma(X_m, m \geqslant n)$-measurable if and only if $Z = Z' \circ \theta_n$ where $Z' \in \mathscr{F}$.

Exercise 2.16. A sequence $\{X_n\}_{n\geqslant 0}$ of random variables defined on $(\Omega, \mathscr{F}, P)$ is a Markov chain of order r if for every $B \in \mathscr{E}$ and integer n,

$$P[X_{n+1} \in B \mid \sigma(X_m, m \leqslant n)] = P[X_{n+1} \in B \mid \sigma(X_m, n - r + 1 \leqslant m \leqslant n)].$$

Prove that the sequence of random variables $Y_n = (X_{n+1}, \ldots, X_{n+r-1})$ with range in E^r is a Markov chain in the ordinary sense.

Exercise 2.17. (1) Suppose that for every pair $m \leqslant n$ of integers there is a T.P. called $P_{m,n}$ such that $P_{l,m}P_{m,n} = P_{l,n}$. Prove that one can associate with these T.P.'s a *non-homogeneous Markov chain*, that is, a sequence of random variables X_n defined on a space $(\Omega, \mathscr{F}, P)$ and such that for $f \in b\mathscr{E}$

$$E[f(X_n) \mid \sigma(X_k, k \leqslant m)] = P_{m,n}f(X_m) \quad \text{a.s.}$$

(2) In the situation of (1) the *space–time chain* $(X_n \times n)$ with range in $E \times \mathbf{N}$ is a homogeneous Markov chain. ((2) may be solved without previous solution of (1).)

Exercise 2.18. Let X be a homogeneous canonical Markov chain and $(W, \mathscr{A}, Q)$ a probability space. Let U be a random variable defined on W and endow the space $\bar{\Omega} = \Omega \times W$ with the σ-algebras

$$\mathscr{G}_n = \sigma(X_0, X_1, \ldots, X_n, U), \qquad \mathscr{G} = \sigma\left(\bigcup_{n=0}^{\infty} \mathscr{G}_n\right)$$

and the probability measures $\bar{P} = P \otimes Q$. Prove that the sequence $\{\bar{X}_n\}$

defined by $\bar{X}_n(\omega, w) = X_n(\omega)$, is still a homogeneous Markov chain with the same T.P. with respect to the σ-algebras $\mathscr{G}_n$.

Exercise 2.19. If $\{X_n\}$ is a homogeneous Markov chain, then $E[f(X_n) \mid X_m] = P_{n-m} f(X_m)$ a.s., but the converse is false as the following example shows. Set $E = \{1, 2, \ldots, N\}$ and define $\Omega = \Omega_1 \cup \Omega_2$ where $\Omega_1 = E$ and Ω_2 is the set of permutations of E. Define a probability measure P on Ω by setting with (obvious notation)

$$P(\omega_1) = N^{-2}, \qquad P(\omega_2) = (1 - N^{-1})\,(N!)^{-1}.$$

Define further random variables X_n, $1 \leqslant n \leqslant N$, with values in E by $X_n(\omega_1) = \omega_1$, $X_n(\omega_2) = i$ if i is the nth number in the permutation ω_2.

If we define a T.P. on E by $P(i, j) = N^{-1}$, prove that

$$P[X_n = i \mid X_{n-1} = j] = P(i, j), \quad n \geqslant 2,$$

although for $N \geqslant 3$, the sequence $\{X_n\}$ is not a Markov chain. By using the space $\Omega \times \Omega \times \Omega \cdots$ build an example where $\{X_n\}$ is indexed by $\mathbf{N}$.

Exercise 2.20. The following chain may be seen as describing the evolution of the size of a population where, at each generation, the random number of offsprings of the individuals are independent and equally distributed. It is called the Galton–Watson chain.

The state space is the set $\mathbf{N}$ and the conditional probability $P[X_{n+1} = x \mid \mathscr{F}_n]$ is equal to the law of the sum of X_n independent equidistributed random variables of law $p = (p(x), x \in \mathbf{N})$. The T.P. is thus given by $P(x, y) = p^{*x}(y)$.

(1) Prove that for every n there exists a probability measure p_n on E such that $P_n(x, y) = p_n^{*x}(y)$.

[*Hint*: Begin by proving that the mapping φ from Ω^x into Ω defined by

$$X_n(\varphi(\omega_1, \ldots, \omega_x)) = X_n(\omega_1) + \cdots + X_n(\omega_x)$$

sends the product measure $P_1 \otimes P_1 \otimes \cdots \otimes P_1$ to P_x. This is the mathematical formulation of the following intuitive fact: the chain starting at x behaves as the sum of x independent chains starting at 1.]

(2) Prove that the generating functions $g_n(t) = \sum_{x \in E} t^x p_n(x)$ are given by the recurrence formula

$$g_{n+1}(t) = g_n(g(t)) = g(g_n(t)),$$

where g is the generating function of the measure p. Compute for P_1 the mean value and variance of X_n as functions of those of X_1 which are assumed to exist.

3. Stopping times. Strong Markov property

Definition 3.1. A *stopping time* of the canonical Markov chain X is a random variable defined on $(\Omega, \mathscr{F})$ with range in $\mathbf{N} \cup \{\infty\}$ and such that for every integer n the event $\{T = n\}$ is in $\mathscr{F}_n$. The family $\mathscr{F}_T$ of events $A \in \mathscr{F}$ such that for every n, $\{T = n\} \cap A \in \mathscr{F}_n$ is called the σ-algebra associated with T.

It is easily verified that $\mathscr{F}_T$ is indeed a sub-σ-algebra of $\mathscr{F}$. The constant random variables are stopping times, and if $T(\omega) = n$ for every $\omega \in \Omega$, then $\mathscr{F}_T = \mathscr{F}_n$. The stopping times thus appear as generalizations of the ordinary times. The following examples of stopping times are basic.

Definition 3.2. For $A \in \mathscr{E}$, we call first *hitting time* of A and first *return time* of A the random variables defined by

$$T_A(\omega) = \inf\{n \geqslant 0 : X_n(\omega) \in A\},$$

$$S_A(\omega) = \inf\{n > 0 : X_n(\omega) \in A\},$$

where in both cases the infimum of the empty set is understood to be $+\infty$.

It is readily checked that both variables are stopping times. For example

$$\{T_A = n\} = \bigcap_{m=0}^{n-1} \{X_m \in A^c\} \cap \{X_n \in A\} \in \mathscr{F}_n.$$

In the same way the random variable

$$\zeta(\omega) = \inf\{n \geqslant 0 : X_n(\omega) = \Delta\}$$

is a stopping time called the *death-time* of X. If P is markovian, ζ is a.s. equal to $+\infty$.

Definition 3.3. With each stopping time T we associate the following objects:

(i) The random variable X_T is defined by setting

$$X_T(\omega) = X_n(\omega) \quad \text{if } T(\omega) = n, \qquad X_T(\omega) = \Delta \quad \text{if } T(\omega) = +\infty.$$

It gives the position of the chain at time T, and the reader will easily check that $X_T \in \mathscr{F}_T/\mathscr{E}_\Delta$.

(ii) The point transformation θ_T on Ω is defined by setting

$$\theta_T(\omega) = \theta_n(\omega) \quad \text{if } T(\omega) = n, \qquad \theta_T(\omega) = \omega_\Delta \quad \text{if } T(\omega) = +\infty,$$

where ω_Δ is the trajectory $\{\Delta, \Delta, \ldots, \Delta, \ldots\}$ of Ω. It is easily seen that $\theta_T \in \mathscr{F}/\mathscr{F}$ and that

$$X_n(\theta_T(\omega)) = X_{n+m}(\omega) \quad \text{on } \{T = m\}, \qquad X_n(\theta_T(\omega)) = \Delta \quad \text{on } \{T = \infty\}.$$

Proposition 3.4. *Let S and T be two stopping times; then the mapping $S + T \circ \theta_S : \omega \to S(\omega) + T(\theta_S(\omega))$ is a stopping time.*

Proof. We have

$$\{S + T \circ \theta_S = n\} = \bigcup_{p \leqslant n} \{S = p\} \cap \{T \circ \theta_S = n - p\}.$$

The event $\{S = p\}$ is in $\mathscr{F}_p \subset \mathscr{F}_n$ and on $\{S = p\}$ we have $\theta_S = \theta_p$; hence

$$\{T \circ \theta_S = n - p\} = \{T \circ \theta_p = n - p\} = \theta_p^{-1}(\{T = n - p\}) \in \mathscr{F}_n$$

since $\{T = n - p\} \in \mathscr{F}_{n-p}$.

Intuitively one should think of stopping times as the first time some physical event occurs and of $\mathscr{F}_T$ as containing the information on the chain up to time T when this event occurs. The time $S + T \circ \theta_S$ is the first time where the "event T" occurs after the "event S" has occurred. For instance if $A \in \mathscr{E}$, $n + T_A \circ \theta_n$ is the first hitting time of A after time n; in particular $S_A = 1 + T_A \circ \theta_1$. If $A, B \in \mathscr{E}$, then $T_A + T_B \circ \theta_{T_A}$ is the first time the chain hits B after having hit A.

Despite its simplicity the following result is basic. It implies in particular that if one starts to observe the chain at a stopping time, the ensuing process is still a homogeneous Markov chain with the same T.P. (cf. Exercise 3.14).

Theorem 3.5 (Strong Markov property). *For every real positive random variable Z on $(\Omega, \mathscr{F})$, starting measure ν and stopping time T,*

$$E_\nu[Z \circ \theta_T \mid \mathscr{F}_T] = E_{X_T}[Z] \quad P_\nu\text{-a.s.}$$

By convention the two members vanish on $\{X_T = \Delta\}$.

The right member is the composite mapping of $\omega \to X_T(\omega)$ and $x \to E_x[Z]$.

Proof. We have to show that for $A \in \mathscr{F}_T$,

$$\int_A Z \circ \theta_T \cdot 1_{\{X_T \neq \Delta\}} \, \mathrm{d}P_\nu = \int_A E_{X_T}[Z] \, \mathrm{d}P_\nu,$$

or equivalently

$$\sum_{n \geqslant 0} \int_{A \cap \{T=n\}} Z \circ \theta_T \cdot 1_{\{X_n \neq \Delta\}} \, \mathrm{d}P_\nu = \sum_{n \geqslant 0} \int_{A \cap \{T=n\}} E_{X_T}[Z] \, \mathrm{d}P_\nu.$$

It suffices therefore to prove that for every $n \geqslant 0$ and $B \in \mathscr{F}_n$,

$$\int_B Z \circ \theta_n \cdot 1_{\{X_n \in E\}} \, \mathrm{d}P_\nu = \int_B E_{X_n}[Z] \, \mathrm{d}P_\nu,$$

but this is precisely the Markov property of Proposition 2.13.

Definition 3.6. With each stopping time T we associate a new T.P., denoted P_T, by setting for $B \in \mathscr{E}$,

$$P_T(x, B) = P_x[X_T \in B].$$

It is easily seen that if $f \in \mathscr{E}_+$, then

$$P_T f(x) = E_x[f(X_T)].$$

We recall that by our conventions $f(X_T) = 0$ on $\{X_T = \Delta\}$. We further notice that if $T = n$ a.s., then $P_T = P_n$. Finally if $A \in \mathscr{E}$, we write P_A instead of P_{T_A}; this T.P. is called the *balayage operator* associated with A.

Before we state our next result, we introduce the following notation. Let $A \in \mathscr{E}$; define I_A to be the operator of multiplication by 1_A, $I_A f(x) = f(x)$ if $x \in A$, $I_A f(x) = 0$ if $x \notin A$. We may now give for P_A the following analytical expression.

Proposition 3.7. *For* $A \in \mathscr{E}$,

$$P_A = \sum_{n \geqslant 0} (I_{A^c} P)^n I_A.$$

Proof. Let $B \in \mathscr{E}$; using eq. (2.2), we have

$$P_A(x, B) = 1_{A \cap B}(x) + \sum_{n=1}^{\infty} P_x[X_0 \notin A, \ldots, X_{n-1} \notin A, X_n \in A \cap B]$$

$$= 1_{A\cap B}(x) + \sum_{n=1}^{\infty} 1_{A^c}(x) \int P(x, \mathrm{d}x_1)\, 1_{A^c}(x_1) \int P(x_1, \mathrm{d}x_2)\, 1_{A^c}(x_2) \cdots$$

$$\int P(x_{n-1}, \mathrm{d}x_n)\, 1_{B\cap A}(x_n),$$

which is the desired result.

Clearly the measures $P_A(x, \cdot\,)$ vanish outside A. Furthermore if $x \in A$, then $P_A(x, \cdot\,) = \varepsilon_x$; indeed $x \in A$ implies $P_x[X_0 = x] = 1$, hence $T_A = 0$ P_x-a.s. and consequently $X_{T_A} = x$ P_x-a.s. In the same way, if $x \notin A$, then $T_A = S_A$ P_x-a.s. and $P_A(x, \cdot\,) = P_{S_A}(x, \cdot\,)$.

We close this section by applying the strong Markov property to compute the products of operators P_T. We first notice that if S and T are two stopping times then $X_T(\theta_S(\omega)) = X_{S+T\circ\theta_S}(\omega)$, which we write simply $X_T \circ \theta_S = X_{S+T\circ\theta_S}$.

Proposition 3.8. *Let S, T be two stopping times; then $P_S P_T = P_{S+T\circ\theta_S}$.*

Proof. Let $f \in b\mathscr{E}_+$. We have, applying Definition 3.6,

$$P_S P_T f(x) = E_x[P_T f(X_S)] = E_x[E_{X_S}[f(X_T)]]$$

$$= E_x[E_x[f(X_T \circ \theta_S) \mid \mathscr{F}_S]] = E_x[f(X_T \circ \theta_S)] = P_{S+T\circ\theta_S} f(x).$$

In particular, we get $P_{S_A} = PP_A$, which implies

$$P_{S_A} = P\left(\sum_{n\geqslant 0} (I_{A^c}P)^n\, I_A\right).$$

We set

$$\Pi_A = I_A P_{S_A} = I_A P\left(\sum_{n\geqslant 0} (I_{A^c}P)^n\, I_A\right) = I_A\left(\sum_{n\geqslant 0} (PI_{A^c})^n\right) PI_A.$$

The interpretation of the kernel Π_A is given in Exercise 3.13.

Exercise 3.9. Let $A, B \in \mathscr{E}$; can $T_{A\cup B}$ and $T_{A\cap B}$ be expressed as functions of T_A and T_B? Given $A \subset B$, compare T_A and T_B and prove that $P_B P_A = P_A$.

Exercise 3.10. Let S and T be two stopping times.

(1) Prove that $\inf(S, T)$ and $\sup(S, T)$ are stopping times. In particular $S \wedge n$ is a stopping time for every n. Moreover $\mathscr{F}_{\inf(S,T)} = \mathscr{F}_S \cap \mathscr{F}_T$.

(2) If $\Gamma \in \mathscr{F}_S$, then $\Gamma \cap \{S \leqslant T\} \in \mathscr{F}_T$. If $S \leqslant T$, then $\mathscr{F}_S \subset \mathscr{F}_T$.

(3) Prove that the event $\{S \leqslant T\}$ is in $\mathscr{F}_S \cap \mathscr{F}_T$;

(4) Prove that the conditional expectations $E[\ |\mathscr{F}_S]$ and $E[\ |\mathscr{F}_T]$ commute and that their product is $E[\ |\mathscr{F}_{\inf(S,T)}]$.

Exercise 3.11. Let φ be a function mapping $\mathbf{N}$ into itself. The random variable $\varphi(T)$ is a stopping time for every stopping time T, if and only if φ enjoys the following property: there is an integer k, which may be $+\infty$, such that

$$\varphi(n) \geqslant n \quad \text{for } n \leqslant k, \qquad \varphi(n) = k \quad \text{for } n > k.$$

If φ is now a function mapping $\mathbf{N}^d$ into $\mathbf{N}$, then $\varphi(T_1, T_2, \ldots, T_d)$ is a stopping time for any d-uple of stopping times T_i if and only if the functions obtained from φ by fixing $d-1$ variables enjoy the above property.

Exercise 3.12. (1) Let $A \in \mathscr{E}$; prove that $T_A = n + T_A \circ \theta_n$ on $\{T_A \geqslant n\}$. Let $A, B \in \mathscr{E}$; has $T_A + T_B \circ \theta_{T_A}$ the same property as T_A?

(2) Let T be a stopping time such that $T = n + T \circ \theta_n$ on $\{T > n\}$. Prove that

$$1_{[T>n]} = 1_{[T>n-1]} \cdot 1_{[T>1]} \circ \theta_{n-1}.$$

(3) Let Y be the sequence of random variables defined by

$$Y_n(\omega) = X_n(\omega) \quad \text{if } T(\omega) > n, \qquad Y_n(\omega) = \Delta \quad \text{if } T(\omega) \leqslant n.$$

Prove that Y is a homogeneous Markov chain with respect to the σ-algebras $\mathscr{F}_{T\wedge n}$ for every probability measure P_x on $(\Omega, \mathscr{F})$. When $T = T_A$ we can take A^c as its state space; prove then that its T.P. is equal to $I_{A^c}PI_{A^c}$. The chain Y is said to be obtained by *killing* X *at time* T_A.

Exercise 3.13. Let $A \in \mathscr{E}$ and define inductively the times

$$T_1 = S_A,\ T_2 = T_1 + S_A \circ \theta_{T_1}, \ldots, T_n = T_{n-1} + S_A \circ \theta_{T_{n-1}}, \ldots .$$

(1) Prove that the times T_n are the successive times at which the chain returns in A.

(2) Prove that the sequence $\{Y_n = X_{T_n}\}$ is a homogeneous Markov chain with respect to the σ-algebras $\mathscr{F}_{T_n}$ with T.P. Π_A. This chain is called the chain *induced* on A or the *trace chain* on A.

Exercise 3.14. Let T be a finite stopping time. Prove that the sequence $Y = \{X_{T+n}\}_{n\geqslant 0}$ is a homogeneous Markov chain with transition probability

P with respect to the σ-algebras $\mathscr{F}_{T+n}$. What can be said if T is not finite?

Exercise 3.15. *Stopped chains.* If T is a stopping time of X prove that the sequence $Y = \{X_{T\wedge n}\}_{n\geqslant 0}$ is a homogeneous Markov chain if and only if $T = T_A$ for a set $A \in \mathscr{E}$. In that case write down the T.P. of the chain Y.

Exercise 3.16. Let T be a stopping time and G a real function on $\Omega \times \Omega$ measurable with respect to $\mathscr{F}_T \otimes \mathscr{F}$. Prove that for any ν,

$$E_\nu[G(\omega, \theta_T(\omega)) \mid \mathscr{F}_T] = \int_\Omega P_{X_{T(\omega)}}(\mathrm{d}\omega')\, G(\omega, \omega') \quad P_\nu\text{-a.s.}$$

for $\omega, \omega' \in \Omega$.

[*Hint*: Begin with $G(\omega, \omega') = \varphi(\omega)\,\psi(\omega')$, where $\varphi \in \mathscr{F}_T$.]

Exercise 3.17. Let T be a stopping time and S a random variable $\mathscr{F}_T$-measurable and $\geqslant T$ a.s. Prove that, for $f \in \mathscr{E}_+$,

$$E_\nu[f \circ X_S \mid \mathscr{F}_T]\,(\omega) = P_{S(\omega)-T(\omega)}(X_T, f) \quad P_\nu\text{-a.s.}$$

[*Hint*: Use the preceding exercise with $G(\omega, \omega') = f(X_{S(\omega)}(\omega'))$.]

Exercise 3.18. A random variable L defined on $(\Omega, \mathscr{F})$, with range in $\mathbf{N} \cup \{\infty\}$, is called a *death time* if $L \circ \theta = (L-1)^+$, that is,

$$L \circ \theta = L - 1 \quad \text{if } L \geqslant 1, \qquad L \circ \theta = 0 \quad \text{if } L = 0.$$

(1) Let $A \in \mathscr{E}$; prove that the last hitting time of A, namely

$$L_A = \sup\{n \geqslant 0, X_n \in A\},$$

where the supremum of the empty set is taken equal to zero, is a death time. The death time of the chain is a death time.

(2) Prove that $X_L \circ \theta = X_L$ on $\{0 < L\}$ if L is a death time.

(3) If L and L' are death times, then $L \vee L'$ and $L \wedge L'$ are death times; $(L-n)^+$ is a death time for every n.

4. Random walks on groups and homogeneous spaces

In this section we define an important family of Markov chains, which will occur frequently in the sequel both for their intrinsic interest and to provide examples about general results.

In the sequel G is an LCCB group and we call $\mathscr{G}$ the σ-algebra of its Borel sets. The elements of G will be denoted $g, g', h, \ldots$ and the inverse of g by g^{-1}. The unit element is denoted e.

Definition 4.1. A *right (left) random walk on* G is a Markov chain with state space $(G, \mathcal{G})$ and transition probability $\varepsilon_g * \mu$ $(\mu * \varepsilon_g)$, where μ is a probability measure on $(G, \mathcal{G})$ which is called the *law* of the random walk.

For the right random walk we have therefore $P(g, A) = \varepsilon_g * \mu(A)$. For $h \in G$ we set $hA = \{hg: g \in A\}$; we then have $P(hg, hA) = P(g, A)$. *The right random walk is invariant under left translations* (see Exercise 1.15). Of course on an abelian group there is only one random walk of law μ.

Proposition 4.2. *Let X be the right random walk of law μ; then for every P_ν the random variables $Z_n = X_{n-1}^{-1}X_n$, $n \geqslant 1$, are independent and equidistributed with law μ.*

Proof. Let f_i, $i = 1, 2, \ldots, p$ be a finite collection of bounded Borel functions. We have

$$E_\nu\left[\prod_{i=1}^{p} f_i(Z_i)\right] = E_\nu\left[\prod_{i=1}^{p-1} f_i(Z_i)\, E_\nu[f_p(X_{p-1}^{-1}\, X_p) \mid \mathcal{F}_{p-1}\right]$$

$$= E_\nu\left[\prod_{i=1}^{p-1} f_i(Z_i)\, E_{X_{p-1}}[f_p(X_0^{-1}\, X_1)]\right].$$

For every g in G we have $E_g[f(X_0^{-1}\, X_1)] = \mu(f)$, so that we get inductively

$$E_\nu\left[\prod_{i=1}^{p} f_i(Z_i)\right] = \prod_{i=1}^{p} \mu(f_i) = \prod_{i=1}^{p} E_\nu[f_i(Z_i)],$$

which is the desired result.

Remarks 4.3. The random variable X_n is thus a.s. equal to the product $X_0\, Z_1 \cdots Z_n$, where the Z_i are independent with law μ. In particular X_n is P_e-a.s. equal to the product of n independent equidistributed random variables. Of course there is a left-handed version of these results and the left random walk may be written $Z_n\, Z_{n-1} \cdots Z_1\, X_0$. The invariance by translation is also obvious from this relation.

We assume now that G operates on the left on an LCCB space M; let $\mathcal{M}$ be the σ-algebra of Borel subsets of M. We may associate with μ a T.P. on $(M, \mathcal{M})$ by setting (cf. Example 1.4 (ii)), for $x \in M$, $A \in \mathcal{M}$,

$$P(x, A) = \int_G 1_A(gx)\, \mu(\mathrm{d}g).$$

We are going to show an interesting way of constructing a Markov chain with transition probability P.

Let X be the left canonical random walk of law μ on $\boldsymbol{G}$. We set $\bar{\Omega} = \boldsymbol{M} \times \Omega$, $\bar{\mathscr{F}} = \mathscr{M} \otimes \mathscr{F}$, $\bar{\mathscr{F}}_n = \mathscr{M} \otimes \mathscr{F}_n$, and if ν is a probability measure on $(\boldsymbol{M}, \mathscr{M})$ we call $\bar{P}_\nu$ the probability measure $\nu \otimes P_e$ on $(\bar{\Omega}, \bar{\mathscr{F}})$. Next, for $\bar{\omega} = (x, \omega)$ we set $Y_0(\bar{\omega}) = x$ and

$$Y_n(\bar{\omega}) = X_n(\omega)\, Y_0(\bar{\omega}) = X_n(\omega)\, x.$$

Proposition 4.4. *The sequence $\{Y_n\}_{n\geq 0}$ is a homogeneous Markov chain with respect to the σ-algebras $\bar{\mathscr{F}}_n$ for any probability measure $\bar{P}_\nu$. Its transition probability is equal to P.*

Proof. The σ-algebra $\bar{\mathscr{F}}_n$ is generated by the rectangles $\Lambda \times \Gamma$ where $\Lambda \in \mathscr{M}$ and $\Gamma \in \mathscr{F}_n$. For A in $\mathscr{M}$ we have

$$\begin{aligned}\int_{\Lambda\times\Gamma} 1_A(Y_{m+n})\, \mathrm{d}\bar{P}_\nu &= \int_\Lambda \nu(\mathrm{d}x) \int_\Gamma 1_A(X_{m+n}(\omega)\, x)\, \mathrm{d}P_e(\omega)\\ &= \int_\Lambda \nu(\mathrm{d}x) \int_\Gamma E_e[1_A(X_{m+n}x) \mid \mathscr{F}_n]\, \mathrm{d}P_e\\ &= \int_\Lambda \nu(\mathrm{d}x) \int_\Gamma (\mu^m * \varepsilon_{X_n})\, (\tilde{A})\, \mathrm{d}P_e,\end{aligned}$$

where $\tilde{A} = \{g: gx \in A\}$; but $(\mu^m * \varepsilon_{X_n})\, (\tilde{A}) = (\mu^m * \varepsilon_{X_n}x)\, (A)$, so that finally

$$\int_{\Lambda\times\Gamma} 1_A(Y_{m+n})\, \mathrm{d}\bar{P}_\nu = \int_{\Lambda\times\Gamma} (\mu^m * \varepsilon_{Y_n})\, (A)\, \mathrm{d}\bar{P}_\nu,$$

which is the desired conclusion.

Remark 4.5. The chain Y thus constructed is not the canonical chain associated with P, but it is sometimes useful to know that one can construct a chain associated with P in the above way, for instance when one must deal simultaneously with the random walks on $\boldsymbol{G}$ and on $\boldsymbol{M}$.

In the preceding discussion one could not use the right random walk instead of the left random walk unless $\boldsymbol{G}$ operates on $\boldsymbol{M}$ on the right. We shall, however, show that under an additional hypothesis one may associate with right random walks some interesting random walks on (left) homogeneous spaces of $\boldsymbol{G}$. We begin with a result of more general scope.

Let X be a Markov chain on $(E, \mathscr{E})$ and σ a measurable mapping from $(E, \mathscr{E})$ onto a space $(E', \mathscr{E}')$ such that, for every $A' \in \mathscr{E}'$,

$$P(x, \sigma^{-1}(A')) = P(x', \sigma^{-1}(A') \quad \text{if } \sigma(x) = \sigma(x'), \tag{4.1}$$

and moreover such that for $A \in \mathscr{E}$, $\sigma(A) \in \mathscr{E}'$. We set

$$X'_n = \sigma(X_n), \qquad P'(x', A) = P(\sigma^{-1}(x'), \sigma^{-1}(A)),$$

where $\sigma^{-1}(x')$ is any point in $\sigma^{-1}(\{x'\})$. Thanks to the second property of σ, it may be checked that P' is a T.P. on E', and we have

Proposition 4.6. *The sequence* $\{X'_n\}_{n\geqslant 0}$ *is a homogeneous Markov chain on* $(E', \mathscr{E}')$ *with transition probability* P'.

Proof. Let $A' \in \mathscr{E}'$ and P_ν be a probability measure on the space Ω of the chain X; we have

$$P_\nu[X'_{m+n} \in A' \mid \mathscr{F}_n] = P_\nu[\sigma(X_{m+n}) \in A' \mid \mathscr{F}_n] = P_\nu[X_{m+n} \in \sigma^{-1}(A') \mid \mathscr{F}_n]$$
$$= P_n[X_m, \sigma^{-1}(A')] = P'_n[X'_m, A'],$$

since it is easily seen from eq. (4.1) that

$$P'_n(x', A') = P_n(\sigma^{-1}(x'), \sigma^{-1}(A')).$$

Remark and Examples 4.7. Here also the chain X'_n is not the canonical chain associated with P', since the random variables X'_n are defined on the space Ω of the chain X. This proposition allows us to construct new chains from already known ones. Let us call *symmetric* a random walk such that $\mu = \hat{\mu}$ where $\hat{\mu}$ is the image of μ by the mapping $x \to x^{-1}$. Let X be a symmetric random walk on $\mathbf{R}$ or $\mathbf{Z}$. Then we may apply the above result with the map $\sigma : x \to |x|$. We then say that X' is obtained from X by reflection at the point 0.

Now let $\boldsymbol{G}$ be a group and $\boldsymbol{K}$ a compact sub-group of $\boldsymbol{G}$; let σ denote the canonical, continuous and open mapping $\boldsymbol{G} \to \boldsymbol{G}/\boldsymbol{K}$. If $x \in \boldsymbol{G}/\boldsymbol{K}$, $\sigma^{-1}(x)$ is of the form $g\boldsymbol{K}$ and the T.P. of the right random walk of law μ on $\boldsymbol{G}$ satisfies the conditions of Proposition 4.6 if μ is $\boldsymbol{K}$-invariant, that is to say that $\varepsilon_k * \mu = \mu$ for every $k \in \boldsymbol{K}$ or equivalently that $\mu = m_{\boldsymbol{K}} * \mu'$, where $m_{\boldsymbol{K}}$ is the normalized Haar measure for $\boldsymbol{K}$ and μ' any probability measure on $\boldsymbol{G}$. The image of the right random walk on $\boldsymbol{G}$ by σ is then a homogeneous Markov chain on $\boldsymbol{G}/\boldsymbol{K}$. When $\boldsymbol{G}$ is a semi-simple Lie group with finite centre and $\boldsymbol{K}$ a maximal compact sub-group, we thus obtain random walks on the Riemannian symmetric space $\boldsymbol{G}/\boldsymbol{K}$.

We could have taken the image of the left random walk by σ, but the well-known equivariance properties of σ would then imply that

$$\sigma(Z_n Z_{n-1} \cdots Z_1 X_0) = Z_n Z_{n-1} \cdots Z_1 \sigma(X_0)$$

and we would be in the general situation of Proposition 4.4.

Exercise 4.8. A *sub-random walk* is a Markov chain on a group $\boldsymbol{G}$ with T.P. $\varepsilon_x * \mu$ (or $\mu * \varepsilon_x$) but with $\mu(\boldsymbol{G}) < 1$.

Prove, using the same notation as in Proposition 4.2, that the random variables of any finite collection $Z_{n_1}, \ldots, Z_{n_k}$ are independent on the set $\{\zeta > n_k\}$.

Exercise 4.9. (1) Let X be a right random walk of law μ and T a stopping time. Prove, after restricting the probability space to $\Omega_T = \{T < \infty\}$, that the variables $X_T^{-1} X_{T+n}$ are independent of $\mathscr{F}_T$ and that $Y = X_{T+n}$ is still a right random walk of law μ.

[*Hint*: See Exercise **3.14.**]

(2) We assume that the underlying group is **R** or **Z** and we define inductively the following sequence of random variables

$$T_0(\omega) = 0,$$

$$T_1(\omega) = \inf\{n \geqslant 1 : X_n(\omega) > X_0(\omega)\},$$

$$T_p(\omega) = \inf\{n > T_{p-1} : X_n(\omega) > X_{T_{p-1}}(\omega)\},$$

where the infinum of the empty set is taken equal to $+\infty$. Prove that the variables T_p are stopping times and that the sequences $\{T_n\}$ and $\{X_{T_n}\}$ are sub-random walks. Under which condition are they true random walks?

Exercise 4.10. Let X be a random walk of law μ on $\boldsymbol{G}$ and T_n the nth return time (cf. Exercise **3.13**) to a closed subgroup H. Prove that $Y = \{X_{T_n}\}$ is a sub-random walk on $\boldsymbol{H}$. Give its law as a function of μ and $1_{\boldsymbol{H}}$. Under which condition is it a true random walk?

Exercise 4.11. Let $\boldsymbol{G}$ be a LCCB group which is the semi-direct product of two groups $\boldsymbol{H}$ and $\boldsymbol{K}$. Every element g in $\boldsymbol{G}$ may be written uniquely as a pair (h, k), and using the additive notation in $\boldsymbol{K}$ we have $(h, k)(h', k') = (hh', hk' + k)$. (Example: $\boldsymbol{G}$ is the group of rigid motions of the euclidean plane, $\boldsymbol{H}$ is the group of rotations and $\boldsymbol{K}$ the group of translations.) Let $X_n = (H_n, K_n)$ be a left random walk on $\boldsymbol{G}$; then H_n is a random walk on $\boldsymbol{H}$ and K_n the Markov chain induced (cf. Proposition 4.4) by X_n on $\boldsymbol{K}$. (We recall that $\boldsymbol{K}$ is an invariant subgroup of $\boldsymbol{G}$ and therefore that $\boldsymbol{G}$ operates on $\boldsymbol{K}$ in a natural way.)

Exercise 4.12. One can study the random walks on the spaces $(W, \mathscr{A}, P)$ equal to the infinite product $(\boldsymbol{G}, \mathscr{G}, \mu)^{\mathbf{N}}$. We let $\{Z_n\}_{n\geq 0}$ be the coordinate mappings, which are clearly independent and equidistributed.

(1) The probability measure P_g on $(\Omega, \mathscr{F})$ is the image of P by the mapping from W to Ω

$$(g_1, g_2, \ldots, g_n, \ldots) \to (g, gg_1, gg_1g_2, \ldots, gg_1, g_n, \ldots)$$

(2) Prove that P is invariant under finite permutations of the coordinates in W.

(3) An event $A \in \mathscr{A}$ is said to be symmetric if it is invariant under finite permutations of coordinates. Prove that the family of symmetric events is a σ-algebra and that if A is symmetric then either $P(A) = 0$ or $P(A) = 1$. This is the so-called *zero-or-one law for symmetric events.* Finally prove that the events in $\bigcap_{n=1}^{\infty} \sigma(Z_m, m \geqslant n)$ are symmetric.

[*Hint*: To prove the zero-or-one law, approximate sets in $\mathscr{A}$ by rectangles C depending on the first n coordinates and use the permutation σ exchanging 1 and $n+1$, 2 and $n+2, \ldots, n$ and $2n$.

5. Analytical properties of integral kernels

This section should be omitted at first reading. Its purpose is to collect some results which will be useful later and which will then be referred to. Since these results deal mainly with compactness, we recall:

Definition 5.1. A linear and continuous operator from a Banach space into another Banach space is said to be *compact* if it maps bounded sets onto relatively compact ones. A kernel on $(E, \mathscr{E})$ is said to be *compact* if it maps the unit ball $\mathscr{U}$ of $b\mathscr{E}$ into a relatively compact set in $b\mathscr{E}$. A compact kernel is thus a bounded kernel.

Let us also recall that an operator with finite-dimensional range is compact and that every compact operator is the uniform limit (cf. ch. 6, Definition 3.1) of operators with finite dimensional range. Finally the adjoint of a compact operator is itself compact. We characterize below the compact kernels.

Proposition 5.2. *If* $(E, \mathscr{E})$ *is separable, a compact kernel is an integral kernel.*

Proof. Let N be a compact kernel; its adjoint N^* is a compact endomorphism of $b\mathscr{E}^*$. But since $b\mathscr{M}(\mathscr{E})$ is closed in $b\mathscr{E}^*$ and invariant by N, the operator N^*, which is equal to N operating on the left, is a compact endomorphism of $b\mathscr{M}(\mathscr{E})$. As a result the set $\{N(x, \cdot): x \in E\}$ is relatively compact in $b\mathscr{M}(\mathscr{E})$, and consequently there is a probability measure λ on $\mathscr{E}$ such that for every x in E one has $N(x, \cdot) \ll \lambda$. The proposition then follows from the following lemma.

Lemma 5.3. *If* $(E, \mathscr{E})$ *is separable, for every probability measure* λ *on* $\mathscr{E}$ *and every proper kernel* N *on* $(E, \mathscr{E})$ *there exists a function* $(x, y) \to n(x, y)$ *which is* $\mathscr{E} \otimes \mathscr{E}$*-measurable and such that* N *is the sum of the integral kernel with basis* λ *associated with* n *and of a kernel* N^1 *such that all measures* $N^1(x, \cdot)$ *are singular with respect to* λ.

The function n is called a *density* of N with respect to λ. It is unique in the sense that if n' is another function with the same properties, then for every x, we have $n(x, \cdot) = n'(x, \cdot)$ λ-a.s. Actually, what the lemma says is that one can choose $n(x, \cdot)$ within the equivalence class of the Radon–Nikodym derivative $dN(x, \cdot)/d\lambda$ in such a way that the resulting bivariate function n is $\mathscr{E} \otimes \mathscr{E}$-measurable.

Proof. By hypothesis there exists a sequence of finite partitions $\mathscr{P}^n$ of E such that $\mathscr{P}^{n+1}$ is a refinement of $\mathscr{P}^n$ and that $\mathscr{E}$ is generated by $\bigcup_0^\infty \mathscr{P}^n$. A point y in E belongs to one and only one set in $\mathscr{P}^n$, which we call E_y^n. Our notation will not distinguish between $\mathscr{P}^n$ and $\sigma(\mathscr{P}^n)$.

Let ν be a probability measure on $\mathscr{E}$; the functions f_n defined by setting

$$f_n(y) = \nu(E_y^n)/\lambda(E_y^n) \quad \text{if } \lambda(E_y^n) > 0,$$

$$f_n(y) = 0 \qquad\qquad\qquad \text{if } \lambda(E_y^n) = 0,$$

form a positive martingale relative to the σ-algebras $\mathscr{P}^n$ and the probability measure λ, hence converge λ-a.s. to a limit f_∞ as n tends to $+\infty$. For every $A \in \bigcup_0^\infty \mathscr{P}^n$ we have by Fatou's lemma

$$\int_A f_\infty \, d\lambda \leqslant \underline{\lim} \int_A f_n \, d\lambda \leqslant \nu(A),$$

and since any set in $\mathscr{E}$ is the limit $(\lambda + \nu)$-a.s. of sets in $\bigcup_0^\infty \mathscr{P}^n$, this inequality holds for all sets in $\mathscr{E}$. Let f be a function such that

$$\int_A f \, d\lambda \leqslant \nu(A), \quad A \in \mathscr{E};$$

then, as is easily seen, $E[f \mid \mathscr{P}^n] \leqslant f_n$ λ-a.s. and therefore by passing to the limit we get $f \leqslant f_\infty$ λ-a.s. The function f_∞ being thus the largest (up to equivalence) function with this property, is equal to the Radon–Nikodym derivative of ν with respect to λ.

Let now N be a bounded kernel; we apply the above discussion to the probability measure $P(x, \cdot) = N(x, \cdot)/N(x, E)$. The functions f_n defined by setting

$$f_n(x, y) = P(x, E_y^n)/\lambda(E_y^n) \quad \text{if } \lambda(E_y^n) > 0, \qquad f_n(x, y) = 0 \quad \text{if } \lambda(E_y^n) = 0,$$

are $\mathscr{E} \otimes \mathscr{E}$-measurable; their superior limit as n goes to infinity is $\mathscr{E} \otimes \mathscr{E}$-measurable and provides the desired function n. If N is merely proper, we begin by multiplying it on the right by a strictly positive function h such that Nh is bounded (see ch. 2, Proposition 1.14). The property may still be extended to some non-positive kernels.

Corollary 5.4. *If $(E, \mathscr{E})$ is separable, then every proper kernel N such that $N(x, \cdot) \ll \lambda$ for every x in E is an integral kernel.*

The following results will go in the converse direction. We start with a preliminary result which is a generalization of Egoroff's theorem.

Proposition 5.5. *Let H be a set of finite functions in $\mathscr{E}$, compact and metrizable for the topology of pointwise convergence; for every $\varepsilon > 0$ there exists a set $F \in \mathscr{E}$ such that $\lambda(F^c) < \varepsilon$ and $I_F(H)$ is compact for the topology of uniform convergence.*

Proof. By hypothesis there exists a sequence $\{x_n\}_{n \geqslant 0}$ of points in E which separates H and a countable subset H_1 of H which is dense in H for the topology of pointwise convergence. We set

$$h_n = \sup\{(f - g) : f, g \in H \quad \text{and } |f(x_k) - g(x_k)| < 2^{-n} \text{ for } k \leqslant n\}.$$

Since H_1 is dense in H, the supremum may be obtained by considering only the functions of H_1; as a result h_n is in $\mathscr{E}$. Each function h_n is finite and moreover $h_{n+1} \leqslant h_n$ and $\inf_n h_n = 0$.

For every sequence $\{g_p\} \subset H$ which converges pointwise to $g \in H$, we have the following property: for every n there exists a p_0 such that $p \geqslant p_0$ implies $|g - g_p| \leqslant h_n$. It follows that for a set $F \in \mathscr{E}$, such that $\{1_F h_n\}$ converges uniformly to zero, the set $1_F H$ is compact for the topology of uniform convergence. The desired result then follows from Egoroff's theorem.

Theorem 5.6. *Let N be an integral kernel with basis λ on $(E, \mathscr{E})$ and such that $N1$ is finite; then there exists an increasing sequence of sets A_n in $\mathscr{E}$ such that:*

(i) *for every $\varepsilon > 0$ there is an integer n such that $\lambda(A_n^c) < \varepsilon$;*

(ii) *the kernels $I_{A_n} N$ are compact.*

Proof. The image $N(\mathscr{U})$ is equal to the image by N of the unit ball in $L^\infty(\lambda)$, which is compact and metrizable for the weak* topology $\sigma(L^\infty(\lambda), L^1(\lambda))$. Since N is a continuous operator with respect to this topology and the topology of pointwise convergence, the set $N(\mathscr{U})$ is compact and metrizable for the topology of pointwise convergence. By Proposition 5.5 there exists a sequence $\{A_n\}$ of sets in $\mathscr{E}$ with the required properties and such that the sets $1_{A_n} N(\mathscr{U})$ are compact for the topology of uniform convergence – hence such that the operators $I_{A_n} N$ are compact.

The preceding results are useful in many situations. In this book we shall use them to prove quasi-compactness properties. We are now going to state in a topological setting some properties which may be used to the same end.

We assume below that E is an LCCB space and $\mathscr{E}$ the σ-algebra of Borel sets.

Definition 5.7. A sub-markovian kernel N on $(E, \mathscr{E})$ is said to be

(i) *Feller* if the map $x \to \varepsilon_x N$ from E to $b\mathscr{M}(\mathscr{E})$ is continuous for the strict topology on $b\mathscr{M}(\mathscr{E})$, in other words if $Nf \in C(E)$ whenever $f \in C(E)$;

(ii) *strong Feller* if the same map is continuous for the weak-star topology $\sigma(b\mathscr{M}(\mathscr{E}), b\mathscr{E})$, in other words if $Nf \in C(E)$ whenever $f \in b\mathscr{E}$;

(iii) *strong Feller in the strict sense* if the same map is continuous for the norm topology on $b\mathscr{M}(\mathscr{E})$.

Plainly each of these conditions is more stringent than the previous one. The convolution kernels provide examples of Feller kernels, and even of strong Feller kernels, whenever the relevant measure is absolutely continuous. Examples of kernels strongly Feller in the strict sense will be obtained as by-products of the following results.

Proposition 5.8. *The following two conditions are equivalent:*
(i) *the kernel N is strong Feller in the strict sense;*
(ii) *the image by N of bounded sets of $C(E)$ are compact for the topology of uniform convergence on compact sets.*

Proof. If f is in the unit ball of $C(E)$ then for every pair (x, y) in $E \times E$,

$$|Nf(x) - Nf(y)| \leqslant \|N(x, \cdot) - N(y, \cdot)\|;$$

thus functions Nf, f in the unit ball of $C(E)$, are equi-uniformly continuous on compact sets and (ii) follows from (i).

The converse follows at once from the relation

$$\|N(x, \cdot) - N(y, \cdot)\| = \sup\{|Nf(x) - Nf(y)| : f \in C(E), \|f\| \leqslant 1\}$$

and the equi-continuity on compact sets of the functions $Nf (f \in C(E), \|f\| < 1)$.

Remark. The image by N of the bounded sets of $b\mathscr{E}$ are also compact for the topology of uniform convergence on compact sets.

Theorem 5.9. *The product of two strong Feller kernels is strong Feller in the strict sense.*

Proof. The theorem follows immediately from the preceding proposition and the next two lemmas, in which N is a strong Feller kernel.

Lemma 5.10. *From every sequence $\{g_n\}$ of functions in the unit ball $\mathscr{U}$ of $b\mathscr{E}$, one can extract a sub-sequence $\{g_n'\}$ such that $\{Ng_n'\}$ is pointwise convergent.*

Proof. Let $\{x_n\}$ be a countable dense subset of E, and set $\lambda = \sum_n 2^{-n} N(x_n, \cdot)$. The measures $N(x, \cdot)$ are all absolutely continuous with respect to λ. Indeed if f is λ-negligible, the continuous function Nf vanishes at all points x_n, hence everywhere.

Let now $\{g_n'\}$ be a subsequence of $\{g_n\}$ convergent in the sense of $\sigma(L^\infty(\lambda), L^1(\lambda))$. (We indulge in the usual confusion between functions and their equivalence classes.) Then the sequence $\{Ng_n'\}$ converges pointwise.

Lemma 5.11. *Let $\{g_n\}$ be a sequence of functions in $\mathscr{U}$ converging pointwise to a function g; then the sequence $\{Ng_n\}$ converges to Ng uniformly on every compact set.*

Proof. It suffices to prove the lemma for the case $g = 0$. Set $h_n = \sup_{m \geqslant n} |g_n|$; as $|Ng_n| \leqslant Nh_n$ it suffices to show that $\{Nh_n\}$ converges to zero uniformly on compact sets. Since the functions Nh_n are continuous and decrease to zero we conclude the proof by applying Dini's lemma.

As an application, let us give the following result. The reader may refer to ch. 2 § 6 for the definition of a resolvent.

Proposition 5.12. *If the kernels $\{V_\alpha\}_{\alpha>0}$ of a submarkovian resolvent are strong Feller, then they are strong Feller in the strict sense.*

Proof. Let $\beta > \alpha$; the map $x \to \varepsilon_x V_\alpha V_\beta$ is continuous for the norm topology by Theorem 5.9, and the maps $x \to \varepsilon_x V_\beta$ converge uniformly to zero whenever $\beta \to \infty$, since $||\varepsilon_x V_\beta|| < \beta^{-1}$. The result thus follows from the relation

$$V_\alpha = (\beta - \alpha) V_\alpha V_\beta + V_\beta.$$

Exercise 5.13. If E is an LCCB space, $N1$ is continuous, and $N(C_K(E)) \subset C(E)$, then N is Feller.

Exercise 5.14. If $\mathbf{G}$ is a group, prove that the convolution kernels $\varepsilon_x * \mu$ on $\mathbf{G}$ are Feller. If μ is absolutely continuous with respect to the Haar measure, then $\varepsilon_x * \mu$ is strong Feller. Prove by an example that it is not always strong Feller in the strict sense.

Exercise 5.15. Let M and N be two kernels on the separable space $(E, \mathscr{E})$; prove that there exists an $\mathscr{E} \otimes \mathscr{E}$-measurable function f and a kernel N^1 such that for every x in E, the measure $N^1(x, \cdot)$ is singular with respect to $M(x, \cdot)$ and

$$N(x, dy) = f(x, y) M(x, dy) + N^1(x, dy).$$

Exercise 5.16. Let N be a bounded kernel taking both positive and negative values. Prove that the map N^+ defined by $N^+(x, \cdot) = (N(x, \cdot))^+$, is also a kernel. As a result, if we set $\alpha(x) = ||N(x, \cdot)||$ the function α is $\mathscr{E}$-measurable.
[*Hint*: See ch. 6 §2.]

Exercise 5.17. Prove the following extension to Lemma 5.3. If there is a probability measure ν and a family of sets E_n increasing to E, such that

$N(\cdot\,, E_n) < \infty$ ν-a.s. for every n, then there exist a function n and a kernel N^1 such that

$$N(x, \cdot\,) = n(x, \cdot\,)\,\lambda + N^1(x, \cdot\,)$$

for ν-almost every x.

CHAPTER 2

POTENTIAL THEORY

Except in a short section devoted to resolvents, our basic data will be a transition probability P on $(E, \mathscr{E})$ and the associated canonical Markov chain.

1. Superharmonic functions and the maximum principle

Definition 1.1. The Poisson equation with second member g is the equation

$$(I - P) f = g.$$

A finite function of $\mathscr{E}$ will be called *harmonic* if it is a solution of the Poisson equation without second member ($g = 0$). A function f in $\mathscr{E}_+$ is called *superharmonic* if $f \geqslant Pf$ everywhere on E.

The positive harmonic functions are superharmonic. The function 1 is always superharmonic; it is harmonic if P is markovian. If X is a right random walk of law μ on a group $\boldsymbol{G}$, the function f is harmonic if and only if

$$f(g) = \int_{\boldsymbol{G}} f(gh)\, \mu(\mathrm{d}h).$$

Other examples will be found in the exercises. Let us mention without proof that the harmonic functions for random walks on symmetric spaces are harmonic in the classical sense, that is they are solutions of Laplace equations on these spaces. In particular, the classical harmonic functions in the unit disc are the harmonic functions of suitable Markov chains.

Proposition 1.2. *The set of superharmonic functions is a left lattice and a convex cone. If $\{f_n\}_{n \geqslant 0}$ is a sequence of superharmonic functions, then $\underline{\lim}_n f_n$ is superharmonic.*

The set of harmonic functions is a vector space. The set of bounded harmonic functions is a Banach subspace of $\mathrm{b}\mathscr{E}$.

Proof. Given that f, g are superharmonic and a, b are in $\mathbf{R}_+$, the reader can easily check that $af + bg$ and $f \wedge g$ are superharmonic. The other properties in the statement are also straightforward.

The subject of potential theory, which we are about to develop in this and the next section, is the study of the cone of superharmonic functions, which will be henceforth denoted by $\mathscr{S}$.

Definition 1.3. The *potential kernel* of P, or of X, is the kernel

$$G = \sum_{n=0}^{\infty} P_n = I + P + P_2 + \cdots .$$

The reader can easily check that G is indeed a kernel, but will observe that the measures $G(x, \cdot\,)$ are not always σ-finite. (Take $P = I$ for instance, but less trivial examples will be seen later on.)

We recall (ch. 1, Definition 1.1) that the kernel G is said to be *proper* if E is the increasing limit of a sequence E_n of sets in $\mathscr{E}$ such that the functions $G(\cdot, E_n)$ are bounded. The probabilistic significance of this condition will be seen in due course.

The kernel G satisfies the identity

$$G = I + PG = I + GP. \tag{1.1}$$

On the other hand, using ch. 1, eq. (2.3), we may write, for $A \in \mathscr{E}$,

$$G(x, A) = E_x\left[\sum_{n=0}^{\infty} 1_{\{X_n \in A\}}\right]. \tag{1.2}$$

Let f be in $\mathscr{E}$; if the function Gf is defined it will be called the *potential of the function* f. If $f = 1_A$, then $Gf(x) = G(x, A)$ is called the potential of the set A, and by eq. (1.2) appears as the mean number of times the chain started at x hits the set A. One sees easily that the potentials of positive functions are superharmonic and we have

Theorem 1.4 (Riesz decomposition theorem). *A finite superharmonic function f has a unique representation as a sum $f = Gg + h$, where Gg is a potential and h a harmonic function.*

Proof. By hypothesis the sequence of function $P_n f$ decreases, and thus admits a positive limit $h = \lim_n P_n f$. We have

$$Ph = P\left[\lim_n P_n f\right] = \lim_n P_{n+1} f = h;$$

hence h is harmonic.

Next, $g = f - Pf$ is a positive function, and

$$\sum_0^N P_n g = f - P_{N+1} f;$$

passing to the limit as N tends to infinity yields $Gg = f - h$, which is the desired representation.

Suppose now that f has another decomposition of the same kind for another pair, say (g', h'). Applying P to both sides of the equality $Gg + h = Gg' + h'$, we get $Gg - g + h = Gg' - g' + h'$; it follows that $g = g'$, hence that $h = h'$ and the proof is complete.

Remark 1.5. If $g \in \mathscr{E}_+$ and Gg is finite, then it is a solution of the Poisson equation

$$(I - P) f = g.$$

If f is another finite solution, it is a superharmonic function which can be written $f = Gg + h$, where $h = \lim_n P_n f$ is harmonic and positive. Thus Gg appears as the *smallest positive solution of the Poisson equation with second member g.*

Corollary 1.6. *A superharmonic function dominated by a finite potential is a potential. A positive harmonic function dominated by a finite potential vanishes identically.*

Proof. By the Riesz decomposition theorem it suffices to show the second sentence in the statement. Let h be harmonic and $h \leqslant Gg < \infty$; for every n we have $h = P_n h \leqslant P_n Gg$ and since this sequence converges to zero as n tends to infinity, the proof is complete.

Theorem 1.7. *If the kernel G is proper every superharmonic function is the increasing limit of a sequence of finite potentials.*

Proof. Let us resume the usual notation and set $g_n = G(n 1_{E_n})$; the sequence g_n is a sequence of finite potentials increasing everywhere to $+\infty$. If f is superharmonic, the functions $f \wedge g_n$ are finite potentials and increase to f.

The following two results will hint at the probabilistic significance of superharmonic functions.

Proposition 1.8. *A function f is superharmonic (bounded harmonic) if and only if the sequence $\{f(X_n)\}$ of random variables is a supermartingale (bounded martingale) with respect to the σ-algebras $\mathscr{F}_n$ for any probability measure P_ν.*

If $f = Gg$ is a finite potential, the supermartingale $\{Gg(X_n)\}$ converges to zero almost-surely.

Proof. For $m \geqslant n$ and f superharmonic, the Markov property implies that

$$E_\nu[f(X_m) \mid \mathscr{F}_n] = E_\nu[f(X_{m-n} \circ \theta_n) \mid \mathscr{F}_n]$$
$$= E_{X_n}[f(X_{m-n})] = P_{m-n} f(X_n) \leqslant f(X_n), \quad P_\nu\text{-a.s.}$$

Conversely, integrating the inequality $f(X_0) \geqslant E_x[f(X_1) \mid \mathscr{F}_0]$ with respect to P_x yields $f(x) \geqslant Pf(x)$.

Let now $Z = \underline{\lim}_n\, Gg(X_n)$, then $\lim_n Gg(X_n) = Z$ a.s. (ch. 0, Theorem 2.1). By Fatou's lemma, for every x in E, we have

$$E_x[Z] \leqslant \lim_n E_x[Gg(X_n)] = \lim_n P_n Gg(x) = 0.$$

The positive random variable Z is thus equal to zero almost-surely, which completes the proof.

One of the golden rules in the study of Markov processes is to replace, as much as possible, constant times by stopping times and other kinds of random times. Although this is not as important in the case of Markov chains we will stress the idea by proving the following characterization of superharmonic functions which is little else than the stopping theorem for supermartingales.

Proposition 1.9. *A function f is superharmonic if and only if $P_S f \geqslant P_T f$ for any pair (S, T) of stopping times such that $S \leqslant T$.*

Proof. The sufficiency is obvious by making $S = 0$ and $T = 1$. To prove the necessity, let us first consider two finite stopping times S and T such that $S \leqslant T$ and $T - S$ is at most equal to 1. Set $A_p = \{S = p\} \cap \{T = p + 1\}$; this event is clearly in $\mathscr{F}_p$ and by the supermartingale property of Proposition 1.8, for any $x \in E$ we have

$$\int_{A_p} f(X_S)\, dP_x = \int_{A_p} f(X_p)\, dP_x \geqslant \int_{A_p} f(X_{p+1})\, dP_x = \int_{A_p} f(X_T)\, dP_x.$$

If we set $B_p = \{S = T = p\}$ it follows that

$$P_S f(x) = \sum_{p=0}^{\infty}\left(\int_{A_p} f(X_S)\, dP_x + \int_{B_p} f(X_S)\, dP_x\right)$$

$$\geqslant \sum_{p=0}^{\infty}\left(\int_{A_p} f(X_T)\, dP_x + \int_{B_p} f(X_T)\, dP_x\right) = P_T f(x).$$

For S and T as in the statement we consequently have

$$P_S f(x) \geqslant P_{(S+1)\wedge T} f(x) \geqslant \cdots \geqslant P_{(S+n)\wedge T} f(x) \geqslant \cdots.$$

As n tends to infinity, $f(X_{(S+n)\wedge T})$ converges to $f(X_T)$ on $\{T < \infty\}$; since on $\{T = \infty\}$ we have $f(X_T) = 0$, it follows that

$$f(X_T) \leqslant \underline{\lim}\, f(X_{(S+n)\wedge T})$$

and by Fatou's lemma $P_T f(x) \leqslant \underline{\lim}\, P_{(S+n)\wedge T} f(x) \leqslant P_S f(x)$.

In ch. 1, §3 we saw how one can associate with X other Markov chains, in particular by killing X at time T_A for $A \in \mathscr{E}$. The T.P. of this new chain is $I_{A^c} P I_{A^c}$; we shall denote by G^A its potential kernel, so that we have

$$G^A = \sum_{n=0}^{\infty} I_{A^c}(P I_{A^c})^n = \sum_{n=0}^{\infty} (I_{A^c} P)^n I_{A^c}.$$

By reasoning in the same way as in ch. 1, Proposition 3.7 the reader can prove without difficulty that

$$G^A f(x) = E_x\left[\sum_{n<T_A} f(X_n)\right];$$

for $f = 1_B$, this is precisely the mean number of times that X hits B before hitting A.

The following result generalizes the formula $G = I + PG$. It could be proved by computations using the analytical expressions of G, G^A, P_A as functions of P and I_A; a simple way of doing this will be introduced in §2.

Proposition 1.10. *For any $A \in \mathscr{E}$,*

$$G = G^A + P_A G.$$

In particular, if g vanishes on A^c, then $P_A Gg = Gg$.

Proof. Let g be in $\mathscr{E}_+$; by applying the strong Markov property, we get

$$P_A G\, g(x) = E_x[G\, g(X_{T_A})] = E_x\left[\sum_{n=0}^{\infty} P_n\, g(X_{T_A})\right]$$

$$= E_x\left[E_{X_{T_A}}\left[\sum_{n=0}^{\infty} g(X_n)\right]\right] = E_x\left[\sum_{n=0}^{\infty} g(X_{n+T_A})\right]$$

$$= E_x\left[\sum_{n=T_A}^{\infty} g(X_n)\right].$$

Since

$$E_x\left[\sum_{n=0}^{\infty} g(X_n)\right] = E_x\left[\sum_{n<T_A} g(X_n)\right] + E_x\left[\sum_{n=T_A}^{\infty} g(X_n)\right],$$

we have the desired result.

The second sentence comes from the fact that $G^A 1_A = 0$.

We are now going to use the foregoing result to prove the very important

Theorem 1.11 (complete maximum principle). *Let f be such that Gf is well defined, g be in $\mathscr{E}_+$ and a be a constant $\geqslant 0$, such that*

$$Gf(x) \leqslant Gg(x) + a$$

for $x \in \{f > 0\}$. Then $Gf \leqslant Gg + a$ everywhere on E.

Proof. Set $A = \{f > 0\}$. If $Gf^+ \leqslant Gg + Gf^- + a$ on A, since the measures $P_A(x, \cdot\,)$ vanish outside A, we have

$$P_A Gf^+ \leqslant P_A Gg + P_A Gf^- + aP_A 1$$

everywhere on E, and by Proposition 1.10,

$$Gf^+ - G^A f^+ \leqslant Gg + Gf^- + a.$$

Since $G^A f^+ = 0$ by definition of G^A, the proof is complete. The reader may even prove as an exercise that one gets $Gf \leqslant Gg + a - f^-$ everywhere.

Corollary 1.12. *The kernel G satisfies the following reinforced maximum principle: for every $a \geqslant 0$ and every pair (f, g) of functions in $\mathscr{E}_+$ such that*

$$Gf(x) \leqslant a + Gg(x) - g(x) \quad \textit{for all } x \in \{f > 0\},$$

we have $Gf \leqslant a + Gg - g$ *everywhere.*

Proof. Since $Gg - g = GPg$, the corollary follows easily from the complete maximum principle.

For our later needs we shall lay down the following

Definition 1.13. A kernel G (not necessarily the potential kernel of a Markov chain) is said to satisfy the *complete maximum principle* (*reinforced maximum principle*) if it satisfies the condition of Theorem 1.11 (1.12).

If, in Theorem 1.12, we replace $Gg + a$ by Gg (by a) we say that G satisfies the *domination principle* (the *maximum principle*).

Clearly a kernel which satisfies the complete principle also satisfies the two last principles. We shall see later that a kernel which satisfies the reinforced principle satisfies the complete principle (see Exercise 2.15).

We now give a first application of the maximum principles by showing that for the potential kernel of a Markov chain one can replace "bounded" by "finite" in the definition of "proper".

Proposition 1.14. *A positive kernel N is proper if and only if there is a function f strictly positive everywhere such that Nf is bounded.*

The potential kernel G of a Markov chain is proper if and only if there is a function f strictly positive such that Gf is finite or equivalently if and only if E is the union of a sequence E_n of sets in $\mathscr{E}$ such that the potentials $G(\cdot\,, E_n)$ are finite.

Proof. For $f > 0$, the sets $E_n = \{f > 1/n\}$ increase to E and $N(\cdot\,, E_n) \leqslant nNf(\cdot)$. Conversely, if $N(\cdot, E_n) \leqslant M_n$, the function $f = \sum_1^\infty M_n^{-1} 2^{-n} 1_{E_n}$ is strictly positive and Nf is bounded.

In the same way, the existence of $f > 0$ with $Gf < \infty$ implies the existence of sets E_n with finite potentials $G(\cdot, E_n)$. Set

$$H_{nm} = \{x \in E_n : G(x, E_n) \leqslant m\};$$

the space E is the union of the sets H_{nm}, and since $H_{nm} \subset E_n$, we have

$$G(\cdot\,, H_{nm}) \leqslant G(\cdot\,, E_n) \leqslant m$$

on H_{nm} and therefore everywhere, by the maximum principle. The kernel G is thus proper, and by the first half of the proof there exists a function $f > 0$ such that Gf is bounded; this completes the proof.

Exercise 1.15. Prove that there is equality in Proposition 1.9, if f is bounded harmonic and S and T are finite.

Exercise 1.16. Let X be the chain of translation on integers ($P(x, \cdot) = \varepsilon_{x+1}(\cdot)$, $x \in \mathbf{Z}$). Compute the kernel G and characterize the harmonic functions and potentials.

Exercise 1.17. Extend the Riesz decomposition theorem to the case where the superharmonic function takes infinite values.

Exercise 1.18. A function h is said to be *harmonic outside* A if $f(x) = Pf(x)$ for all $x \in A^c$. Prove that Gg is harmonic outside $\{g > 0\}$.

Exercise 1.19. If $G(\cdot, E)$ is finite everywhere on E, then every bounded superharmonic function is a potential.

Exercise 1.20. A set $A \subset E$ is said to be *absorbing* if $P(x, A) = 1$ for all x in A. If Gg is a bounded potential, the set $\{Gg = 0\}$ is absorbing and the set $\{Gg > 0\}$ is the union of a sequence of sets with bounded potentials.

Exercise 1.21. *Principle of lower envelope.* The lower envelope of a sequence of finite potentials is a potential.

Exercise 1.22. (1) *Principle of uniqueness of charges.* If (f, g) are in $\mathscr{E}_+$ and $Gf = Gg < \infty$, then $f = g$.

(2) If f and g are two positive functions vanishing outside A and if $Gf = Gg < \infty$ on A, then $f = g$.

Exercise 1.23. Let p_m be the Poisson distribution of parameter m. We define inductively a sequence of random variables: X_1 is a random variable of law p_1, X_2 is a random variable of law $p_{X_1}, \ldots, X_n$ is a random variable of law $p_{X_{n-1}}, \ldots$. Prove that the point 0 is reached with probability 1.

[*Hint*: This is an example of a Galton–Watson chain (ch. 1 §2.20) for which the function $f(n) = n$ is harmonic.]

Exercise 1.24. Assume that E is the set of non-negative integers and that P is given by

$$P(x, x+1) = p_x, \qquad P(x, x-1) = q_x, \qquad P(x, x) = r_x,$$

with $q_0 = 0$, $p_0 > 0$, $p_x > 0$ and $q_x > 0$ for $x > 0$, and $p_x + q_x + r_x = 1$ for every x in E. This chain is called a *Birth and Death* chain. Find all the harmonic functions for this chain.

Exercise 1.25. If G satisfies the complete maximum principle, then for every positive real α the kernel $I + \alpha G$ satisfies the reinforced maximum principle.

Exercise 1.26. Let $f \in \mathscr{S}$ and set $E_f = \{0 < f < \infty\}$. Prove that, by setting

$$P^f(x, A) = \frac{1}{f(x)} \int_A P(x, dy)\, f(y),$$

one defines a transition probability on $(E_f, E_f \cap \mathscr{E})$. Prove that a function g is P^f-superharmonic if and only if there is a P-superharmonic function u such that $u = fg$.

Exercise 1.27. Let f, g be two superharmonic functions; we say that $f < g$ for the proper order of the cone $\mathscr{S}$ if $g - f \in \mathscr{S}$. Prove that $\mathscr{S}$ is a lattice for its proper order.

[*Hint*: Prove first that the least upper bound of two harmonic functions h and h' is

$$\text{l.u.b.}(h, h') = \lim_n P_n[h \vee h'];$$

then if $f = Gg + h$, $f' = Gg' + h'$, prove that

$$\text{l.u.b.}(f, f') = G[g \vee g'] + \text{l.u.b.}(h, h').]$$

2. Reduced functions and balayage

This section is devoted to the notion of reduced functions and to some applications. We give two different proofs of the balayage theorem, the first one to hint at the probabilistic aspect of the theorem, the second one of more analytical and general character. We introduce thereafter a family of kernels which will be very important in the sequel.

Theorem 2.1 (balayage theorem). *Let f be superharmonic and A a set in $\mathscr{E}$. The function $P_A f$ is the smallest superharmonic function which dominates f on A. It is called the reduced function of f on A.*

Proof. By Proposition 1.9 we have $P_A f \leqslant f$. Moreover the function $P_A f$ is superharmonic; indeed

$$PP_A f = P_{1+T_A \circ \theta_1} f = P_{S_A} f$$

and since $T_A \leqslant S_A$, we conclude again from Proposition 1.9.

Finally, let $h \in \mathscr{S}$ and $h \geqslant f$ on A; since $P_A(x, \cdot)$ vanishes outside A, we have

$$h \geqslant P_A h \geqslant P_A f.$$

The proof is thus complete.

Remark. If f is a potential, $P_A f$ is also a potential, by Corollary 1.6. It is the potential of a function vanishing outside A which is given in Exercise 2.11.

Let $g \in \mathscr{E}_+$ and define inductively a sequence $\{g_n\}_{n \geqslant 0}$ of functions in $\mathscr{E}_+$, by setting

$$g_0 = g, \qquad g_n = g_{n-1} \vee P g_{n-1}, \quad n \geqslant 1;$$

this sequence is clearly increasing, hence converges pointwise to a limit $Rg \geqslant 0$.

Theorem 2.2. *The function Rg is the smallest superharmonic function which dominates g. Moreover, $PRg = Rg$ on $\{g < Rg\}$ and $Rg = g$ if and only if g is superharmonic.*

Proof. Passing to the limit in the definition of g_n as a function of g_{n-1} yields that $Rg = Rg \vee PRg$, which proves that Rg is superharmonic. Obviously, $Rg \geqslant g$, and for $h \in \mathscr{S}$ with $h \geqslant g$, we have $h \geqslant Ph \geqslant Pg$; hence $h \geqslant g \vee Pg = g_1$, and, inductively, $h \geqslant g_n$ for every n. It follows that $h \geqslant Rg$, which gives the first conclusion in the statement.

We may also consider the sequence $\{g_n'\}$ defined inductively by

$$g_0' = g, \qquad g_n' = g \vee Pg_{n-1}', \quad n \geqslant 1;$$

this sequence also increases and thus converges to a limit $R'g \leqslant Rg$. Passing to the limit we get also $R'g = g \vee PR'g$; thus $R'g$ is superharmonic, and since it dominates g we get $Rg = R'g$. Consequently $Rg = g \vee PRg$, which implies that $PRg = Rg$ on $\{g < Rg\}$.

Remark. Theorem 2.1 is, in a sense, but a particular case of Theorem 2.2. Indeed, if $f \in \mathscr{S}$ and $A \in \mathscr{E}$, let us apply the second of the above schemes to $g = 1_A f$. We have

$$g \vee Pg = 1_A f \vee P\, 1_A f = I_{A^c} P\, I_A f + 1_A(f \vee P\, I_A f),$$

and since $PI_A f \leqslant Pf \leqslant f$, we have $g \vee Pg = I_{A^c} P\, I_A f + 1_A f$. It follows inductively that

$$g_n' = I_A f + \sum_{0<p\leqslant n} (I_{A^c} P)^p\, I_A f,$$

and consequently that $Rg = P_A f$. But in the particular setting of Theorem 2.1, we have a means of computing the reduced function, namely the kernel P_A, which has no counterpart in the general case. Moreover Theorem 2.1 stresses the probabilistic aspect of the reduced function.

The above results allow us to return to the Maximum Principle and give a characterization of superharmonic functions which will be useful in ch. 10. Another proof of the sufficiency is to be found in Exercise 6.12.

Theorem 2.3. *If G is proper, a function $f \in \mathscr{E}_+$ is superharmonic if and only if it satisfies the following property: for any function g in $\mathscr{E}$ such that Gg is well defined, the inequality $Gg \leqslant f$ holds everywhere whenever it holds on $\{g > 0\}$.*

Proof. Since $P_A f \leqslant f$ for $f \in \mathscr{S}$ and $A \in \mathscr{E}$, the necessity is proved in exactly the same way as was proved Theorem 1.11.

Conversely, begin with a function f majorized by a finite potential; then the reduced function Rf is also majorized by this potential; hence it is a potential, namely, $Rf = Gk$, where $k = Rf - P(Rf)$. But since $Rf = P(Rf)$ on $\{f < Rf\}$, we have $f = Gk$ on $\{k > 0\}$; hence, by the maximum property, $f \geqslant Gk$ everywhere. It follows that $f = Rf$; that is, f is superharmonic.

Next suppose f arbitrary and let Gg be a finite potential. The function $f \wedge Gg$ is majorized by a finite potential and satisfies the property of the statement because f and Gg do so. Thus $f \wedge Gg$ is superharmonic. The kernel G being proper, we may apply this to the functions $f \wedge Gg_n$, where Gg_n is a sequence of finite potentials converging everywhere to $+\infty$.

In the course of the proof of the balayage theorem, we saw that for $f \in \mathscr{S}$ the function $P_A f$ is superharmonic and also that $PP_A f = P_{S_A} f \leqslant f$, in other words, that superharmonic functions are also superharmonic for the chain induced on A (ch. 1, Exercise 3.13). We are going to generalize this result, but we must first introduce a family of kernels which will be very useful in several parts of this book. We recall that the unit ball of $b\mathscr{E}$ is denoted by $\mathscr{U}$: thus $\mathscr{U}_+$ is the set of measurable functions h such that $0 \leqslant h \leqslant 1$.

Definition 2.4. With each function h of $\mathscr{U}_+$ we associate the kernel

$$U_h = \sum_{n\geqslant 0} (PI_{1-h})^n P = \sum_{n\geqslant 0} P(I_{1-h}P)^n$$

where the kernel I_f is defined by $I_f g = fg$ or $I_f(x, \cdot) = f(x)\varepsilon_x$. When $h = 1_A$ we write U_A instead of U_h, and if $h = c \in [0, 1]$ we write U_c instead of U_h.

By reasoning in a way similar to ch. 1, Proposition 3.7, it is easily seen that, for $f \in \mathscr{E}_+$,

$$U_h f(x) = E_x\left[\sum_{n\geqslant 1} (1 - h(X_1))\,(1 - h(X_2)) \cdots (1 - h(X_{n-1}))\, f(X_n)\right],$$

$$U_A f(x) = E_x\left[\sum_{n=1}^{S_A} f(X_n)\right].$$

These kernels will be very important in the sequel, for the following reasons. On the one hand, all useful kernels may be easily expressed as functions of the U_h's; for instance

$$\Pi_A = I_A U_A I_A, \qquad P_A = I_A + I_{A^c} U_A I_A, \qquad P = U_1, \qquad G = I + U_0.$$

On the other hand, these kernels are very easy to handle, thanks to the following

Proposition 2.5 (resolvent equation). *For $h, k \in \mathscr{U}_+$ and $h \leqslant k$,*

$$U_h = \sum_{n\geqslant 0} (U_k I_{k-h})^n U_k = \sum_{n\geqslant 0} U_k (I_{k-h} U_k)^n,$$

which implies

$$U_h = U_k + U_h I_{k-h} U_k = U_k + U_k I_{k-h} U_h.$$

Proof. This is an immediate consequence of ch. 1, Exercise 1.18, which we solve rapidly below.

For every integer n,

$$(N(A + B))^n N = \sum_{\substack{k, n_0, \ldots, n_k \\ k + n_0 + \ldots + n_k = n}} ((NA)^{n_0} N)\, B\, ((NA)^{n_1} N)\, B \cdots B\, ((NA)^{n_k} N)$$

by summing over n and interchanging the order of summations, which is possible since all the kernels are positive, we obtain

$$\sum_{n\geqslant 0} (N(A+B))^n N = \sum_{k\geqslant 0} \sum_{n_0,\dots,n_k\in \mathbf{N}^{k+1}} ((NA)^{n_0}N)\, B\cdots B((NA)^{n_k}N)$$
$$= \sum_{k\geqslant 0} (SB)^k S.$$

Remark. This result says that one may compute U_h as a function of U_k in the same way as as a function of $P = U_1$. We notice also that $U_h \geqslant U_k$.

Proposition 2.6. *If $f \in \mathscr{S}$ and h, k are functions in $\mathscr{U}_+$ such that $h \leqslant k$, then*

$$U_h I_h f \leqslant U_k I_k f \leqslant f.$$

In particular $U_h(h) \leqslant 1$ for all h in $\mathscr{U}_+$ and $U_h(h)$ increases with h.

Proof. Let us first assume that $U_k I_k f \leqslant f$, which implies that

$$U_k I_h f + (U_k I_{k-h})\, U_k I_k f \leqslant U_k I_k f \leqslant f.$$

This is, for $p = 1$, what the inequality

$$\sum_{n<p} (U_k I_{k-h})^n\, U_k I_h f + (U_k I_{k-h})^p\, U_k I_k f \leqslant U_k I_k f \leqslant f \tag{2.1}$$

reduces to. Assume this inequality to be true up to p, and multiply through to the left by $U_k I_{k-h}$; then adding through $U_k I_h f$ yields

$$U_k I_h f + \sum_{n<p} (U_k I_{k-h})^{n+1}\, U_k I_h f + (U_k I_{k-h})^{p+1}\, U_k I_k f$$
$$\leqslant U_k I_h f + U_k I_{k-h} f = U_k I_k f \leqslant f.$$

Thus eq. (2.1) is true for all p. If we let p tend to infinity in eq. (2.1), it follows by the resolvent equation of Proposition 2.6, that

$$U_h I_h f \leqslant U_k I_k f \leqslant f.$$

Now if $k = 1$, then $U_k I_k f = Pf \leqslant f$, and by the preceding discussion $U_h I_h f \leqslant f$ for all $h \in \mathscr{U}_+$. It follows that the hypothesis with which we started the proof always holds, and consequently the proof is complete.

If we had started with f harmonic and positive, since the term $\lim (U_k I_{k-h})^p\, U_k I_k f$ is not bound to vanish (cf. below), we would still only get an inequality in the above result. In the same way $P_A f = f$ is not always true, as is seen from the example of the function 1, when $P1 = 1$ and $P_A 1$ is not equal to 1. We have, however,

Lemma 2.7. *If h is harmonic and positive, and if $U_A I_{A^c} h < \infty$, then $P_A h = P_{S_A} h = h$.*

Proof. The resolvent equation, $U_A = P + U_A I_{A^c} P$, yields

$$U_A h = h + U_A I_{A^c} h,$$

and by substracting $U_A I_{A^c} h$ from both members we get

$$P_{S_A} h = U_A I_A h = h,$$

and also $P_A h = I_A h + I_{A^c} U_A I_A h = h$.

We shall now use the reduced functions to give a characterization of potentials. We assume G to be proper.

Proposition 2.8. *A finite superharmonic function f is a potential if and only if $\lim_n P_{A_n} f = 0$ for every decreasing sequence $\{A_n\}_{n \geqslant 0}$ of sets in $\mathscr{E}$ such that $\bigcap A_n = \emptyset$.*

Proof. Let f be a potential; since $T_{A_n} \leqslant T_{A_{n+1}}$, Proposition 1.9 shows that the sequence $g_n = P_{A_n} f$ decreases; let g denote its limit. Since by Theorem 2.1, $Pg_n = g_n$ on A_n^c, an application of Lebesgue's theorem yields $Pg = g$. By Corollary 1.6, it follows that g vanishes identically.

Conversely, we suppose that f is not a potential and we call h the harmonic part of its Riesz decomposition. Let E_n be an increasing sequence of sets such that $E = \bigcup E_n$ and $G(x, E_n) < \infty$ for every integer n and each x in E. Then set $A_n^c = E_n \cap \{h \leqslant n\}$; we have $U_{A_n} I_{A_n^c} h \leqslant U_0 I_{A_n^c} h \leqslant n\, G\, 1_{E_n} < \infty$. Therefore $P_{A_n} h = h$ for every n, and it follows that $\lim_n P_{A_n} f \geqslant \lim_n P_{A_n} h = h \neq 0$ although $\bigcap A_n = \emptyset$.

Exercise 2.8. Let f be a finite superharmonic function. For every x in E, the set function L defined on $\mathscr{E}$ by $L(A) = P_A f(x)$ is strongly sub-additive, that is

$$L(A \cup B) + L(A \cap B) \leqslant L(A) + L(B).$$

[*Hint*: Use the results in ch. 1, Exercise 3.9.]

Exercise 2.9. Let f be superharmonic and finite and $\{S_n\}$ a sequence of stopping times decreasing to the stopping time S; prove that

$$P_S f = \lim_n P_{S_n} f.$$

In particular, if $\{A_n\}$ is an increasing sequence of sets with union A, then $\lim_n P_{A_n} 1 = P_A 1$. What can be said if $\{A_n\}$ is decreasing?

Exercise 2.10. (1) Let f be bounded from below and such that $Pf \leqslant f$ on the set A^c. Prove that the sequence $\{f(X_{T_A \wedge n})\}$ of random variables is a supermartingale for every P_ν and derive therefrom that $f \geqslant P_A f$.

[*Hint*: Use the stopped chain of ch. 1, Exercise 3.15. The inequality $f \geqslant P_A f$ may actually be shown directly.]

(2) Assume further that $P_A 1 = 1$ and prove the following Minimum principle:

$$f \geqslant \inf\{f(x),\ x \in A\}.$$

Exercise 2.11. For $A \in \mathscr{E}$, we define a kernel H_A by setting for $f \in \mathscr{E}_+$

$$H_A f(x) = 1_A(x)\, E_x \left[\sum_{n=0}^{S_A - 1} f(X_n) \right].$$

(1) Prove that

$$H_A = I_A \sum_{n \geqslant 0} (PI_{A^c})^n = I_A + I_A U_A I_{A^c}.$$

In particular $H_A f$ vanishes outside A.

(2) Prove that $GH_A = G - G^A = P_A G$ (cf. Proposition 1.10), and derive that if Gg is a potential, then $P_A Gg$ is the potential of the function $H_A g$ which vanishes outside A (the reader will see in the next section other interpretations of H_A).

Exercise 2.12 (Dirichlet's problem). Let $A \in \mathscr{E}$ and $f \in \mathscr{E}_+$; prove that $P_A f$ is the smallest function equal to f on A and harmonic outside A.

Exercise 2.13. For $h \in \mathscr{U}_+$, show that

$$U_h(h)\,(x) = 1 - E_x \left[\prod_{n=1}^{\infty} (1 - h(X_n)) \right].$$

Derive therefrom that $U_h(h) \leqslant 1$ and that $U_h(h)$ increases with h. If P is markovian and h is bounded away from zero, then $U_h(h) = 1$. This last result may fail to be true if one of the two conditions is not verified; give examples.

Exercise 2.14. For g in $\mathscr{E}_+$, prove that the sequence $\{Rg(X_n)\}$ is for every P_ν, the smallest supermartingale which majorizes the sequence $\{g(X_n)\}$.

Exercise 2.15. Let N be a kernel on $(E, \mathscr{E})$ satisfying the reinforced maximum principle and let $\mathscr{B}$ be the space of bounded functions f such that $N|f|$ is bounded.

(1) If $h \in \mathscr{B}$ is such that $Nh \geqslant 0$, then $Nh \geqslant h$. Derive that $Nf = Nf'$ with $f, f' \in \mathscr{B}$ implies $f = f'$.

(2) Prove that N satisfies the complete maximum principle for the functions in $\mathscr{B}_+$. (This exercise is solved in ch. 10.)

Exercise 2.16. Prove that the kernel U_0 satisfies the complete maximum principle if it is proper, and derive (cf. Exercise 1.25) that G satisfies the reinforced maximum principle.

Exercise 2.17. (1) Let L be a death time (ch. 1, Exercise 3.18). Prove that $\varphi(x) = P_x[L > 0]$ is superharmonic. Express its Riesz decomposition. Under what conditions is φ a potential? In particular if $L = L_A$, what condition must A satisfy in order that φ be a potential?

(2) For every probability measure P_ν on Ω, the sequence $\{Y_n\}_{n \geqslant 0}$ defined by

$$Y_n(\omega) = X_n(\omega) \quad \text{if } n < L(\omega),$$

$$Y_n(\omega) = \Delta \qquad \text{if } n \geqslant L(\omega),$$

is a homogeneous Markov chain with transition probability P^φ (notation of Exercise 1.26) and initial measure $\varphi\,\nu$. (Notice that this measure may have a total mass less than one.)

Exercise 2.18. Show that $\mathscr{S}$ is a lattice for the ordinary order.

[*Hint*: For the l.u.b. use reduced functions.]

3. Equilibrium, invariant events and transient sets

The following definition is basic.

Definition 3.1. An event $\Gamma \in \mathscr{F}$ is called *invariant* if $\theta^{-1}(\Gamma) = \Gamma$. The class of invariant events is a σ-algebra denoted $\mathscr{I}$. A random variable Z is $\mathscr{I}$-measurable if and only if $Z \circ \theta = Z$; such a random variable is called *invariant*. Two invariant random variables Z and Z' are said to be *equivalent* if $P_\nu[Z \neq Z'] = 0$ for every probability measure ν.

Proposition 3.2. *If P is markovian, the formula*

$$h(x) = E_x[Z]$$

sets up a one-to-one and onto correspondence between equivalence classes of bounded invariant random variables and bounded harmonic functions. Moreover

$$\lim_n h(X_n) = Z \quad a.s.$$

Proof. If Z is invariant, the function $E.[Z]$ is harmonic, as is seen by a straightforward application of the Markov property. Conversely, if h is bounded and harmonic, then $\{h(X_n)\}$ converges almost-surely to the random variable $Z = \overline{\lim}_{n\to\infty} h(X_n)$ which is obviously invariant. By Lebesgue's theorem we have

$$h(x) = E_x[Z]$$

for every x in E. The correspondence is clearly one-to-one.

If P is not markovian, the function $E.[Z]$ is superharmonic but not harmonic at points x such that $P_x[\zeta = 1] > 0$. The reader will look at Exercise 3.10 for what can be said in this case.

Definition 3.3. For $A \in \mathscr{E}$, set

$$R(A) = \left\{\omega : \overline{\lim_{n\to\infty}}\, 1_A(X_n(\omega)) = 1\right\} = \bigcap_{n=0}^{\infty} \{n + S_A \circ \theta_n < \infty\}.$$

The event $R(A)$ is the set of trajectories ω which hit A infinitely often; it is clearly an invariant event. We further set $h_A = P.[R(A)]$; this function is bounded and harmonic and $h_A(x)$ is the probability that the chain started at x hits A infinitely many times. This function has the following potential theoretic interpretation.

Definition 3.4. Let A be in $\mathscr{E}$; the superharmonic function $P_A 1$ is called the *equilibrium potential* of A.

This superharmonic function is actually a potential only in some "classical" cases, as is seen from the following result, which specifies its Riesz decomposition.

Proposition 3.5. *For any set A in $\mathscr{E}$, we have*

$$P_A 1 = G g_A + h_A,$$

where g_A is the probability that the chain never returns to A ($g_A(x) = 0$ for for $x \notin A$ and $g_A(x) = P_x[S_A = \infty]$ for $x \in A$). Moreover

$$\lim_n P_{X_n}[S_A < \infty] = \lim_n P_A 1(X_n) = \lim_n h_A(X_n) = 1_{R(A)} \quad a.s.$$

Proof. By definition, $P_A 1(x) = P_x[T_A < \infty]$. By the Riesz decomposition theorem, 1.6, we have $P_A 1(x) = G g_A(x) + h_A(x)$, where h_A is harmonic and equal to $\lim_n P_n P_A 1(x)$. But by ch. 1, Proposition 3.8,

$$P_n P_A 1(x) = P_x[n + T_A \circ \theta_n < \infty] = P_x\left[\bigcup_{m=n}^{\infty} \{X_m \in A\}\right].$$

Consequently,

$$h_A(x) = P_x\left[\bigcap_{n=0}^{\infty} \bigcup_{m=n}^{\infty} \{X_m \in A\}\right] = P_x\left[\overline{\lim_{n\to\infty}} \{X_n \in A\}\right],$$

which is the probability that X hits A infinitely many times.

By Theorem 1.4, we have also

$$g_A(x) = P_A 1(x) - P P_A 1(x) = P_x[T_A < \infty] - P_x[S_A < \infty].$$

By ch. 1, Proposition 3.8 it follows that g_A vanishes on A^c, and is equal to $1 - P_x[S_A < \infty] = P_x[S_A = \infty]$ on A.

The second sentence follows at once from Proposition 1.8 and Proposition 3.2 and the fact that $P_{X_n}[S_A < \infty] = P_{X_n}[T_A < \infty]$ a.s. for $n \geqslant 1$.

Definition 3.6. A set A in $\mathscr{E}$ will be called *transient* if $P_A 1$ is a potential, hence if $h_A \equiv 0$. A set A is called *recurrent* if $h_A \equiv 1$.

A subset of a transient set is a transient set and a finite union of transient sets is transient. Moreover we have the following characterization of transient sets in terms of potentials.

Proposition 3.7. *A set A is transient if and only if there exists a function g vanishing outside A and such that $Gg \geqslant 1$ on A and Gg is finite everywhere on E. In particular, if $G(\cdot, A)$ is everywhere finite then A is transient.*

Proof. The "only if" part is obvious by the foregoing result. For the "if" part, we observe that since Gg is finite we have $\lim Gg(X_n) = 0$ a.s. and therefore $\lim 1_A(X_n) = 0$ which is the desired result.

This result permits us to understand the probabilistic significance of the condition "G is proper"; it means that E is the union of an increasing sequence of transient sets.

It is essential to point out that a set may be neither recurrent nor transient. If P is not markovian no set is recurrent, not even the set E. If P is markovian we have the following result which describes an interesting situation occurring for large classes of random walks as well as for other chains studied later on.

Proposition 3.8. *If P is markovian, the following three statements are equivalent:*
(i) *The bounded harmonic functions are constant;*
(ii) *The σ-algebra $\mathscr{I}$ is trivial (up to equivalence);*
(iii) *Every set is either recurrent or transient.*

Proof. The equivalence of (i) and (ii) is an immediate consequence of Proposition 3.2. Since, in addition, $R(A)$ is in $\mathscr{I}$, these conditions imply (iii). It remains to show that (iii) implies (ii).

Let $\Gamma \in \mathscr{I}$ and put $A = \{x: P_x[\Gamma] > a\}$, where $a \in]0, 1[$. We know that $1_\Gamma = \lim_n P_{X_n}[\Gamma]$ a.s.; if A is recurrent, then $\Gamma = \Omega$ a.s., and if A is transient, then $\Gamma = \emptyset$ a.s.

Exercise 3.9. Prove that a set A is recurrent if and only if $U_A(1_A) = 1$ everywhere on E. Prove that if A is not transient, there are points x for which $h_A(x)$ is arbitrarily close to 1. Prove the even better result: if $h_A \leqslant a < 1$ on A, then $h_A \equiv 0$.

Exercise 3.10. (1) If $\Gamma = \{\zeta = \infty\}$, prove that $\Gamma \vartriangle \theta^{-1}(\Gamma) = \emptyset$ a.s.

(2) If P is not markovian there is a one-to-one correspondence between bounded harmonic functions h and equivalence classes of bounded invariant random variables Z which vanish on Γ^c. This correspondence is given by

$$Z = \lim_n h(X_n), \qquad h(x) = E_x[Z].$$

In particular if $\zeta < \infty$ a.s., the only bounded harmonic function is the zero function.

Exercise 3.11. (1) For $A \in \mathscr{E}$ and every integer n prove that

$$P_A - P_n P_A = \sum_{m<n} P_m(I_A - \Pi_A),$$

and derive that if A is transient then $P_A = G(I_A - \Pi_A)$.

(2) If f is superharmonic and A is transient, the superharmonic function $P_A f$ is a potential. If G is proper and $\{A_n\}$ is an increasing sequence of transient sets whose union is E, then $f = \lim_n P_{A_n} f$.

Exercise 3.12. (1) Pick $a \in]0, 1[$ and set $A = \{x: P_x[\Gamma] > a\}$, where Γ is an invariant event. Prove that $\Gamma = \lim\{X_n \in A\}$ a.s.

(2) A set $A \in \mathscr{E}$ is called *regular* if $\lim\{X_n \in A\}$ exists almost surely. Prove that the collection Σ of regular sets is a sub-algebra of $\mathscr{E}$. Transient sets are regular, and the class of transient sets is an ideal in Σ.

(3) Two regular sets A and A' are said to be equivalent if $A \triangle A'$ is transient. If A and A' are equivalent, then $P_x[R(A)] = P_x[R(A')]$ for all x in E.

[*Hint*: Prove that $R(A) \triangle R(A') \subset R(A \triangle A')$.]

(4) Prove that the Boolean algebras of equivalence classes of regular sets and invariant events are isomorphic.

[*Hint*: Associate with A the event $\lim\{X_n \in A\}$.]

(5) Prove that if $f \in b\mathscr{E}$ is such that $\lim f(X_n)$ exists a.s. if and only if f is the uniform limit of simple functions on Σ. This exercise is solved in ch. 7 §1.

Exercise 3.13. (1) Let m be a positive measure on $(E, \mathscr{E})$. The following two statements are equivalent:

(i) If $\Gamma \in \mathscr{I}$, then either $P_\nu[\Gamma] = 1$ for every $\nu \ll m$ or $P_\nu[\Gamma] = 0$ for every $\nu \ll m$;

(ii) If $A \in \mathscr{E}$, then either $P.[R(A)] = 1$ m-a.e. or $P.[R(A)] = 0$ m-a.e.

(2) These conditions imply:

(iii) the bounded harmonic functions are constant m-a.e. Prove by an example that the converse may fail to be true. Prove that the converse is true if m satisfies the following condition: if $m(A^c) = 0$, then $P_\nu[R(A)] = 1$ for all $\nu \ll m$.

4. Invariant and excessive measures

In the last three sections, P was viewed as operating on the right on functions. We are now going to let it operate on the left on measures and sketch a potential theory relative to measures. We thereafter introduce the important notion of Markov chains in duality.

Definition 4.1. A σ-finite measure m is said to be *invariant* by P, or P-invariant, if $mP = m$. We shall also say that m is invariant for the chain X, or simply invariant if there is no risk of confusion.

Examples. The left (right) Haar measure on a group $\boldsymbol{G}$ is invariant for the left (right) random walks on $\boldsymbol{G}$. If P is the T.P. defined in ch. 1, Example 1.4(iii), the measure m is P-invariant if and only if $\theta(m) = m$, namely if m is invariant with respect to θ, in which case θ is said to be a *measure-preserving* point transformation of $(E, \mathscr{E}, m)$.

The probabilistic significance of an invariant measure is the following: even if the positive measure m is unbounded, one can define a measure P_m on $(\Omega, \mathscr{F})$ by setting

$$P_m[\Gamma] = \int_E P_x[\Gamma]\, dm(x).$$

If the measure m is invariant, the chain X is then *stationary* for the measure P_m: for all integers $n_1, n_2, \ldots, n_k$ and n and sets $A_1, \ldots, A_k \in \mathscr{E}$,

$$P_m[X_{n_1} \in A_1, \ldots, X_{n_k} \in A_k] = P_m[X_{n+n_1} \in A_1, \ldots, X_{n+n_k} \in A_k],$$

which may also be expressed by saying that P_m is invariant with respect to the shift operator θ.

Definition 4.2. A positive σ-finite measure m is said to be *P-excessive* if $mP \leqslant m$. We also say that m is excessive for X, or simply excessive when there is no fear of mistake.

Let ν be a positive measure such that the measure νG is σ-finite; we then say that ν is a charge and νG is an excessive measure called the *potential measure* of the charge ν. If G is proper, all bounded measures are charges.

If m is excessive (invariant) then αm, where $\alpha \in \mathbf{R}_+$, is still excessive (invariant). In the sequel, when we state that an excessive (invariant) measure is unique, we mean unique up to multiplication. Moreover the sum of two excessive (invariant) measures is an excessive (invariant) measure; thus the set of excessive (invariant) measures is a convex cone. As for superharmonic functions, one can state a Riesz decomposition theorem.

Proposition 4.3. *Every excessive measure may be uniquely written as the sum of an invariant measure and of a potential measure.*

Proof. If $f \in b\mathscr{E}_+$, the sequence of numbers $mP_n(f)$ decreases to a limit $\nu(f)$ which defines an invariant measure, as is easily seen. The proof then follows the same pattern as for functions.

As in the case of functions, we shall see that the restriction to a set A of an excessive measure is excessive for the chain induced on A, namely $mP \leqslant m$ implies $(1_A m)\Pi_A \leqslant 1_A m$, or equivalently $mI_A U_A I_A \leqslant mI_A$. This is shown by the following.

Proposition 4.4. *If m is excessive and h, k are in $\mathscr{U}_+$ and such that $h \leqslant k$, then*

$$mI_h U_h \leqslant mI_k U_k \leqslant m.$$

The proof is the same as for Proposition 2.6 with right multiplication instead of left multiplications. As in the case of functions, if m is invariant, $1_A m$ is not bound to be Π_A-invariant (cf. Exercise 4.11). In the next chapter, we shall find sufficient conditions implying this result.

Let us now prove the result analogous to Theorem 2.1.

Theorem 4.5 (balayage theorem). *Let m be an excessive measure. The measure mH_A, where $H_A = I_A + I_A U_A I_{A^c}$, is the smallest excessive measure which dominates m on A.*

Proof. Clearly $mH_A = m$ on A, and by Proposition 4.4, $mH_A \leqslant m$; moreover, by the resolvent equation of Proposition 2.5, we have

$$\begin{aligned} mH_A P &= mI_A P + mI_A U_A I_{A^c} P \\ &= mI_A P + mI_A (U_A - P) = mI_A U_A. \end{aligned}$$

But Proposition 4.4 implies that

$$mI_A U_A = mI_A U_A I_A + mI_A U_A I_{A^c} \leqslant mI_A + mI_A U_A I_{A^c},$$

so that $mH_A P \leqslant mH_A$, which proves that mH_A is excessive.

Finally if m' is excessive and $m'I_A \geqslant mI_A$, then $m' \geqslant m'H_A \geqslant mH_A$ and the proof is completed.

The kernel H_A was introduced and studied in Exercise 2.11. It is called the *co-balayage operator* relative to A, because it acts with respect to measures as P_A acts with respect to functions, as one can see by comparing on the one hand Theorems 2.1 and 4.5 and on the other hand Exercises 2.11 and 4.12. This duality between the operators P_A and H_A appears in an even more striking form under the "duality hypothesis" that we are going to introduce now.

Definition 4.6. Two kernels N and $\hat{N}$ on $(E, \mathscr{E})$ are said to be *in duality* relative to the positive σ-finite measure m, if for every pair (f, g) of functions in $\mathscr{E}_+$,

$$\int_E Nf\, g\, \mathrm{d}m = \int_E f\, \hat{N}g\, \mathrm{d}m,$$

which will also be written $\langle Nf, g\rangle_m = \langle f, \hat{N}g\rangle_m$, or even $\langle Nf, g\rangle = \langle f, \hat{N}g\rangle$ when the relevant measure m cannot be mistaken.

Two Markov chains X and $\hat{X}$ are said to be in duality relative to m if their T.P.'s are in duality relative to m. One of the two chains is said to be the *dual* or *reversed* chain of the other one. The objects defined in terms of $\hat{X}$ will be designated by the symbol ˆ and the prefix "co-". For example, a function co-superharmonic is a function which is superharmonic for $\hat{P}$, the T.P. of $\hat{X}$, and the set of co-superharmonic functions is denoted by $\hat{\mathscr{S}}$.

Some easy remarks and examples are collected in Exercise 4.14. In particular, by making $g = 1$ ($f = 1$) in the definition it is seen that m is excessive (co-excessive). More generally, we must emphasize the duality relationship between superharmonic functions and co-excessive measures.

The reader will notice that, without any special hypothesis, there may be several T.P.'s in duality with a given T.P. Their relationship is settled in the following.

Lemma 4.7. *If $(E, \mathscr{E})$ is separable and if each of the two transition probabilities $\hat{P}$ and $\hat{P}'$ is in duality with P relative to m, then there is an m-negligible set N such that $\hat{P}(x, \cdot) = \hat{P}'(x, \cdot)$ for all x in N^c.*

Proof. The duality hypothesis implies at once that for every $f \in b\mathscr{E}_+$, we have $\hat{P}f = \hat{P}'f$ m-a.e. We can therefore find an m-negligible set N such that if x is in N^c, then $\hat{P}(x, A_n) = \hat{P}'(x, A_n)$ for every set A_n in a countable algebra generating $\mathscr{E}$, hence $\hat{P}(x, \cdot) = \hat{P}'(x, \cdot)$.

Definition 4.8. Any T.P. which is in duality with P relative to m is called a *modification* of $\hat{P}$.

Proposition 4.9. *Let X and $\hat{X}$ be in duality relative to m; then for every $h \in \mathscr{U}_+$ the kernels U_h and $\hat{U}_h$ are in duality relative to m.*

Proof. One must prove that, for $f, g \in b\mathscr{E}_+$,

$$\int_E \left(\sum_{n\geqslant 0} (PI_{1-h})^n Pf\right) g \, dm = \int_E f \left(\sum_{n\geqslant 0} (\hat{P}I_{1-h})^n \hat{P}g\right) dm,$$

but this follows from repeated applications of the formula

$$\int_E (PI_{1-h}) Pf \, g \, dm = \int_E f \, (\hat{P}I_{1-h}) \hat{P}g \, dm,$$

which in turn is an obvious consequence of the duality formula of Definition 4.6.

Corollary 4.10. *For $A \in \mathscr{E}$, the kernels H_A and $\hat{P}_A$ and the transition probabilities Π_A and $\hat{\Pi}_A$ are in duality.*

Proof. Since the operator I_A is its own dual, the corollary follows at once from Proposition 4.9 and the expression of the kernels as functions of U_A. For example,

$$\begin{aligned} \langle H_A f, g\rangle_m &= \langle I_A f, g\rangle_m + \langle I_A U_A I_{A^c} f, g\rangle_m \\ &= \langle f, I_A g\rangle_m + \langle f, I_{A^c} \hat{U}_A I_A g\rangle_m = \langle f, \hat{P}_A g\rangle_m. \end{aligned}$$

Exercise 4.11. Prove by an example that in Proposition 4.4, one may have $m\Pi_A < 1_A m$, even if m is invariant.

[*Hint*: One can use the chain of translation on integers.]

Exercise 4.12 (potential theoretic properties). (1) If λ and ν are two charges such that $\lambda G = \nu G$, then $\lambda = \nu$.

(2) If ν is a charge, the measure $\nu' = \nu P_A$ is the unique measure carried by A and such that: (i) $\nu' G = \nu G$ on A, (ii) $\nu' G \leqslant \nu G$ everywhere.

(3) If ν is carried by A and $\nu G \leqslant \lambda G$ on A, then $\nu G \leqslant \lambda G$ everywhere. [*Hint*: Use the operator H_A.]

Exercise 4.13 (capacities). (1) For $A \in \mathscr{E}$, prove that

$$(I - P_{S_A})\, G I_A = I_A.$$

(2) Let m be an excessive measure for P and f, g two functions in $\mathscr{E}_+$ vanishing outside A and such that $Gf \leqslant Gg$ on A. Prove that

$$\int f\, dm \leqslant \int g\, dm.$$

(3) Prove that for $A, B \in \mathscr{E}$, we have

$$P_{A \cup B} 1 + P_{A \cap B} 1 \leqslant P_A 1 + P_B 1,$$

[*Hint*: See Exercise 2.8.]

(4) We call *capacity* (relative to m) of the transient set A the number

$$C(A) = \int_A P_x[S_A = \infty]\, m(dx).$$

Prove that $C(A) \leqslant C(B)$ if $A \subset B$, and that for any pair (A, B) of transient sets we have

$$C(A \cup B) + C(A \cap B) \leqslant C(A) + C(B).$$

(5) Let $\mathscr{L}$ be the subset of $\mathscr{E}_+$ of functions g vanishing outside A and such that $Gg \leqslant 1$ on A. Prove that

$$C(A) = \sup_{g \in \mathscr{L}} \int_A g\, dm.$$

(6) Assume furthermore that X is in duality with $\hat{X}$ relative to m, which is then also co-excessive, and prove that if A is transient for both chains then $C(A) = \hat{C}(A)$.

Exercise 4.14. Let P and $\hat{P}$ be two T.P.'s in duality relative to the measure m.

(1) Prove that m is both excessive and co-excessive. If E is countable and $\mathscr{E}$ discrete, then the empty set is the only set of m-measure zero; the duality hypothesis is equivalent to the requirement that

$$m(x)\, P(x, y) = m(y)\, \hat{P}(y, x)$$

for every pair (x, y) of points in E.

(2) A right (left) random walk of law μ on a group G in duality relative to the right (left) Haar measure on G with the right (left) random walk of law $\hat{\mu}$, where $\hat{\mu}$ is the image of μ by the map $x \to x^{-1}$.

(3) If $f \in \mathscr{S}$, the measure $f\,m$ is co-excessive. If f is harmonic, the measure $f\,m$ is co-invariant.

Exercise 4.15. (1) If m is P-excessive then P maps the functions m-a.e. bounded into functions m-a.e. bounded, and thus induces a positive contraction on $L^\infty(E, \mathscr{E}, m)$.

[*Hint*: Use the weaker condition $mP \ll m$. Solution in ch. 4, Proposition 1.3.]

(2) Solve the same problem with $L^1(m)$ instead of $L^\infty(m)$. By using Holder's inequality for the measures $P(x, \cdot\,)$, prove that for every $p \in [1, \infty]$ the transition probability P induces on $L^p(m)$ a positive contraction (one may also use the Riesz convexity theorem). If P and $\hat{P}$ are in duality relative to m, the operators induced by P on $L^1(m)$ and by $\hat{P}$ on $L^\infty(m)$ are adjoint operators.

Exercise 4.16. Let P and $\hat{P}$ be in duality relative to m and let f be finite and P-superharmonic. Then, with the notation of Exercise 1.26, P^f is in duality with $\hat{P}$ relative to $f\,m$.

Exercise 4.17. (1) If P is in duality with itself with respect to m, that is if

$$\int Pf\, g\, dm = \int f\, Pg\, dm$$

for every pair (f, g) of functions in $\mathscr{E}_+$, then P induces on $L^2(m)$ a self-adjoint operator of norm less than or equal to one (see Exercise 4.15 above). There exists therefore a resolution of the identity $E(\mathrm{d}\lambda)$ in $L^2(m)$ such that

$$P_n = \int_{-1}^{+1} \lambda^n\, E(\mathrm{d}\lambda).$$

In particular, if E is countable,

$$m(x)\, P_n(x, y) = \int_{-1}^{+1} \lambda^n\, \mu_{x,y}(\mathrm{d}\lambda),$$

where $\mu_{x,y}$ is the measure defined by $\mu_{x,y}([-1, \lambda[) = \langle E[-1, \lambda[y, x\rangle$. (In the last formula x and y stand for the elements $1_{\{x\}}$ and $1_{\{y\}}$ of $L^2(m)$.)

(2) The space E still being countable, it is assumed that $G(x, y) < \infty$ for every pair (x, y) of points in E. Prove that the infinite matrix

$$\Gamma(x, y) = m(x)\, G(x, y) = m(y)\, G(y, x)$$

is positive definite. Derive that the space $\mathscr{V}$ of functions with finite supports may be endowed with a structure of pre-Hilbert space, by setting

$$(f, g) = \sum_{x,y} \Gamma(x, y)\, f(x)\, g(y) = \langle f, Gg\rangle_m.$$

Let $\mathscr{V}_A$ be the subspace of $\mathscr{V}$ of functions with support in A; prove that the orthogonal projector proj_A on $\mathscr{V}_A$ satisfies the relation

$$\text{proj}_A f = P_A f = (f\, m)\, H_A/m,$$

where the last term is the Radon–Nikodym derivative of $(f\, m)\, H_A$ with respect to m.

(3) If $g_A(x) = 1_A(x)\, P_x[S_A = \infty]$ set $C(A) = m(g_A)$. (See Exercise 4.13(4).) Prove that on the set of functions g of $\mathscr{V}_A$ such that $m(g) = 1$, the quantity (g, g) reaches its infimum $C(A)^{-1}$ for the unique function $g_A/C(A)$.

Exercise 4.18. (1) Let M be a homogeneous space of a group G admitting a measure ν invariant by G, that is $\varepsilon_g * \nu = \nu$ for all $g \in G$. Prove that ν is invariant for all random walks induced on M. If M is compact, every random walk on M induced by a random walk of law μ on G has at least one invariant probability measure [*Hint*: A cluster point of the sequence $n^{-1} \sum_{m=1}^{m=n} \mu^m * \nu'$, where ν' is any probability measure on M] and the set of invariant probability measures is a vaguely compact convex set.

(2) If ν is an invariant probability measure for the induced random walk, then for every function $\varphi \in L^\infty(\nu)$, the function

$$f(g) = \int_M \varphi(gx)\, \nu(\mathrm{d}x)$$

is μ-harmonic.

Exercise 4.19 (Continuation of Exercise 1.24). (1) Prove that there exists one and only one excessive measure m such that P is in duality with itself with respect to m. Compute this measure and find under which condition it is an invariant measure.

(2) Using the notation and results of Exercise 4.17, prove that there are

polynomial functions $Q_k(\lambda)$ and a unique measure μ such that for every pair (k, l) of integers,

$$\mu_{k,l}(\mathrm{d}\lambda) = Q_k(\lambda)\, Q_l(\lambda)\, \mu(\mathrm{d}\lambda),$$

and therefore

$$m(k)\, P_n(k, l) = \int_{-1}^{+1} \lambda^n Q_k(\lambda)\, Q_l(\lambda)\, \mu(\mathrm{d}\lambda).$$

Exercise 4.20. (1) If E is an LCCB and P maps the functions of C_K into continuous functions, every limit in the vague topology of excessive Radon measures is itself an excessive measure.

(2) If P is strong Feller, every excessive measure is a Radon measure.

5. Randomized stopping times and filling scheme

This section may be skipped without hampering the understanding of the sequel. It aims at giving another insight into the connections between Potential theory and the probabilistic and ergodic properties of Markov chains. In particular, we are going to describe a procedure of construction of reduced measures thus furthering the results in §§2 and 4. This procedure, known as the *filling scheme* has often been used in the theory of Markov chains and, although we shall not do so, we could use it to derive some of the results in the following chapters; some exercises in this and some further sections will present the alternative methods based on the filling scheme.

We begin with a generalization of the notion of stopping time.

Definition 5.1. A *randomized stopping time* T (abbreviated R.S.time T) of the canonical Markov chain X is a map from $\bar{\Omega} =]0, 1] \times \Omega$ to $\mathbf{N} \cup \{+\infty\}$ such that

(i) for each $u \in]0, 1]$, the map $T(u, \cdot\,)$ is a stopping time;

(ii) for each $\omega \in \Omega$, the map $T(\,\cdot\,, \omega)$ is decreasing and left continuous.

Notice that an ordinary stopping time T may be turned into a R.S.time by setting $T(u, \cdot\,) = T$ for each u.

Let U be the random variable defined on $\bar{\Omega}$ by $U(u, \omega) = u$ and set $\mathscr{G}_n = \sigma(X_0, \ldots, X_n, U)$, $\mathscr{G} = \sigma(\bigcup_0^\infty \mathscr{G}_n)$. We leave to the reader the task of showing that a R.S.time is a stopping time relative to the family $\mathscr{G}_n$, that is $\{T = n\} \in \mathscr{G}_n$ for every n. For every ν on $(E, \mathscr{E})$ we now define a probability measure $\bar{P}_\nu$ on $(\bar{\Omega}, \mathscr{G})$ by setting $\bar{P}_\nu = m \otimes P_\nu$ where m is the Lebesgue measure on

]0, 1]. We will still use X_n for the map $(u, \omega) \to X_n(\omega)$ and X for the collection of all X_n's. For $\bar{P}_\nu$, the random variable U is independent of X and X is still a homogeneous Markov chain with transition probability P with respect to the σ-algebras $\mathscr{G}_n$ (see ch. 1, Exercise 2.18). We have thus enlarged the probability space on which X is defined and a R.S.time is a stopping time in the enlarged setting.

As for ordinary stopping times we may define P_T by setting, for $f \in \mathscr{E}_+$,

$$P_T f(x) = \bar{E}_x[f(X_T)] = E_x\left[\int_0^1 f(X_{T_u})\, du\right] = \int_0^1 P_{T_u} f(x)\, du,$$

where $T_u = T(u, \cdot)$. A R.S.time thus appears as a "mixture" of ordinary stopping times. From Proposition 1.9 it is now clear that if f is superharmonic, then $P_T f \leqslant f$, a fact which will be important below.

Definition 5.2. A *balayage sequence* is a sequence (r_n, s_n), $n \geqslant 0$, of pairs of bounded measures such that for every $n > 0$,

$$r_n + s_n = s_{n-1} P.$$

The measure $r_0 + s_0$ is called the *initial measure* of the sequence and $\sum_0^\infty r_n$ its *terminal measure*.

With a R.S.time T and a starting measure λ we may associate a balayage sequence by setting for $A \in \mathscr{E}$ and $n \geqslant 0$,

$$r_n(A) = \bar{P}_\lambda[(X_n \in A) \cap (T = n)], \qquad s_n(A) = \bar{P}_\lambda[(X_n \in A) \cap (T > n)].$$

If we think of λ as a mass moved by the chain and stopped at time T, r_n is the mass stopped at time n, whereas s_n is the mass still moving at time n. Conversely we have

Proposition 5.3. *For any balayage sequence (r_n, s_n) there is a R.S.time T such that $\sum_0^\infty r_n = (r_0 + s_0) P_T$.*

Proof. Set $\lambda = r_0 + s_0$ and for every n define α_n to be a function in $\mathscr{U}_+$ such that $s_n = \alpha_n(r_n + s_n)$; such a function exists by the Radon–Nikodym Theorem. We now define a sequence $\{M_n\}$ of random variables on $\bar{\Omega}$ by

$$M_n = \alpha_0(X_0)\, \alpha_1(X_1) \cdots \alpha_n(X_n).$$

This sequence is decreasing, bounded by 1 and $M_n \in \mathscr{G}_n$ for every n. We clearly define a R.S.time by

$$T(u, \cdot) = \inf\{n: M_n < u\}$$

if there is at least one such n and $T = \infty$ otherwise; we have $\{T > n\} = \{M_n \geqslant u\}$ so that for every ν, $\bar{P}_\nu[\{T > n\} \mid X_0, \ldots, X_n] = M_n$.

Let now (r'_n, s'_n) be the balayage sequence associated with T by using λ as starting measure; we have

$$\begin{aligned} s'_n(A) &= \bar{E}_\lambda[1_A(X_n)\, 1_{[T>n]}] \\ &= \bar{E}_\lambda[1_A(X_n)\, \bar{E}_\lambda[1_{[T>n]} \mid X_0, \ldots, X_n]] \\ &= \bar{E}_\lambda[1_A(X_n)\, M_n] = \bar{E}_\lambda[1_A(X_n)\, \alpha_n(X_n)\, 1_{[T>n-1]}] \\ &= \bar{E}_\lambda[1_A(X_n)\, \alpha_n(X_n)\, 1_{[T=n]}] + \bar{E}_\lambda[1_A(X_n)\, \alpha_n(X_n)\, 1_{[T>n]}] \\ &= \int_A \alpha_n\, dr'_n + \int_A \alpha_n\, ds'_n. \end{aligned}$$

We thus have proved that $s'_n = \alpha_n(r'_n + s'_n)$ for every $n \geqslant 0$. Since $r_0 + s_0 = r'_0 + s'_0 = \lambda$, it follows that $s_0 = s'_0$ hence $r_0 = r'_0$ and inductively that $r_n + s_n = r'_n + s'_n$ hence $s_n = s'_n$ and in turn $r_n = r'_n$. The proof is complete.

The sequence $\{\alpha_n\}$ exhibited in the previous proof allows us to give an analytical expression for P_T which generalizes that of P_A.

Proposition 5.4. *For any* $f \in \mathscr{E}_+$,

$$\lambda P_T f = \sum_0^\infty \int f\, dr_n = \sum_0^\infty \lambda I_{\alpha_0} P I_{\alpha_1} P I_{\alpha_2} P \cdots P I_{\alpha_{n-1}} P I_{1-\alpha_n} f.$$

Proof. The first equality is clear. To prove the second we shall compute r_n as a function of P and the α_n's. By the definition of α_n we have $s_n = \alpha_n(s_{n-1}P) = s_{n-1} P I_{\alpha_n}$; it follows that

$$r_n = s_{n-1}P - s_n = s_{n-1}P - s_{n-1} P I_{\alpha_n} = s_{n-1} P I_{1-\alpha_n}.$$

From these formulas we easily derive that

$$r_n = s_0 P I_{\alpha_1} P \cdots P I_{\alpha_{n-1}} P I_{1-\alpha_n}$$

and $s_0 = \lambda I_{\alpha_0}$ whence the result follows.

Remark. The above reasonings may be performed with any sequence $\{\alpha_n\}$ of functions of $\mathscr{U}_+$ and any initial measure λ. With a constant sequence we get some of the operators previously described; more specifically if $\alpha_0 = 1$, $\alpha_n = 1 - h$ for $n \geqslant 1$, we get $P_T = U_h I_h$ and for $\alpha_n = 1_{A^c}$, $n \geqslant 0$, we get P_A.

It is clear that for a given starting measure λ, different R.S.times may have the same balayage sequence. In particular, if in Proposition 5.3 we start with the sequence associated with a time T, the time constructed is not necessarily equal to T. But all these times have the same measure λP_T. We shall now turn to the converse problem: two probability measures λ and ν being given, under what condition does there exist a R.S.time T such that $\nu = \lambda P_T$ or equivalently a balayage sequence with λ and ν as initial and terminal measures?

If T is a R.S.time we have already pointed out that $P_T f \leqslant f$ for any superharmonic function f. If $\nu = \lambda P_T$, we therefore get $\lambda(f) \geqslant \nu(f)$ for every $f \in \mathscr{S}$. This necessary condition turns out to be also sufficient.

Two bounded positive measures (λ, ν) being given we shall now describe a particular fashion of constructing a balayage sequence which will enable us to solve the above problem. We define inductively (λ_n, ν_n) by

$$\lambda_0 = (\lambda - \nu)^+, \qquad \nu_0 = (\lambda - \nu)^-,$$

$$\lambda_n = (\lambda_{n-1}P - \nu_{n-1})^+, \qquad \nu_n = (\lambda_{n-1}P - \nu_{n-1})^-.$$

We also put $\sigma_n = \sum_0^n \lambda_k$ and $\sigma = \sum_0^\infty \lambda_k$.

The following interpretation explains the title of this section: ν is viewed as a "hole" and a heap of dirt is dumped with distribution λ; λ_0 is the part of the dirt which does not fall in the hole whereas ν_0 is the remaining hole. The dirt λ_0 is then moved with the random "shovel" P and λ_1 is again this part of the dirt which does not fall in the remaining hole, and so on and so forth. We shall find the conditions under which the hole is entirely filled.

Lemma 5.5. (i) *For every n, we have $\lambda_{n+1} - \nu_{n+1} = \lambda_n P - \nu_n$;*

(ii) *for every n, the measures λ_n and μ_n are mutually singular;*

(iii) *the sequence $\{\nu_n\}$ decreases to a limit ν_∞ and if we define $r_0 = \nu - \nu_0$, $r_n = \nu_{n-1} - \nu_n$ for $n > 0$, the sequence (r_n, λ_n) is a balayage sequence with initial measure λ.*

(iv) *for every n, $\lambda_{n+1} \leqslant \lambda_n P$, so that for any superharmonic function f, the sequence of numbers $\{\lambda_n(f)\}$ is decreasing.*

Proof. Straightforward.

To say that the hole is filled is to say that $\nu_\infty = 0$; in any case there is a R.S.time T associated with (r_n, λ_n) which is such that $\lambda P_T = \sum_0^\infty r_n = \nu - \nu_\infty$. As already observed we thus have $\lambda(f) \geqslant \langle \nu - \nu_\infty, f\rangle$ for every superharmonic function f. Let us denote by $\mathrm{b}\mathscr{S}$ the set of bounded superharmonic functions and write $\nu \prec \lambda$ if $\nu(f) \leqslant \lambda(f)$ for every $f \in \mathrm{b}\mathscr{S}$; with this notation $\nu - \nu_\infty \prec \lambda$. We moreover have the following

Proposition 5.6. *Let $A \in \mathscr{E}$ be such that $\nu_\infty(A^c) = 0$ and $\lambda_n(A) = 0$ for every n, then $\nu_\infty = (\nu - \lambda) P_A$, and for $f \in \mathrm{b}\mathscr{S}$,*

$$\nu_\infty(f) = \sup\{\langle \nu - \lambda, g\rangle; g \in \mathrm{b}\mathscr{S}, g \leqslant f\}.$$

Proof. Since $\lambda_n(A) = 0$, we have $\lambda_n PP_A = \lambda_n P_A$ for every n, and therefore

$$(\lambda_{n+1} - \nu_{n+1}) P_A = (\lambda_n - \nu_n) P_A = \cdots = (\lambda - \nu) P_A. \tag{5.1}$$

On the other hand, it follows from Lemma 5.5(iv), that for $g \in \mathrm{b}\mathscr{E}_+$,

$$\lambda_n(P_A g) = \lambda_n(I_{A^c} P_A g) \leqslant \lambda_{n-1}(P I_{A^c} P_A g)$$

and inductively $\lambda_n(P_A g) \leqslant \lambda((P I_{A^c})^n P_A g)$ which goes to zero as n tends to infinity by the analytical expression of $P_A g$ (ch. 1, Proposition 3.7). Passing to the limit in eq. (5.1) we thus get the first equality of the statement.

To prove the second, observe that if $g \leqslant f$ then

$$\langle \nu - \lambda, g\rangle = \langle \nu_\infty, g\rangle + \langle \nu - \nu_\infty - \lambda, g\rangle \leqslant \langle \nu_\infty, g\rangle$$

by the remark preceding the statement, hence $\langle \nu - \lambda, g\rangle \leqslant \langle \nu_\infty, f\rangle$. On the other hand, the equality obtains for $g = P_A f$, which completes the proof.

Theorem 5.7. *The following three statements are equivalent:*

(i) $\nu_\infty = 0$;

(ii) *there exists a R.S.time T such that $\nu = \lambda P_T$;*

(iii) $\nu \prec \lambda$.

Proof. If $\nu_\infty = 0$, then $\nu = \sum_0^\infty r_n = \lambda P_T$ where T is the time associated with (r_n, λ_n) by Proposition 5.3. Thus (i) implies (ii).

The fact that (ii) implies (iii) was noted earlier. Finally, to prove that (iii) implies (i), we use the first formula in Proposition 5.6. Since $P_A 1 \in \mathrm{b}\mathscr{S}$, (iii) implies that $\nu_\infty(1) \leqslant 0$ which entails that $\nu_\infty = 0$.

Let us observe that there is no uniqueness property for the statement (ii) above; in other words there may exist several R.S.times T such that $\nu = \lambda P_T$. If G is proper these stopping times have a common mathematical expectation; indeed if $\mathscr{B}_+$ is the set of functions $f \in \mathscr{U}_+$ such that Gf is bounded

$$(\lambda - \nu)\, Gf = \bar{E}_\lambda \left[\sum_0^{T-1} f(X_n) \right]$$

for any $f \in \mathscr{B}_+$ and therefore $\bar{E}_\lambda[T] = \sup\{(\lambda - \nu)\, Gf; f \in \mathscr{B}_+\}$. When this expectation is finite it is natural to ask if among these R.S.times there is one with minimal variance or more generally such that $\bar{E}_\lambda[\varphi(T)]$ is minimum for a convex function φ on $\mathbf{N}$. This will be tackled in Exercise 5.15.

With G still proper the filling scheme also provides a method for constructing reduced measures.

Proposition 5.8. *If G is proper, the measure $\nu_\infty G$ is the smallest excessive measure larger than $\alpha = (\nu - \lambda)\, G$. Moreover $\sigma = \nu_\infty G - \alpha$ and in particular if $\nu \prec \lambda$, the measure σ is a positive solution of the Poisson equation $\sigma = \sigma P + (\lambda - \nu)$.*

Proof. We may write $\alpha = (\nu_\infty + (\nu - \nu_\infty) - \lambda)\, G$ and we have already observed before Proposition 5.6 that $(\nu - \nu_\infty)\, G \leqslant \lambda G$ when G is proper, so that $\alpha \leqslant \nu_\infty G$. Let now ξ be an excessive measure such that $\alpha \leqslant \xi \leqslant \nu_\infty G$. By the Riesz decomposition theorem, there is a bounded measure θ such that $\xi = \theta G$ and clearly $\langle \nu - \lambda, f\rangle \leqslant \langle \theta, f\rangle$ for $f \in \mathrm{b}\mathscr{S}$.

If now A is a set as in Proposition 5.6, we have for $f \in \mathrm{b}\mathscr{S}$,

$$\langle \nu - \lambda, P_A f\rangle \leqslant \langle \theta, P_A f\rangle \leqslant \langle \theta, f\rangle,$$

and therefore $\nu_\infty \prec \theta$; consequently $\nu_\infty G \leqslant \theta G$, hence $\nu_\infty = \theta$ which completes the proof of the first sentence. To prove the second, notice that the equation $\lambda_n - \nu_n = \lambda_{n-1} P - \nu_{n-1}$ entails upon summation that

$$\sigma_{n+1} = \sigma_n P + (\lambda - \nu) + \nu_{n+1};$$

passing to the limit yields

$$\sigma = \sigma P + (\lambda - \nu) + \nu_\infty,$$

and applying G through to the right of this equality, we get

$$\sigma = (\lambda - \nu)\, G + \nu_\infty G.$$

The second part of the statement is now obvious.

When G is not proper the above reasonings break down. These topics will be taken up in ch. 6, Exercise 6.10 in the case when G is no longer proper. The reader may already solve Exercise 5.11 in this section.

Exercise 5.9. In the setting of Proposition 5.4 prove that the following three properties are equivalent:

(i) T is $\bar{P}_\lambda$-a.s.finite;

(ii) $\lim_n s_n(E) = 0$;

(iii) $\lim_n M_n = 0$ P_λ-a.s.

Exercise 5.10. With the notation of Proposition 5.4 prove that the measure $\sigma = \sum_0^\infty s_n$ is equal to $\sum_0^\infty \lambda I_{\alpha_0} P I_{\alpha_1} \cdots P I_{\alpha_n}$.

Exercise 5.11. If $\nu \prec \lambda$ and if there is a positive σ-finite solution ξ to the Poisson equation $\xi = \xi P + \lambda - \nu$, then the measure σ provided by the filling scheme is the minimal positive σ-finite solution. This is true even if G is not proper.

[*Hint*: Prove inductively that $\sigma_n \leqslant \xi$ for every n.]

Exercise 5.12. If G is proper, the filling scheme (λ_n, ν_n) for (λ, ν) may be defined inductively by the conditions

(i) $\lambda_n \wedge \nu_n = 0$;

(ii) $(\nu_0 - \lambda_0)\, G = (\nu - \lambda)\, G$;

(iii) $(\nu_{n+1} - \lambda_{n+1})\, G = (\nu_n - \lambda_n)\, GP \vee (\nu_n - \lambda_n)\, G$.

[*Hint*: Use a set A_n such that $\lambda_n(A_n^c) = \nu(A_n) = 0$ and prove that the right-hand side of (iii) is equal to $(\nu_n - \lambda_n P)\, G$.]

Exercise 5.13. A function $g \in \mathscr{E}_+$ is called *subharmonic* if $Pg \geqslant g$. If $\nu(1) = \lambda(1)$ and $\nu = \lambda P_T$ where T is associated with (λ, ν) by the filling scheme, then for all ν-integrable subharmonic functions g the following three statements are equivalent:

(i) $\nu_n(g) \geqslant \lambda_n(g)$ for every n;

(ii) $\lim_n \lambda_n(g) = 0$;

(iii) for $\bar{P}_\lambda$, the submartingale $\{g(X_{T \wedge n})\}$ is equi-integrable.

Exercise 5.14. In the general situation of the filling scheme prove that the following two conditions are equivalent:

(i) $\lim_n \lambda_n(1) = 0$, in other words the "dirt" is entirely used up;

(ii) $\lambda(g) \leqslant \nu(g)$ for every bounded subharmonic function g.

Exercise 5.15. Let (λ, ν) be such that $\lambda G \geqslant \nu G$ where G is proper.

(1) Prove that one defines unambiguously a balayage sequence (t_n, t_n') by setting

$$\lambda = t_0 + t_0', \qquad t_n' = (t_n + t_n') \wedge (t_n'' G - \nu G)$$

where $t_n'' = t_0 + t_1 + \cdots + t_n + t_n'$. In particular $t_n'' G \geqslant \nu G$ for each n.

[*Hint*: For the last fact use the equality $t_{n+1}'' G = t_n'' G - t_n'$.]

(2) Prove that $t_n \wedge t_m' = 0$ for $m > n$. The sequence (r_n, λ_n) of the filling scheme has the opposite property: $r_n \wedge \lambda_m = 0$ for $n > m$.

[*Hint*: Prove that $t_n \wedge (t_{n+1}'' G - \nu G) = 0$ and write $t_n \wedge t_m' = t_n \wedge (t_m'' - t_{m+1}'') G$.]

(3) Prove that $\sum_0^n r_k \leqslant \sum_0^n t_k$ [*Hint*: Proceed by induction, using the formula $\sum_0^n r_k = (\sum_0^n (\lambda_k + r_k) - \nu)^+$.] and derive that $r_n'' G \geqslant t_n'' G$ with $r_n'' = r_0 + r_1 + \cdots + r_n + \lambda_n$. Then show that (t_n, t_n') has terminal measure ν.

(4) Let φ be a positive function on $\mathbf{N}$ such that $\varphi(n+1) - \varphi(n) \geqslant \varphi(n) - \varphi(n-1)$ and let S (resp. T) be the R.S.time associated with (t_n, t_n') (resp. (r_n, λ_n)). Suppose that $\varphi(0) = 0$ and set $\varphi(-1) = 0$; prove that

$$\bar{E}_\lambda[\varphi(S)] = \sum_0^\infty (\varphi(n+1) - \varphi(n-1))\,(t_n'' - \nu)\, G(E).$$

Deduce therefrom that

$$\bar{E}_\lambda[\varphi(T)] \geqslant \bar{E}_\lambda[\varphi(S)].$$

6. Resolvents

In this section we digress from our main subject to introduce another situation which also gives rise to a potential theory and which is related to Markov processes with continuous time parameters. Some results relevant to this situation may be obtained by mere translation of results about transition probabilities. Numerous exercises in the following chapters will be designed to this end.

Definition 6.1. A *sub-markovian resolvent* on $(E, \mathscr{E})$ is a family $\{V_\alpha, \alpha > 0\}$ of positive kernels such that

(i) $\alpha V_\alpha(x, E) \leqslant 1$ for each x in E and $\alpha > 0$;

(ii) $V_\alpha - V_\beta = (\beta - \alpha)\, V_\alpha V_\beta = (\beta - \alpha)\, V_\beta V_\alpha$ for $\alpha, \beta > 0$.

Equation (ii) is called the *resolvent equation*. The reader will notice its resemblance to the equation of Proposition 2.5 and should also refer to Exercise 6.11.

For $f \in \mathscr{E}_+$, it follows from the resolvent equation that the map $\alpha \to V_\alpha f$ is decreasing and that we may thus define

$$V_0 f = \sup_\alpha V_\alpha f = \lim_{\alpha \to 0} V_\alpha f. \tag{6.1}$$

If (f_n) is an increasing sequence of functions with limit f, by interchange of increasing limits we get immediately

$$V_0 f = \lim_n V_0 f_n;$$

this shows that V_0 is a positive kernel on $(E, \mathscr{E})$ which satisfies the identities

$$V_0 = V_\alpha + \alpha V_0 V_\alpha = V_\alpha + \alpha V_\alpha V_0. \tag{6.2}$$

In the remainder of this section we consider only *proper resolvents*, that is, resolvents such that V_0 is a proper kernel.

For each $\alpha > 0$, the kernel $P^\alpha = \alpha V_\alpha$ is a T.P. on E. We shall call X^α the canonical Markov chain associated with P^α.

Definition 6.2. A function f in $\mathscr{E}_+$ is called *supermedian* (*invariant*) for the resolvent $\{V_\alpha\}$ if it is superharmonic (harmonic) for every chain X^α, in other words if

$$\alpha V_\alpha f \leqslant f \; (\alpha V_\alpha f = f) \quad \text{for each } \alpha > 0.$$

Proposition 6.3. *If $f \in \mathscr{E}_+$, then $V_0 f$ is supermedian. If $V_0 f$ is finite and if h is positive, invariant and smaller than $V_0 f$, then h vanishes identically.*

Proof. The first sentence is an obvious consequence of eq. (6.2). If $h \leqslant V_0 f$, then eq. (6.2) again yields

$$h = \alpha V_\alpha h \leqslant \alpha V_\alpha V_0 f = V_0 f - V_\alpha f$$

for every $\alpha > 0$. As α tends to zero, $V_0 f - V_\alpha f$ converges to zero and consequently $h = 0$.

Proposition 6.4. *The potential kernel G^α of X^α is equal to $I + \alpha V_0$.*

Proof. Since $I + \alpha V_0$ is a proper kernel, it suffices to show that

$$\alpha V_0 f = \sum_{n=1}^{n=\infty} (\alpha V_\alpha)^n f$$

for every $f \in \mathscr{E}_+$ such that $V_0 f$ is finite. By applying k times the identity of eq. (6.2), we get

$$\alpha V_0 f = \sum_{n=1}^{n=k-1} (\alpha V_\alpha)^n f + (\alpha V_\alpha)^k V_0 f;$$

if k tends to infinity, the term $(\alpha V_\alpha)^k V_0 f$ decreases to an invariant function which is smaller than $V_0 f$, and hence vanishes identically. The proof is then easily completed.

Remark. The kernels G^α are thus also proper kernels.

We now state the complete maximum principle for V_0 which, as in §2, is a characterization of supermedian functions.

Theorem 6.5. *The function $g \in \mathscr{E}_+$ is supermedian if and only if whenever $V_0 f$ is well defined and $V_0 f \leqslant g$ on $\{f > 0\}$, then $V_0 f \leqslant g$ everywhere.*

Proof. We suppose that for $x \in \{f > 0\}$ we have

$$V_0 f^+(x) \leqslant g(x) + V_0 f^-(x).$$

For every integer n we set $f_n = f^+ \wedge n$; we, a fortiori, have

$$V_0 f_n(x) \leqslant g(x) + V_0 f^-(x), \quad x \in \{f > 0\},$$

and consequently, for every $\alpha > 0$,

$$G^\alpha f_n(x) \leqslant \alpha g(x) + G^\alpha f^-(x) + n, \quad x \in \{f > 0\}.$$

By Theorem 2.3, applied to the chain X^α, we get

$$G^\alpha f_n \leqslant \alpha g + G^\alpha f^- + n$$

everywhere on E, and upon dividing by α and letting α tend to infinity, it follows that $V_0 f_n \leqslant g + V_0 f^-$. The proof of the necessity is completed by letting n tend to infinity.

Conversely, let $f > 0$ be such that $V_0 f$ be bounded, and set $g_n = g \wedge nf$, $h_n = \alpha(g_n - \alpha V_\alpha g_n)$. The potential $V_0 h_n$ is well defined and therefore

$$g \geqslant \alpha V_\alpha g_n = V_0 h_n \quad \text{on } \{g - \alpha V_\alpha g_n > 0\},$$

hence a fortiori on $\{g_n - \alpha V_\alpha g_n > 0\} = \{h_n > 0\}$. The hypothesis then implies that for every n, $g \geqslant \alpha V_\alpha g_n$ everywhere, and by passing to the limit, we obtain $g \geqslant \alpha V_\alpha g$. Since this is true for all α, the function g is supermedian.

Exercise 6.6. Let $\{V_\alpha, \alpha > 0\}$ be a resolvent and β a real number > 0. Prove that the family of kernels $V'_\alpha = V_{\alpha+\beta}$ is a proper resolvent.

Exercise 6.7 (Resolvent of linear brownian motion). Prove that the family of convolution kernels V_α defined on $\mathbf{R}$ by the densities

$$v_\alpha(x, y) = (2\alpha)^{-1/2} \exp\{-(2\alpha)^{-1/2} |y - x|\}$$

is a resolvent.

Exercise 6.8. The integral kernels defined on $\mathbf{R}$ by the densities

$$\begin{aligned} v_\alpha(x, y) &= v^{-1} \exp\{-\alpha(y - x)/v\} \quad \text{if } y > x, \\ &= 0 \qquad\qquad\qquad\qquad\quad \text{if } y \leqslant x, \end{aligned}$$

form a proper resolvent.

Exercise 6.9. If h is a positive finite function such that $V_\beta h$ is finite for all $\beta > 0$ and if there is an α such that $\alpha V_\alpha h = h$, then h is invariant.

Exercise 6.10. If $f \in \mathscr{E}_+$, then for every x in E the function $\alpha \to V_\alpha f(x)$ is decreasing, right continuous and continuous on every open interval on which it is finite.

Exercise 6.11. (1) Let $\{V_\alpha, \alpha > 0\}$ be a resolvent and h a function in $b\mathscr{E}_+$. Pick $\alpha > \|h\|$ and define

$$V_h = \sum_{n \geqslant 0} (V_\alpha I_{\alpha-h})^n V_\alpha.$$

Prove that V_h does not depend on α and that for $h \leqslant k$ we have

$$V_h = \sum_{n \geqslant 0} (V_k I_{k-h})^n V_k = \sum_{n \geqslant 0} V_k (I_{k-h} V_k)^n,$$

and

$$V_h = V_k + V_k I_{k-h} V_h = V_k + V_h I_{k-h} V_k.$$

These equations contain the resolvent equation, since for $h = \beta$ we have $V_h = V_\beta$.

[*Hint*: Use the chains X^α for which $V_h = \alpha^{-1} U_{h\alpha^{-1}}$; in particular for $h \leqslant 1$, we have $V_h = U_h$, where this kernel is computed for X^1.]

(2) The family of kernels $\tilde{V}_\alpha = V_{\alpha h} I_h$ is a resolvent.

Exercise 6.12. (1) Let P be a T.P. on $(E, \mathscr{E})$. Prove that one defines a resolvent by setting

$$V_\alpha = (1+\alpha)^{-1} \sum_{n \geqslant 0} ((1+\alpha)^{-1} P)^n.$$

Prove that $V_0 = G$.

(2) Assume that G is a proper kernel, and prove that a function is superharmonic for P if and only if it is supermedian for V_α. Then derive Theorem 2.3 from Proposition 6.5.

Exercise 6.13. Consider a proper resolvent $\{V_\alpha, \alpha > 0\}$.

(1) Prove that, if $\alpha < \beta$, the superharmonic functions for X^β are superharmonic for X^α.

(2) Let f be supermedian, and for $A \in \mathscr{E}$ let $P^\alpha_A f$ be the reduced function of f relative to X^α. Prove that the following limit

$$R_A f = \lim_{\alpha \to \infty} P^\alpha_A f$$

exists, and defines the smallest supermedian function which dominates f on A.

(3) Prove that $R_A f \leqslant f$, $R_A f = f$ on A and if f, g are supermedian and $f \leqslant g$ on A, then $R_A f \leqslant R_A g$. If $\{f_n\}$ is an increasing sequence of supermedian functions with limit f, then $R_A f = \lim R_A f_n$.

(4) If h is a positive function vanishing outside A, then $R_A V_0 h = V_0 h$. The operator $R_A V_0$ is given by a kernel. Is the same thing true for the operator R_A?

(5) If f, g are supermedian, then $R_A(f + g) = R_A f + R_A g$.

(6) If $\{A_n\}$ is an increasing sequence of sets with union A, then $R_A f = \lim_n R_{A_n} f$.

Exercise 6.14 (excessive functions). (1) If f is supermedian, the function $\alpha \to \alpha V_\alpha f(x)$ is increasing for each x in E. We call f *excessive* for the given resolvent if in addition $\lim_{\alpha \to \infty} \alpha V_\alpha f = f$. Prove that the potentials $V_0 g$, $g \in \mathscr{E}_+$, are excessive functions.

(2) Let f be supermedian and set $f^* = \lim_{\alpha \to \infty} \alpha V_\alpha f$. Show that f^* is an excessive function and that $f^* \leqslant f$. Show that the set $K = \{f^* < f\}$ on which f^* differs from f is of *potential zero*, that is $V_\alpha(\cdot, K) = 0$ for one (hence for all) $\alpha > 0$. Show that f^* is the greatest excessive function bounded above by f.

(3) Let f be a supermedian function and define, with the notation of Exercise 6.13, $(R_A f)^* = \lim_{\alpha \to \infty} \alpha V_\alpha R_A f$. Prove that $(R_A f)^*$ is excessive and that it is the smallest supermedian function which majorizes f on A up to a set of potential zero.

Exercise 6.15. A σ-finite positive measure m is *invariant* (*supermedian*) for a resolvent $\{V_\alpha\}$ if $\alpha m V_\alpha = m$ ($\alpha m V_\alpha \leqslant m$) for all $\alpha > 0$. It is *excessive* if it is supermedian and if in addition $\lim_{\alpha \to 0} \alpha m V_\alpha = m$.

(1) Work out for supermedian measures results similar to those in Exercise 6.13. For A in $\mathscr{E}$ and m supermedian, set

$$mL_A = \lim_{\alpha \to \infty} mH_A^\alpha.$$

(2) Prove that the operator $V_0 L_A$ is given by a kernel and that $V_0 L_A = R_A V_0$. (Compare with Exercise 2.11.)

Exercise 6.16. If E is locally compact and if $\lim_{\alpha \to \infty} \alpha V_\alpha(x, \cdot) = \varepsilon_x$ in the vague topology, show that the supermedian Radon measures are excessive measures.

CHAPTER 3

TRANSIENCE AND RECURRENCE

In this chapter we describe the basic classification of Markov chains according to their asymptotic behaviour. We start with countable state space where the situation is particularly nice and then proceed to more general state spaces.

1. Discrete Markov chains

Definition 1.1. A Markov chain will be called *discrete* if the state space E is countable and $\mathscr{E}$ is the discrete σ-algebra on E.

Remark. If E is an uncountable set and $\mathscr{E}$ is the discrete σ-algebra, then for every T.P. on $(E, \mathscr{E})$, every point $x \in E$ is contained in the countable absorbing set $\{y: G(x, y) > 0\}$. It suffices therefore to study countable state spaces, justifying the above terminology. In the same way, in the case of random walks on a discrete group, the identity is contained in a countable absorbing subgroup. In the sequel, the term *discrete group* will therefore be understood to mean countable and discrete.

In this section, we shall deal only with discrete chains, and shall use the notational device of ch. 1, §1.4, namely, we shall not distinguish between the point x and the set $\{x\}$; when μ is a measure we write $\mu(y)$ in place of $\mu(\{y\})$, and in particular $P(x, y)$ for $P(x, \{y\})$. We shall also abbreviate $S_{\{x\}}$ (resp. $R(\{x\})$, $h_{\{x\}}$) to S_x (resp. $R(x)$, h_x) and speak of the state x instead of the point x to follow a time-honoured custom.

Definition 1.2. A state x is said to be *transient* if the set $\{x\}$ is transient; otherwise it is said to be *recurrent*.

The attention of the reader is drawn to the fact that x may be recurrent without the set $\{x\}$ being recurrent in the sense of §3 in ch. 2.

Proposition 1.3. *The following four properties are equivalent*:

(i) *x is recurrent*;

(ii) $G(x, x) = +\infty$;

(iii) $P_x[S_x < \infty] = 1$;

(iv) $h_x(x) = 1$.

Proof. If $G(x, x) < +\infty$ then x is transient by ch. 2 §3.7, so (i) implies (ii). From Proposition 3.5 in ch. 2, we may write

$$1 = P_{\{x\}}1(x) = G(x, x)\, P_x[S_x = \infty] + h_x(x).$$

If $G(x, x) = +\infty$ we thus must have $P_x[S_x = \infty] = 0$, hence (ii) implies (iii). The same formula shows that (iii) implies (iv) and finally that (iv) implies (i) is obvious.

We observe that for any x in E, either $h_x(x) = 0$ and x is transient or $h_x(x) = 1$ and x is recurrent. We also have

Corollary 1.4. *The following three properties are equivalent*:

(i) *x is transient*;

(ii) $P_x[S_x = \infty] > 0$;

(iii) $G(x, x) < +\infty$ *and in that case* $G(x, x) = (P_x[S_x = \infty])^{-1}$.

Proposition 1.5. *If the state x is recurrent, then for every $y \in E$, $y \neq x$, either $h_y(x) = P_x[S_y < \infty] = 0$ or $h_y(x) = P_x[S_y < \infty] = 1$; if*

$$C_x = \{y: h_y(x) = 1\},$$

every state y in C_x is recurrent and $C_y = C_x$.

Proof. By ch. 2 §3.5, we have

$$\lim_n P_{X_n}[S_y < \infty] = \lim_n h_y(X_n) = 1_{R(y)} \quad P_x\text{-a.s.}$$

Since x is recurrent we also have

$$\overline{\lim_n}\, P_{X_n}[S_y < \infty] = P_x[S_y < \infty] \quad P_x\text{-a.s.}$$

and

$$\overline{\lim_n}\, h_y(X_n) = h_y(x) \quad P_x\text{-a.s.}$$

It follows that

$$1_{R(y)} = P_x[S_y < \infty] = h_y(x) \quad P_x\text{-a.s.}$$

As a result either $P_x[S_y < \infty] = h_y(x) = 1$ or $P_x[S_y < \infty] = h_y(x) = 0$.

If $y \in C_x$, then $h_y(x) = 1$ and y cannot be transient; it is thus recurrent. Let z be in C_x; since the chain goes through x and y infinitely often P_x-a.s., we have

$$\lim_n h_z(X_n) = h_z(x) = h_z(y) \quad P_x\text{-a.s.}$$

Consequently $h_z(y) = 1$, so $z \in C_y$ and $C_x \subset C_y$. Therefore $x \in C_y$ and $C_y \subset C_x$. The proof is complete.

We may sum up the preceding results in the following

Theorem 1.6. *There is a unique partition* $(C_i, i \in I)$ *of the set of recurrent states such that each* C_i *is an absorbing set and for every pair* (x, y) *of states within the same class* C_i, *we have* $h_x(y) = 1$. *Moreover* $G(x, y) = +\infty$ *or* 0 *according as* x *and* y *are within the same class or not. The sets* C_i *are called the* recurrence classes *of* X.

Proof. The sets C_i are obtained as the sets C_x where x runs through the set of recurrent states.

The reader is invited to solve Exercise 1.11 which gives another, more elementary, approach to the above results.

Definition 1.7. The chain X will be called *transient* if all the states are transient. In that case $G(x, y) < \infty$ for every pair of states in E. If C is a class of recurrent states, then for the study of X with respect to the probability measures P_x, $x \in C$, we can restrict the state space to C. The chain X will be called *irreducible recurrent* if all the states are recurrent and if there is only one equivalence class. We then have $G(x, y) = \infty$ identically.

We shall end this section by studying irreducible recurrent chains. The following proposition, which will be generalized later, shows that the main concern of the potential theory of ch. 2 is with transient chains. In the recurrent case, we will have to give another definition of potential theory.

Proposition 1.8. *If X is irreducible recurrent the superharmonic and bounded harmonic functions are constant.*

Proof. If f is superharmonic, $f(X_n)$ is a positive supermartingale which converges a.s. But for x and y in E, $f(x)$ and $f(y)$ are limit points of $f(X_n)$ and therefore $f(x) = f(y)$. The same argument works for the bounded harmonic functions.

Theorem 1.9. *If X is irreducible recurrent, there exists a unique (up to multiplication) excessive measure m and this measure is invariant. If x is any point in E, we may define m by*

$$m(y) = H_{\{x\}}(x, y) = U_{\{x\}}(x, y).$$

Proof. Pick a point x in E and set

$$m(y) = E_x\left[\sum_{n<S_x} 1_y(X_n)\right] = H_{\{x\}}(x, y).$$

For every y in E, $0 < m(y) < \infty$; indeed, there is a strictly positive probability of entering y before returning to x, and on the other hand the chain killed at S_x is clearly transient.

Now using the resolvent equation, we have

$$\begin{aligned} mP(y) &= H_{\{x\}}P(x, y) = I_{\{x\}}P(x, y) + I_{\{x\}}U_{\{x\}}I_{\{x\}}P(x, y) \\ &= I_{\{x\}}U_{\{x\}}(x, y) = U_{\{x\}}(x, y). \end{aligned}$$

But for $n = 0$, we have $1_y(X_n) = 0$ P_x-a.s. if $y \neq x$, and $1_y(X_n) = 1$ P_x-a.s. if $y = x$, and on $\{S_x = n\}$ we have also $1_y(X_n) = 0$ if $y \neq x$ and $1_y(X_n) = 1$ if $y = x$; consequently

$$E_x\left[\sum_{n=1}^{S_x} 1_y(X_n)\right] = E_x\left[\sum_{n<S_x} 1_y(X_n)\right],$$

that is exactly $H_{\{x\}}(x, y) = U_{\{x\}}(x, y)$, and it follows that $mP = m$.

To prove that m is the only excessive measure for P we set $\hat{P}(x, y) = m(y)\, P(y, x)/m(x)$. The reader will easily check that $\hat{P}$ is a T.P. in duality with P with respect to m, and that the corresponding chain is irreducible recurrent.

If μ is an excessive measure for P, then $f(\cdot) = \mu(\cdot)/m(\cdot)$ is superharmonic for $\hat{P}$, so that the result follows from Proposition 1.8.

Remarks. (1) The empty set is the only set of m-measure zero.

(2) For two different points x in E we get, by the above process, two proportional values for m.

(3) The T.P. $\hat{P}$ defined in the proof above is the only one to be in duality with P. For an irreducible recurrent chain we may therefore speak freely of its dual chain, which exists and is unique. The two chains play totally symmetric roles, in particular m is invariant for $\hat{P}$ as well as for P.

Corollary 1.10. *If X is irreducible recurrent, either $E_x[S_x] < \infty$ for every x in E and the invariant measure m is bounded, or $E_x[S_x] = \infty$ for every x in E and the invariant measure m is unbounded. In the former case there is an invariant probability measure π which is given by*

$$\pi(x) = 1/E_x[S_x], \quad x \in E.$$

Proof. By the previous result, for every x in E, we have

$$m(E) = \sum_{y \in E} m(y) = U_{\{x\}}(x, E) = H_{\{x\}}(x, E) = E_x[S_x],$$

and since x was arbitrary, this proves the first sentence. To prove the second, observe that $\pi = m/m(E)$, therefore

$$\pi(x) = m(x)/m(E) = U_{\{x\}}(x, x)/E_x[S_x] = 1/E_x[S_x].$$

This last result says that $\pi(x)$ measures the frequency with which the chain goes back to x. In the former case the chain is said to be *positive*, in the latter it is said to be *null*.

The theory of recurrence we have been sketching cannot be carried over in an obvious way to general state spaces. The sequel of this chapter aims at giving some of the possible extensions. Other viewpoints will be developed in the following chapter.

Exercise 1.11. Let $S_1 = S_x, \ldots, S_n = S_{n-1} + S_x \circ \theta_{S_{n-1}}$ the successive random times for which $X_n = x$.

(1) Let Z be an integer-valued and $\mathscr{F}_{S_x}$-measurable random variable. Prove that the random variables $Z \circ \theta_{S_n}$ are independent and equally distributed with respect to the probability measure P_x.

(2) By applying the foregoing to $Z = S_x$ prove that $P_x[S_n < \infty] = (P_x[S_x < \infty])^n$ and derive Proposition 1.3 and Corollary 1.4.

(3) If x is recurrent and $P_x[S_y < \infty] > 0$, prove that y is recurrent and

$P_y[S_x < \infty] > 0$ by the use of Markov property. Derive Proposition 1.5 without using Martingale theory.

Exercise 1.12. Let E be finite and P markovian; prove that there is at least one recurrent state. If there are N states in E and if $P_x[S_y < \infty] > 0$, prove that $P_x[S_y < N] > 0$.

Exercise 1.13 (another proof of Proposition 1.8). (1) Assume the chain to be irreducible recurrent and derive from the equality

$$\left(\sum_{m<n} P_m\right)(f - Pf) = f - P_n f,$$

that every superharmonic function is harmonic. Then prove that if a harmonic function vanishes at one point in E, it vanishes identically, and apply this result to $f - (f \wedge k)$ for every choice of the constant k.

(2) Prove the following converse to Proposition 1.8: if every bounded superharmonic function is constant, then X is irreducible recurrent.

Exercise 1.14 (continuation of ch. 2, Exercise 4.19, birth and death chain). Prove that the chain is either transient or irreducible recurrent and that the latter case occurs if and only if

$$\int_{-1}^{+1} (1 - \lambda)^{-1}\, d\mu(\lambda) = +\infty.$$

Exercise 1.15 (Galton–Watson chain; continuation of ch. 1, Exercise 2.20).

(1) Call σ the smallest solution of the equation $g(s) = s$ in the interval $]0, 1]$. Prove that $\sigma = 1$ if $E_1[X_1] \leqslant 1$ and $\sigma < 1$ if $E_1[X_1] > 1$.

(2) Assume that $0 < p(0) < 1$, the other cases offering neither interest nor difficulty. Show that 0 is an absorbing (hence recurrent) state and that

$$P_x\left[\varliminf_n \{X_n = 0\}\right] = \sigma^x.$$

Interpret this probability with respect to the relevant population. Prove that all the other states are transient and describe the asymptotic behaviour of the chain according to the value of σ. See ch. 2, Exercise 1.23 for a particular case.

(3) Assume further that $E_1[X_1]$ is finite and prove that $\{X_n/(E_1[X_1])^n\}$

is a Martingale which converges a.s. to a random variable Z. Describe completely the asymptotic behavior of X.

Exercise 1.16 (renewal chain). Define a T.P. on the set of positive integers by $P(n, n+1) = p_n$, $P(n, 0) = 1 - p_n$ with $0 < p_n < 1$. Find the condition under which the chain is irreducible recurrent, then compute the invariant measure. Prove that if the chain is transient, there is no invariant measure.

Exercise 1.17 (cyclic classes). Let X be irreducible recurrent and for x in E, call $d(x)$ the greatest common divisor of the set of integers

$$\{n > 0: P_n(x, x) > 0\}.$$

(1) Prove that there is a number d such that $d(x) = d$ for every x in E. If $d = 1$ we say that X is aperiodic, otherwise X is periodic with period d.

(2) If $d > 1$, there exists a partition of E into sets C_i, $i = 1, \ldots, d$, such that

$$P(\cdot\,, C_{i+1}) = 1_{C_i}, \quad i = 1, 2, \ldots, d-1, \qquad P(\cdot\,, C_1) = 1_{C_d}.$$

(3) (The Ehrenfest model of diffusion). N particles are distributed in two containers U_1 and U_2. At every time n a particle is chosen at random and is moved from its container into the other. Define X_n to be the random number of particles in U_1 at time n. Show that the sequence X_n is a Markov chain; write down its T.P. and compute its invariant measure. Prove that the chain is periodic of period 2.

(4) Find the conditions under which the renewal chain of Exercise 1.16 is irreducible recurrent and periodic.

Exercise 1.18. Let X be a chain on $\mathbf{N}$ such that $P_x[T_y < \infty] > 0$ for any pair (x, y) of points in $\mathbf{N}$.

(1) Prove that X is recurrent if and only if there is an integer $N > 0$ and a positive function f such that $Pf \leqslant f$ on $[N, \infty[$ and $\lim_n f(n) = \infty$.

[*Hint*: See ch. 2 §2.10.]

(2) Prove that X is transient if and only if there is a bounded positive function f such that $Pf \leqslant f$ on $[N, \infty[$ for some $N > 0$ and $f(x) < \sup\{f(y) : y < N\}$ for some $x > N$.

2. Irreducible chains and Harris chains

We now return to a general state space $(E, \mathscr{E})$, but we assume henceforth that it is separable.

Definition 2.1. The chain X is said to be *ν-irreducible* if there exists a probability measure ν on $\mathscr{E}$ absolutely continuous with respect to $U_c(x, \cdot)$ for all $x \in E$ and $c \in]0, 1[$.

Clearly if the above condition is satisfied for a measure ν', it is satisfied for a probability measure ν equivalent to ν'; it is satisfied for all numbers $c \in]0, 1[$ as soon as it is satisfied for one of them. This property can also be spelled out as: if $A \in \mathscr{E}$ and $\nu(A) > 0$, then for all x in E there is an integer n such that $P_n(x, A) > 0$. Furthermore if X is ν-irreducible and if we set $\nu' = c\nu U_c$, the reader will easily check that X is still ν'-irreducible, but that ν' has the additional features: (i) $\nu' P \ll \nu'$, (ii) $P_{\nu'}[S_A < \infty] = 0$ if $\nu'(A) = 0$. The significance of condition (i) will be seen in the following chapter.

The purpose of this section is to show that for a ν-irreducible chain either the potential kernel is proper or the chain is recurrent (in a strong sense to be defined later).

Following ch. 1, §5.3 we shall call $p_c(x, y)$ an $\mathscr{E} \otimes \mathscr{E}$-measurable version of the density of $U_c(x, \cdot)$ with respect to ν. The irreducibility of X implies that for all x in E, $p_c(x, \cdot)$ is > 0 ν-a.s.; then $p_c(x, \cdot) + 1_{\{p_c=0\}}(x, \cdot)$ is still a density for $U_c(x, \cdot)$ and is strictly positive everywhere. We shall therefore take p_c to be strictly positive in the sequel.

Proposition 2.2. *There is a function $h_0 \in \mathscr{U}_+$, strictly positive and strictly less than* 1, *and a positive measure m_0 equivalent to ν such that*

$$U_{h_0} \geqslant U_{h_0}(h_0) \otimes m_0.$$

Proof. Let $c < c' < 1$; by the resolvent equation of ch. 2, Proposition 2.5,

$$U_c \geqslant (c' - c)\, U_c U_{c'} \geqslant (c' - c)\, q\nu,$$

where

$$q(x, y) = \int_E p_c(x, z)\, p_{c'}(z, y)\, \nu(\mathrm{d}z).$$

Since p_c is $\mathscr{E} \otimes \mathscr{E}$-measurable and > 0, the functions $f_k(\cdot) = \nu(\{z\colon p_c(\cdot, z) \geqslant 1/k\})$ are measurable and increase to 1 as k tends to infinity.

We can therefore define the integer

$$k(x) = \inf\{k \geqslant 1 : f_k(x) \geqslant \tfrac{3}{4}\}.$$

Since

$$\{x : k(x) = p\} = \{x : f_{p-1}(x) < \tfrac{3}{4}\} \cap \{x : f_p(x) \geqslant \tfrac{3}{4}\}$$

the map $x \to k(x)$ is measurable, and setting $a(x) = 1/k(x)$, we have

$$\nu(\{z : p_c(x, z) \geqslant a(x)\}) \geqslant \tfrac{3}{4}$$

for every x in E. In the same way we can find a function b such that, for every y in E,

$$\nu(\{z : p_{c'}(z, y) \geqslant b(y)\}) \geqslant \tfrac{3}{4}.$$

We then have

$$q(x, y) \geqslant a(x)\, b(y)\, \nu(\{z : p_c(x, z) \geqslant a(x), p_{c'}(z, y) \geqslant b(y)\}) \geqslant \tfrac{1}{2}a(x)\, b(y),$$

hence $U_c \geqslant \frac{1}{2}(c' - c)\, a \otimes b\nu$.

Set now $h_0 = \inf(\frac{1}{2}c, a)$; we have $0 < h_0 < 1$. Also, since $h_0 \leqslant a$ we still have $U_c \geqslant \frac{1}{2}(c' - c)\, h_0 \otimes b\nu$ and since $h_0 \leqslant c$, the resolvent equation implies

$$U_{h_0} \geqslant U_{h_0} I_{c-h_0} U_c \geqslant \tfrac{1}{2}c U_{h_0} U_c$$

$$\geqslant c(c' - c)\, U_{h_0}(h_0 \otimes b\nu)/4 = U_{h_0}(h_0) \otimes \tfrac{1}{4}c(c' - c)\, b\nu.$$

Now set $m_0 = \frac{1}{4}c(c' - c)\, b\nu$; since b is strictly positive m_0 is equivalent to ν and the proof is complete.

We recall from ch. 2 that a set F in $\mathscr{E}$ is said to be *absorbing* for X if $P(x, F) = 1$ for every x in F, or equivalently if 1_{F^c} is superharmonic. The recurrent classes of discrete chains provide examples of absorbing sets.

Theorem 2.3. *For a ν-irreducible chain X, there are only two possibilities:*

(i) *the potential kernel is a proper kernel;*

(ii) *there is an absorbing set F such that $\nu(F^c) = 0$ and a strictly positive function $h_0 \in \mathscr{U}_+$, such that for every x in F*

$$U_{h_0}(h_0)(x) = 1 \quad \textit{and } U_{h_0}(x, \cdot) \geqslant m_0(\cdot).$$

Proof. Let h_0 be the function of Proposition 2.2. According as $U_{h_0}(h_0) = 1$ ν-a.s. or not, we will get the second or the first possibility.

(1) Suppose that $1 - U_{h_0}(h_0)$ is not ν-negligible, which implies that $\langle h_0 m_0, 1 - U_{h_0}(h_0)\rangle = c > 0$. Since $U_{h_0} \geqslant U_{h_0}(h_0) \otimes m_0$, we have

$$(U_{h_0} I_{h_0})^2\, 1 = U_{h_0}(h_0) - U_{h_0} I_{h_0}(1 - U_{h_0}(h_0)) \leqslant (1 - c)\, U_{h_0}(h_0),$$

and inductively

$$(U_{h_0} I_{h_0})^n\, 1 \leqslant (1 - c)^{n-1}\, U_{h_0}(h_0).$$

Now by ch. 2, Proposition 2.5,

$$Gh_0 = h_0 + \sum_{n \geqslant 0} (U_{h_0} I_{h_0})^n\, U_{h_0}(h_0),$$

so that Gh_0 is a bounded function, and it follows from ch. 2, Proposition 1.14 that (i) holds.

(2) If, on the other hand, $U_{h_0}(h_0) = 1$ ν-a.s., the set $F = \{U_{h_0}(h_0) = 1\}$ is such that $\nu(F^c) = 0$, and we claim that it is an absorbing set. The resolvent equation $U_{h_0} = P + PI_{1-h_0} U_{h_0}$ implies

$$1 - U_{h_0}(h_0)\,(x) \geqslant P((1 - h_0)\,(1 - U_{h_0}(h_0)))\,(x);$$

for x in F the first member of this inequality vanishes, hence $P(x, \cdot\,)$ vanishes outside the set $\{(1 - h_0)\,(1 - U_{h_0}(h_0)) = 0\}$, which is equal to F, since $h_0 < 1$ on E. On the other hand, by ch. 2 §2.6, $P(x, E) = 1$, because $U_{h_0}(h_0)\,(x) = 1$ and consequently $P(x, F) = 1$. The set F is thus an absorbing set, and by Proposition 2.3 the proof is complete.

To proceed we will need the following lemma in which B denotes the Banach space of bounded measures ν such that $\nu(E) = 0$.

Lemma 2.4. *The Banach space B is left invariant by any markovian transition probability Q. If the norm Δ of Q as an endomorphism of B is strictly less than one, there exists a unique Q-invariant probability measure Π on $(E, \mathscr{E})$, and for every x in E*

$$\|Q^n(x, \cdot\,) - \Pi(\cdot\,)\| \leqslant 2\Delta^n.$$

Proof. The first sentence of the lemma is straightforward. Let ν be any probability measure on $\mathscr{E}$ and k any integer. The measure $\nu - \nu Q^k$ is in B and its norm is at most equal to 2; thus

$$\|\nu Q^n - \nu Q^{k+n}\| = \|(\nu - \nu Q^k)\, Q^n\| \leqslant 2\Delta^n$$

which shows that the sequence $\{\nu Q^n\}$ is a Cauchy sequence in the space of bounded measures. It has therefore a limit Π which is plainly Q-invariant and such that

$$\|\nu Q^n - \Pi\| \leqslant 2\Delta^n.$$

If ν' is another probability measure on $\mathscr{E}$, we have

$$\|\nu' Q^n - \Pi\| = \|(\nu' - \Pi)\, Q^n\| \leqslant 2\Delta^n;$$

it follows that if ν' is invariant, then $\nu' = \Pi$, hence Π is the only Q-invariant probability measure. Moreover, letting $\nu' = \varepsilon_x$, we obtain the last conclusion in the statement.

In case (i) of Theorem 2.3 the space E is the union of an increasing sequence of transient sets. We now want to show that the case (ii) corresponds to a strong recurrence property on F. Since F is absorbing, the chain started at a point x in F stays in F P_x-a.s.; it is therefore natural to restrict P to F and thus obtain a T.P. on the space $(F, F \cap \mathscr{E})$, the more so in the present instance since ν is carried by F. For notational convenience we will rather assume below that $F = E$ or, in other words, that $U_{h_0}(h_0) = 1$ everywhere on E and $U_{h_0} \geqslant 1 \otimes m_0$. We then get

Theorem 2.5. *There exists a positive, σ-finite, P-invariant measure m such that $m \gg \nu$ and such that $m(A) > 0$ implies*

$$P_x\left[\sum_{n=1}^{\infty} 1_A(X_n) = \infty\right] = 1$$

for every x in E. More generally, we have $U_h(h) = 1$ for every $h \in \mathscr{U}_+$ such that $m(h) > 0$.

Proof. Let λ be a measure in the space B of Lemma 2.4; we have $\lambda U_{h_0} I_{h_0} = \lambda(U_{h_0} I_{h_0} - 1 \otimes h_0 m_0)$, and the norm of the operator $U_{h_0} I_{h_0}$ on B is therefore less than or equal to $1 - m_0(h_0) < 1$. By Lemma 2.4, there exists therefore a probability measure Π that is invariant by $U_{h_0} I_{h_0}$, hence such that $\Pi = \Pi U_{h_0} I_{h_0} \geqslant h_0 m_0$.

Next set $m = (1/h_0)\, \Pi$; the measure m is positive and σ-finite, since h_0 is strictly positive and $m \geqslant m_0$, which implies $m \gg \nu$. Finally for $g \in b\mathscr{E}_+$, using the resolvent equation $U_{h_0} = P + U_{h_0} I_{1-h_0} P$, we get

$$\langle m, Pg\rangle = \langle \Pi, h_0^{-1} Pg\rangle = \langle \Pi U_{h_0} I_{h_0}, h_0^{-1} Pg\rangle$$

$$= \langle \Pi, U_{h_0} Pg \rangle = \langle \Pi, U_{h_0} g - Pg + U_{h_0} I_{h_0} Pg \rangle$$

$$= \langle \Pi, U_{h_0} g \rangle = \langle \Pi, h_0^{-1} g \rangle = \langle m, g \rangle,$$

which proves that m is P-invariant.

We proceed to prove the last conclusion in the statement. If h is not m-negligible, then $h' = h \wedge h_0$ is also non-negligible, and since $U_h(h)$ increases with h (ch. 2, Proposition 2.6), we have $U_{h'}(h') \leqslant U_h(h) \leqslant 1$. It suffices, therefore, to prove the desired result for $h \leqslant h_0$. We then have

$$U_h(h) = U_{h_0}(h) + U_{h_0}((h_0 - h)\, U_h(h)),$$

and since $U_{h_0}(h_0) = 1$, setting $g = 1 - U_h(h)$, we get

$$g = U_{h_0}((h_0 - h)\, g).$$

Consequently $g \leqslant (U_{h_0} I_{h_0})\, g \leqslant \cdots \leqslant (U_{h_0} I_{h_0})^n\, g \leqslant \cdots$, and by Lemma 2.4 we get $g \leqslant \Pi(g)$ everywhere on E. It follows that $g \leqslant U_{h_0}(h_0 - h)\, \Pi(g)$, and integrating with respect to Π yields

$$\Pi(g) \leqslant \Pi(g)\, \langle \Pi, U_{h_0}(h_0 - h) \rangle = (1 - m(h))\, \Pi(g).$$

If $m(h) > 0$, then $\Pi(g) = 0$ and $g = 0$ everywhere.

If A is a set such that $m(A) > 0$, we thus have $U_A(1_A) = P_\cdot[S_A < \infty] = 1$ everywhere on E, so that by ch. 2 §3.5 we have $\mathrm{R}(A) = 1_\Omega$ a.s., which completes the proof.

Definition 2.6. The chain X will be said to be *recurrent in the sense of Harris* if there exists a positive, σ-finite, invariant measure m such that $m(A) > 0$ implies

$$P_x\left[\sum_{n=1}^{\infty} 1_A(X_n) = \infty\right] = 1$$

for all x in E. We shall also say more simply that X is a *Harris chain* or that X is *Harris*. The T.P. of a Harris chain will also be said to be Harris.

This definition makes perfect sense even if $\mathscr{E}$ is not separable. However in the non-separable case a ν-irreducible chain may be non-Harris even though its potential kernel is not proper (see Exercise 2.21). In the sequel we keep on working with separable spaces, leaving as exercises to the reader the task of determining which results extend to the general case.

A Harris chain is clearly irreducible, and conversely we have just shown that an irreducible chain such that the potential kernel is not proper is recurrent in the sense of Harris up to a negligible set. The irreducible recurrent discrete chains are Harris chains, and we shall study other important examples later on.

One of the main subjects of this book will be the study of Harris chains. We start here with a few preliminary results.

The condition in Definition 2.6 is equivalent to $U_A(1_A) = 1$ everywhere on E provided $m(A) > 0$. Actually, it is equivalent to a less stringent condition.

Proposition 2.7. *The following three statements are equivalent:*

(i) *X is recurrent in the sense of Harris with respect to the invariant measure m;*

(ii) *there is a strictly positive function $h_0 \in \mathscr{U}_+$ and a non-zero measure m_0, such that $U_{h_0}(h_0) = 1$ on E and $U_{h_0} \geqslant 1 \otimes m_0$;*

(iii) *there is a non-zero positive measure m_1 such that*

$$P_{\cdot}[S_A < \infty] = U_A(1_A) = 1$$

everywhere on E provided $m_1(A) > 0$.

The measures m_0 and m_1 are then absolutely continuous with respect to m.

Proof. Clearly (i) implies (iii), and we saw that (ii) implies (i). It remains to show that (iii) implies (ii).

If (iii) is satisfied, X is clearly m_1-irreducible. Thus, by Proposition 2.2, there is a strictly positive function $h_0 \in \mathscr{U}_+$ and a measure m_0 equivalent to m_1 such that $U_{h_0} \geqslant U_{h_0}(h_0) \otimes m_0$. We must show that $U_{h_0}(h_0) = 1$. Since h_0 is strictly positive there is a number $\alpha \in]0, 1[$ such that the set $A = \{h_0 \geqslant \alpha\}$ is of positive m_1-measure. Since $h_0 \geqslant \alpha 1_A$, we have $U_{h_0}(h_0) \geqslant U_{\alpha 1_A}(\alpha 1_A)$; but the resolvent equation (ch. 2, Proposition 2.5), applied to the kernels U_A and $U_{\alpha 1_A}$, yields

$$U_{\alpha 1_A} = \sum_{n \geqslant 0} (1-\alpha)^n (U_A I_A)^n U_A,$$

and it follows that

$$U_{\alpha 1_A}(\alpha 1_A) = \alpha \sum_{n \geqslant 0} (1-\alpha)^n (U_A I_A)^{n+1} 1 = \alpha \sum_{n \geqslant 0} (1-\alpha)^n = 1,$$

which completes the proof.

We now generalize the results obtained in §1 for irreducible recurrent discrete chains.

Proposition 2.8. *If X is Harris, the measure m is the only σ-finite P-excessive measure (up to multiplication by a positive scalar).*

Proof. Let λ be a σ-finite excessive measure and h_0 the function of Proposition 2.2. The measure $h_0\lambda$ is σ-finite, and ch. 2, Proposition 4.4 implies that

$$h_0\lambda \geqslant (h_0\lambda)\, U_{h_0} I_{h_0} \geqslant \cdots (h_0\lambda)\, (U_{h_0} I_{h_0})^n \geqslant \cdots,$$

and passing to the limit yields, by Fatou's lemma,

$$h_0\lambda \geqslant \lambda(h_0)\, (h_0 m).$$

This implies $\lambda(h_0) < \infty$; otherwise $h_0\lambda$ would not be σ-finite. Now, since $m(h_0) = 1$, the measures on both sides of this inequality have the same total variation, hence they are equal, and the proof is complete.

We shall now improve ch. 2, Proposition 4.4, and prove that, for Harris chains, $1_A m$ is actually invariant by Π_A when A is not negligible.

Proposition 2.9. *If $f \in \mathscr{U}_+$ and $m(f) > 0$, the T.P. $U_f I_f$ is Harris with invariant measure fm; moreover $m = (fm)\, U_f$. In particular for $A \in \mathscr{E}$ and $m(A) > 0$, the chain induced on A is Harris, the trace $1_A m$ of m on A is Π_A-invariant and $mH_A = m$.*

Proof. Set $Q = U_f I_f$; for $h \in \mathscr{U}_+$, the kernel U_h associated with Q will be denoted by U_h^Q and is equal to

$$U_h^Q = \sum_{n \geqslant 0} (U_f I_f I_{1-h})^n\, U_f I_f = U_{hf} I_f.$$

Consequently, if $m(f 1_A) > 0$, Theorem 2.5 implies that

$$U_A^Q(1_A) = U_{1_A f}(1_A f) \equiv 1.$$

The T.P. Q is thus Harris; since by ch. 2 §4.4, the measure fm is Q-excessive, it is Q-invariant and it is the only Q-excessive measure.

Now, by the resolvent equation, we have

$$(fm)\, U_f P = (fm)\, U_f - (fm)\, P + (fm)\, U_f I_f P;$$

since $(fm)\, P \leqslant mP = m$, the measure $(fm)\, P$ is σ-finite, and by the above discussion the two last summands cancel. The measure $(fm)\, U_f$ is therefore P-invariant, and consequently there is a number k such that $km = (fm)\, U_f$.

If f is integrable, multiplying to the right by f immediately gives $k = 1$. If f is not integrable, let g be an integrable function smaller than f. By the above discussion $(gm)\, U_g = m$; applying (gm) to the left of both sides of the equality $U_g = U_f + U_g(f - g)\, U_f$ gives

$$(gm)\, U_g = (gm)\, U_f + (fm)\, U_f - (gm)\, U_f.$$

Since $U_g \geqslant U_f$, the measure $(gm)\, U_f$ is σ-finite, and by cancellation we obtain

$$(fm)\, U_f = (gm)\, U_g = m.$$

Finally, if $A \in \mathscr{E}$ and $m(A) > 0$, the preceding equality yields $mI_A U_A I_{A^c} = 1_{A^c} m$; hence $mH_A = m$.

Remark. That $mH_A = m$ may be derived from ch. 2 §4.5 and the fact that m is the only excessive measure. This result contains Theorem 1.9.

We now turn to the study of superharmonic functions. The next result is dual to Proposition 2.8 (see ch. 2 §4.14 (3)).

Proposition 2.10. *The bounded harmonic functions of a Harris chain are constant and the superharmonic functions are constant m-a.e. More precisely if f is superharmonic there is a constant C such that $f = C$ m-a.e. and $f \geqslant C$ everywhere.*

Proof. If the superharmonic function f is not constant m-a.e. we can find two numbers $a < b$ such that $m(\{f < a\}) > 0$ and $m(\{f > b\}) > 0$. The sequence $\{f(X_n)\}$ is a positive supermartingale which converges a.s. to a random variable Z. Since by hypothesis, X hits the sets $\{f < a\}$ and $\{f > b\}$ infinitely often, we have $Z < a$ a.s. and $Z > b$ a.s., which is a contradiction. There is thus a constant C such that $f = C$ m-a.e. and $\{f(X_n)\}$ converges to C a.s. Since for every x and n we have $f(x) \geqslant E_x[f(X_n)]$ it follows by Fatou's lemma that $f \geqslant C$ everywhere.

If h is bounded and harmonic, by applying the preceding to $h + \|h\|$ and $\|h\| - h$, we see that h is equal to a constant C everywhere.

This proposition shows that for a Harris chain the σ-algebra $\mathscr{I}$ is a.s. trivial. A much stronger result along this line will be proved in ch. 6. By what was seen in ch. 2 §3, we also see that every set is either recurrent or transient. The following proposition gives a criterion to decide which one a particular set is.

Proposition 2.11. *The following three properties are equivalent:*

(i) *A is recurrent;*

(ii) $m(A) > 0$;

(iii) $P_m[S_A < \infty] > 0$.

Proof. If $m(A) = 0$, for every k we have

$$E_m\left[\sum_0^k 1_A(X_n)\right] = 0$$

and A cannot be recurrent; thus (i) implies (ii). That (ii) implies (iii) is obvious. Finally, if (iii) holds, there exists a number $a > 0$ such that

$$m(\{x: P_x[S_A < \infty] > a\}) > 0.$$

Since X is Harris we get therefore, using ch. 2, Proposition 3.4, that

$$\lim_n P_{X_n}[S_A < \infty] = 1_{R(A)} \geqslant a > 0 \quad \text{a.s.}$$

It follows that $P_x[R(A)] = 1$ for each x in E, that is property (i).

Corollary 2.12. *Every transient set A is contained in a transient set B such that B^c is absorbing.*

Proof. Pick $c \in]0, 1[$ and set $B = A \cup \{x: cU_c(x, A) > 0\}$. Then B^c is absorbing by the Markov property and $m(A) = 0$ implies $cmU_c(A) = 0$ and in turn $m\{x: cU_c(x, A) > 0\} = 0$.

Exercise 2.13. Prove that there is no measure for which the chain of translation on $\mathbf{Z}$ ($P(x, \cdot) = \varepsilon_{x+1}(\cdot)$) is irreducible.

Exercise 2.14. If X is Harris a finite function f is superharmonic if and only if it can be written $f = C + Gg$, where C is a constant and $g \in \mathscr{E}_+$ with $m(g) = 0$.

Exercise 2.14. (1) Using the notation of ch. 1, Lemma 5.3, prove that for any T.P. on a separable state space and any probability measure ν such that $\nu P \ll \nu$,

$$P^1_{n+m}(x, \cdot) \leqslant \int_E P^1_n(x, dy)\, P^1_m(y, \cdot),$$

and derive that $g(x) = \lim_n P_n^1(x, E)$ exists for all x in E.

(2) If X is Harris and if m and ν are not mutually singular, then $g(x) = 0$ for all x in E.

[*Hint*: Show that $1 - g$ is superharmonic.]

Exercise 2.15. (1) If X is irreducible and if there is a bounded invariant measure, then we have the situation (ii) of Theorem 2.3. Give an example to show that we can have (i) with a non-bounded invariant measure.

[*Hint*: See §3.]

(2) Prove that in the situation (i) a measure is invariant if and only if it can be written $\lambda + \lambda I_{h_0} U_{h_0}$, where λ is a positive measure such that $\langle \lambda, h_0\, U_0(h_0)\rangle < \infty$ and $\lambda I_{1-h_0} P = \lambda$.

Exercise 2.16. Give an example in which the set $\{U_{h_0}(h_0) < 1\}$ of Theorem 2.3 is not empty although $P1 = 1$.

[*Hint*: Use a discrete irreducible recurrent chain on the space E and take as new space the sum of E and $\mathbf{N}$. Define P such that $P(n, n + 1) = q_n$ and $P(n, E) = 1 - q_n$, and choose the q_n's suitably.]

Exercise 2.17 (converse to Proposition 2.10). Let X be a Markov chain for which there exists a measure ν such that every bounded superharmonic function is ν-a.s. equal to a constant and larger than this constant everywhere; prove that X is Harris.

Exercise 2.18. If X is Harris and L is a death time (ch. 1, Exercise 3.18 and ch. 2, Exercise 2.17) prove that either $L = \infty$ a.s. or $L = 0$ P_m-a.s.

Exercise 2.19. Resume the notation of ch. 2, Exercise 6.11 and for $h = 1_A$ write $V_h = V_A$.

(1) The resolvent $\{V_\alpha\}$ is said to be (*recurrent in the sense of*) *Harris* if there is a positive measure m_1 such that $V_A(1_A) = 1$ everywhere on E provided $m_1(A) > 0$. Prove that the resolvent is Harris if and only if X^1 is Harris and that X^α is then Harris for each $\alpha > 0$. If the resolvent is Harris, there exists a unique invariant measure m which has the same property as m_1. Furthermore the supermedian functions are constant m-a.e.; if in addition the kernels V_α are absolutely continuous with respect to m, the excessive functions are constant.

(2) The resolvent $\{V_\alpha\}$ is said to be ν-*irreducible* if there is a probability measure ν on $(E, \mathscr{E})$ such that $\nu \ll V_\alpha(x, \cdot)$ for all x in E. Prove that Theorems 2.3 and 2.5 extend to the present situation.

Exercise 2.20. Proposition 2.8 cannot be extended to signed measures. For instance, the measure of density x with respect to the Lebesgue measure dx is invariant for the chain X^1 of the resolvent of Brownian motion; see ch. 2, Exercise 6.7. (However, the reader should be cautious, because this measure is a Radon measure, which is not defined for all Borel subsets of $\mathbf{R}$.) The same example shows that there may exist non-constant unbounded harmonic functions.

Exercise 2.21. The following example shows that the separability assumption in Theorems 2.3 and 2.5 cannot be dropped. The state space is the positive half-line endowed with the σ-algebra generated by countable subsets. Let ν be the probability measure defined by $\nu(A) = 0$ if A is countable and $\nu(A) = 1$ otherwise; pick a sequence of numbers a_n such that $0 < a_n < 1$ and $\prod_0^\infty a_n > 0$ and define the T.P. by

$$P(x, A) = a_n I(x + 1, A) + (1 - a_n)\, \nu(A),$$

where $n \leqslant x < n + 1$.

Exercise 2.22. If X is Harris on a non-separable space, prove that m is still the only excessive measure.

[*Hint*: Use the admissible σ-algebras of ch. 1, Exercise 1.17.]

Exercise 2.23 (theory of R-recurrence). We consider a ν-irreducible chain X on a space $(E, \mathscr{E})$ which is not assumed to be separable.

(1) Prove that there exists a real number $R \geqslant 1$, a ν-null set N and a sequence $\{E_n\}$ of sets increasing to E such that R is the common radius of convergence of the series $\sum_0^\infty r^n P_n(x, A)$ for every $x \notin N$ and every A such that $A \subset E_n$ for n sufficiently large and $\nu(A) > 0$. Moreover this series is convergent for every x and A as above or it is divergent for every x in E and every A as above. If $R = 1$, then $N = \emptyset$; give an example where $N \neq \emptyset$ for $R > 1$. In the first case (second case) the chain is said to be R-transient (R-recurrent).

[*Hint*: One can use the kernels $V_r = \sum_1^\infty r^{n-1} P_n$ which satisfy a resolvent-type equation.]

(2) A finite function $f \in \mathscr{E}_+$ is called r-superharmonic (r-harmonic) if $rPf \leqslant f$ ($rPf = f$) ν-a.e., and a measure μ is r-excessive (r-invariant) if $r\mu P \leqslant \mu$ ($r\mu P = \mu$). Prove that for $r > R$ the r-superharmonic functions vanish ν-a.e. and the only r-excessive measure is the zero measure. Prove that in general an r-superharmonic function which does not vanish ν-a.e. is strictly positive.

(3) Assume from now on that X is R-recurrent and prove that the R-superharmonic functions are R-harmonic and the R-excessive measures are R-invariant.

(4) Let $U_B^R = \sum_0^\infty (RPI_{B^c})^n RP$ and prove that for $\nu(B) > 0$ and f R-harmonic and strictly positive we have

$$f = U_B^R I_B f \quad \nu\text{-a.e.}$$

Deduce therefrom that two R-harmonic functions are proportional ν-a.e.

3. Topological recurrence of random walks

In this section we deal with a right random walk X of law μ on a locally compact separable group $\boldsymbol{G}$. The results will obviously carry over to the case of left random walks.

Definitions 3.1. We denote by T_μ (resp. $\boldsymbol{G}_\mu$) the smallest closed semi-group (resp. closed subgroup) containing the support of μ. We shall say that X (or its law μ) *is adapted* if $\boldsymbol{G}_\mu = \boldsymbol{G}$, *irreducible* if $T_\mu = \boldsymbol{G}$, and *aperiodic* if it is adapted and the support of μ is not contained in a coset of a normal closed subgroup.

Unless otherwise stated, we will always assume that the random walks we are dealing with are adapted. The reason for this is the following. It is necessary and sufficient for a point g to be in T_μ, that for every neighbourhood V of g, there exists an integer n such that $\mu^n(V) = P_e[X_n \in V] > 0$. The group $\boldsymbol{G}$ being separable, we therefore have

$$P_e\left[\bigcap_1^\infty \{X_n \in T_\mu\}\right] = 1,$$

and by translation in $\boldsymbol{G}$, for all $g \in \boldsymbol{G}_\mu$

$$P_g\left[\bigcap_0^\infty \{X_n \in \boldsymbol{G}_\mu\}\right] = 1.$$

The sub-group $\boldsymbol{G}_\mu$ is therefore an absorbing set for X, to which we can restrict ourselves for the probabilistic study of X.

Definition 3.2. The random walk X is said to be *recurrent* if every open set O is recurrent, or in other words if $P_\cdot[R(O)] = 1$ everywhere on $\boldsymbol{G}$. Otherwise it is said to be *transient*.

Theorem 3.3. *A random walk is recurrent if and only if for every neighbourhood V of e,*

$$P_e[R(V)] = P_e\left[\overline{\lim_{n\to\infty}}\, \{X_n \in V\}\right] = 1.$$

Proof. Only the sufficiency needs to be shown. Let R_μ be the set of points g in $\boldsymbol{G}$ such that $P_e[R(V)] = 1$ for every neighbourhood V of g. We take $g \in R_\mu$ and $h \in T_\mu$ and we claim that $h^{-1}g \in R_\mu$.

Let U be a neighbourhood of $h^{-1}g$; hUg^{-1} is then a neighbourhood of e and one can find a neighbourhood V of e such that $V^{-1}V \subset hUg^{-1}$. Since Vh is a neighbourhood of h, there exists an integer $k > 0$ such that the event $A = \{X_k \in Vh\}$ is of positive P_e-probability. For $\omega \in A$, the fact that $X_{n+k}(\omega) \in Vg$ implies

$$X_k^{-1}(\omega)\, X_{n+k}(\omega) \in h^{-1}V^{-1}Vg \subset U;$$

and g being in R_μ, for P_e-almost all ω in A, there exist infinitely many integers n such that $X_{n+k}(\omega) \in Vg$, so that

$$P_e\left[\overline{\lim_{n\to\infty}}\, \{X_k^{-1}\, X_{n+k} \in U\} \cap A\right] = P_e[A].$$

According to ch. 1 §4, the sequence of random variables $\{X_k^{-1}\, X_{n+k}\}$ has the same law as the sequence $\{X_n\}$ and is independent of $A \in \mathscr{F}_k$, hence

$$P_e\left[\overline{\lim_{n\to\infty}}\, \{X_n \in U\}\right] = 1.$$

We have thus shown that $T_\mu^{-1}R_\mu \subset R_\mu$, and the hypothesis implies that R_μ contains at least the point e. On the other hand it is easily seen that R_μ is closed. The set R_μ is therefore a closed sub-group containing T_μ, and consequently $R_\mu = \boldsymbol{G}$, which completes the proof.

We shall prove below that if X is transient, then for every compact subset K in $\boldsymbol{G}$ and every $g \in \boldsymbol{G}$,

$$P_g[R(K)] = P_g\left[\overline{\lim_{n\to\infty}}\, \{X_n \in K\}\right] = 0.$$

We begin by generalizing some of the results obtained for discrete chains.

Proposition 3.4. *The random walk is recurrent if and only if for every neighbourhood V of e, $P_e[S_V < \infty] = 1$.*

Proof. Only the sufficiency needs to be shown. Let d be a left-invariant metric on $\boldsymbol{G}$ and V_r the corresponding open ball of centre e and radius r. Let $0 < r' < r$ be two reals; from the relation

$$\{X_{m+n} \notin V_r\} \cap \{X_m \in V_{r'}\} \subset \{X_m^{-1} X_{m+n} \notin V_{r-r'}\},$$

we deduce, since the sequence $\{X_m^{-1} X_{m+n}\}_{n \geqslant 0}$ has the same law as the sequence $\{X_n\}_{n \geqslant 0}$, that

$$P_e\left[\{X_m \in V_{r'}\} \cap \bigcap_{n \geqslant 1} \{X_{m+n} \notin V_r\}\right] \leqslant P_e\left[\bigcap_{n \geqslant 1} \{X_m^{-1} X_{m+n} \notin V_{r-r'}\}\right]$$
$$= P_e[S_{V_{r-r'}} = \infty] = 0.$$

Let $\{r_k\}$ increase to r; as

$$\{X_m \in V_r\} = \lim_k \{X_m \in V_{r_k}\},$$

passing to the limit in the above inequality yields

$$P_e\left[\{X_m \in V_r\} \cap \bigcap_{n \geqslant 1} \{X_{m+n} \notin V_r\}\right] = 0.$$

If $L = L_{V_r}$ (see ch. 1 §3.18), we then have

$$P_e[L < \infty] = \sum_m P_e[L = m] = \sum_m P_e\left[\{X_m \in V_r\} \cap \bigcap_{n \geqslant 1} \{X_{m+n} \notin V_r\}\right] = 0,$$

and the random walk is recurrent.

The potential kernel $G(g, \cdot)$ of X is a convolution kernel. We shall call $\bar{G}$ the measure, possibly non-σ-finite, such that $G(g, \cdot) = (\varepsilon_g * \bar{G})(\cdot)$, that is $\bar{G} = \sum_0^\infty \mu^n$.

Theorem 3.5. *The random walk X is recurrent if and only if $\bar{G}(O) = +\infty$ for every open neighbourhood O of e.*

Proof. The necessity is obvious. To show the sufficiency let us use the notation of the preceding proof. We have

$$1 \geqslant P_e[L < \infty] = \sum_m P_e\left[\{X_m \in V_r\} \cap \bigcap_{n > 0} \{X_{m+n} \notin V_r\}\right]$$

$$\geqslant \sum_m P_e\left[\{X_m \in V_r\} \cap \bigcap_{n>0} \{X_m^{-1}X_{m+n} \notin V_{2r}\}\right],$$

because $x \in V_r$, $x^{-1}y \notin V_{2r}$ implies $y \notin V_r$; as X_m and $X_m^{-1}X_{m+n}$ are independent,

$$1 \geqslant \bar{G}(V_r)\, P_e[S_{V_{2r}} = \infty].$$

By hypothesis $\bar{G}(V_r) = \infty$; thus $P_e[S_{V_{2r}} = \infty] = 0$, and since r is arbitrary, the random walk X is recurrent.

Corollary 3.6. *If the random walk X is transient, then the potential kernel is a proper kernel, the measures $G(g, \cdot)$ are Radon measures, and the set $\{G(g, \cdot), g \in \mathbf{G}\}$ is vaguely relatively compact.*

Proof. By the preceding result there exists an open neighbourhood O of e such that $\bar{G}(O) < \infty$. Let V be an open neighbourhood of e such that $V^{-1}V \subset O$. If $g \in V$,

$$G(g, V) = \bar{G}(g^{-1}V) \leqslant \bar{G}(V^{-1}V) \leqslant \bar{G}(O),$$

and by the maximum principle $G(\cdot, V) \leqslant \bar{G}(O)$ everywhere. For every compact K of $\mathbf{G}$, there exists a finite sequence $g_1, \ldots, g_n$ of points in K such that the sets $g_i^{-1}V$ form an open covering of K. For all $g \in \mathbf{G}$ we then have

$$G(g, K) \leqslant \sum_{i=1}^{n} G(g, g_i^{-1}V) = \sum_{i=1}^{n} G(g_i g, V) \leqslant n\bar{G}(O),$$

which completes the proof.

Remark. If a random walk is recurrent, the dual random walk is also recurrent. On a non-abelian random walk, the right random walk is recurrent if and only if the left random walk is recurrent.

We end this section by studying the relationship between the topological recurrence of random walks and the recurrence in the sense of Harris. This leads us to consider the important class of laws defined below.

Lemma 3.7. *For a probability measure μ on a group $\mathbf{G}$, the following two properties are equivalent:*

(i) *there exists an integer p such that μ^p is not singular with respect to a right Haar measure m on $\mathbf{G}$.*

(ii) *there exists an integer q such that μ^q dominates a multiple of m on a non-empty open subset of* G.

Proof. That (ii) implies (i) is clear. Conversely if (i) holds, there exists a non-zero function g in $L^\infty(m) \cap L^1(m)$ such that $\mu^p \geqslant gm$. Then $\mu^{2p} \geqslant (g * g)\, m$ and, as $g * g$ is continuous and non-zero, (ii) holds for $q = 2p$.

Definition 3.8. A probability measure μ will be called *spread out* if it satisfies the equivalent conditions of Lemma 3.7. A random walk whose law μ is spread out will also be called *spread out.*

This class is much larger than the class of absolutely continuous probability measures. A probability measure can be singular with respect to Haar measure and yet be spread out (see Exercise 3.16). On a discrete group, all probability measures are spread out.

Let us observe that if μ is spread out, for any c in $]0, 1[$ there is a non-zero g in $\mathscr{L}^1_+(m)$ such that $U_c(x, \cdot\,) \geqslant \varepsilon_x * (gm)$, and therefore for any bounded positive Borel function h such that $m(h) > 0$, the function $U_c h$ will dominate a positive non-zero continuous function. This will be used to prove

Theorem 3.9. *A recurrent random walk is recurrent in the sense of Harris if and only if its law μ is spread out.*

Proof. Since the right Haar measure m is σ-finite, positive and invariant, if X is Harris, its invariant measure must be m. Then $m(A) > 0$ implies

$$\sum_{n=0}^{\infty} 2^{-n}\mu^n(A) = \sum_{n=0}^{\infty} 2^{-n} P_e[X_n \in A] > 0;$$

the measures m and $\sum_0^\infty 2^{-n}\mu^n$ cannot be mutually singular, and μ is therefore spread out.

Conversely, if X is recurrent, $U_c\varphi > 0$ for any non-zero $\varphi \in C_K^+$. Let A be such that $m(A) > 0$; by the above remark, $U_{c'} 1_A$ dominates a non-zero continuous function and by the resolvent equation with $c > c'$, we get $U_c 1_A \geqslant (c - c')\, U_c U_{c'} 1_A > 0$. As a result, X is ν-irreducible for any $\nu \sim m$.

Suppose now that h is a bounded, strictly positive function such that $U_0 h$ is bounded. We then have

$$cU_0 U_c h \leqslant U_0 h < \infty$$

which is impossible since $U_c h$ dominates a non-zero positive continuous function.

As a result, X enjoys property (ii) of Theorem 2.3. There exists an absorbing set F such that $m(F^c) = 0$, and for all $g \in F$ and $A \in \mathscr{E}$ with $m(A) > 0$, $P_g[R(A)] = 1$. The random walk X being translation invariant, all the sets obtained from F by translations have the same property and every point in $\mathbf{G}$ belongs to one of these sets; so for $m(A) > 0$, $P.[R(A)] = 1$ everywhere on $\mathbf{G}$.

Corollary 3.10. *If X is recurrent and spread out, a set is transient if and only if it is of Haar measure zero; otherwise it is recurrent.*

Proof. This corollary follows immediately from Proposition 2.11.

Exercise 3.11. Every random walk on a compact group is recurrent.

Exercise 3.12. Let $\mathbf{G}$ be $\mathbf{R}$ or $\mathbf{Z}$ and assume that

$$\int |x| \, \mathrm{d}\mu(x) < \infty, \qquad \lambda = \int x \, \mathrm{d}\mu(x) \neq 0.$$

Show that the random walk X is transient. If $\lambda > 0$, show that X_n converges a.s. towards $+\infty$.

[*Hint*: Use the strong law of large numbers.]

Exercise 3.13 (Right continuous random walks on $\mathbf{Z}$). (1) Assume that $\mathbf{G} = \mathbf{Z}$ and $\mu(x) = 0$ for $x > 1$. Show that, if, with the notation of the preceding exercise, $\lambda > 0$, then $G(0, x) = \lambda^{-1}$ for $x \geqslant 0$.

[*Hint*: Show first that $G(0, x) = G(0, 0)$ for all $x \geqslant 0$. Then write

$$\lambda = P(0, 1) - \sum_{x=1}^{\infty} \sum_{y=-\infty}^{0} P(x, y)$$

and use the identity

$$\sum_{x=-\infty}^{0} G(0, x) - 1 = \sum_{y} G(0, y) \sum_{n=-\infty}^{0} P(y, n).]$$

(2) Particularize to the Bernoulli case: $\mu(1) = p$, $\mu(-1) = 1 - p = q$ with $p > \frac{1}{2}$, and show that for $x < 0$, $G(0, x) = (p - q)^{-1} (p/q)^x$.

Exercise 3.14. (1) Let X be a transient random walk on $\mathbf{R}$ or $\mathbf{Z}$. Show that $t^{-1}\bar{G}([-t, t])$ is bounded by $2\bar{G}([-1, 1])$ for any integer $t > 0$. Extend the result to $\mathbf{R}^p \oplus \mathbf{Z}^q$.

[*Hint*: Split up $[-t, t]$ into disjoint intervals of length 1 and apply the maximum principle.]

(2) Assume that $\int |x| \, d\mu(x) < \infty$ and $\int x \, d\mu(x) = 0$ and show that the random walk is recurrent. Combine this and Exercise 3.13 to prove that the only recurrent Bernoulli random walk is obtained for $p = q = \frac{1}{2}$.

[*Hint*: From the law of large numbers, there exists an integer n_ε such that $n > n_\varepsilon$ implies $P_e[|X_n| < \varepsilon n] > \frac{1}{2}$; then show that

$$t^{-1} \bar{G}([-t, t]) > \tfrac{1}{2}(\varepsilon^{-1} - n_\varepsilon t^{-1})$$

and finish by using the fact that ε is arbitrary.]

(3) Show that the resolvent of brownian motion (ch. 2 §6.7) is recurrent in the sense of Harris.

Exercise 3.15. Let $\boldsymbol{H}$ be the smallest closed normal sub-group containing the support of $\mu * \hat{\mu}$ (where as usual $\hat{\mu}$ is the law of the reversed random walk). Show that $\boldsymbol{H}$ is the smallest closed normal sub-group such that the support of μ is contained in a coset of $\boldsymbol{H}$ in $\boldsymbol{G}$. Give an example in which $\boldsymbol{H}$ is a proper sub-group of $\boldsymbol{G}$, or in other words in which X is not aperiodic.

The following exercises are designed to give a collection of easy results about spread-out probability measures.

Exercise 3.16. Give examples of recurrent random walks non-recurrent in the sense of Harris. Give examples of transient spread-out random walks which are not irreducible. Give examples of singular and spread-out probability measures.

[*Hint*: For the last one, it is possible to use the nilpotent group of upper triangular matrices, that is $\mathbf{R}^3$ with the multiplication law

$$(x, y, z)\,(x', y', z') = (x + x', y + y', z + z' + xy').$$

Check that Lebesgue measure is still the Haar measure and that a probability measure carried by the plane xOy and thus singular, can be spread out.]

Exercise 3.17. Call S_μ the set of points g such that a power μ^p of μ dominates a multiple of $\boldsymbol{m}$ on a neighbourhood of g.

(1) The set S_μ is non-empty if and only if μ is spread-out; it is an open sub-semigroup of $\boldsymbol{G}$ contained in T_μ. Finally, prove that $S_\mu T_\mu \subset S_\mu$.

(2) For a closed semigroup T the following three conditions are equivalent:

(i) there exists a spread-out probability measure μ such that $T_\mu = T$;
(ii) the interior of T is not empty;
(iii) the Haar measure of T is strictly positive.

(3) If μ is spread-out, aperiodic and irreducible, for any $f \in C_K^+$, there is a real α and a power μ^p such that $\mu^p \geqslant \alpha(fm)$.

Exercise 3.18. If $\mathbf{G}$ is connected, any spread-out probability measure on $\mathbf{G}$ is adapted.

Exercise 3.19. A random walk for which the σ-algebra $\mathscr{I}$ of invariant events is a.s.trivial is spread out.

Exercise 3.20. Let μ be spread out and D_n a set carrying the absolutely continuous component of μ^n and such that D_n^c carries the singular component of μ^n. Show that $\mu^n(D_n)$ tends to one as n tends to infinity and derive the existence of an a.s. finite stopping time T such that $P_T(x, \cdot)$ is absolutely continuous with respect to Haar measure, for all $x \in \mathbf{G}$.

[*Hint*: Take T to be the first time for X_n to be in D_n.]

Exercise 3.21. This exercise is designed to give a proof of Theorem 3.9 which does not rely on the theory of irreducible chains.

(1) Using a countable basis for the topology of $\mathbf{G}$, show that there exists a P_e-negligible set N such that, for $\omega \notin N$ and for every open set O,

$$\sum_{n=1}^{\infty} 1_O(X_n(\omega)) = \infty.$$

(2) Let A be a set such that $m(A) > 0$. Set

$$B(\omega) = \left\{g \in \mathbf{G}: \sum_{n=1}^{\infty} 1_A(g^{-1} X_n(\omega) < \infty\right\},$$

and show that $m(B(\omega)) = 0$. Then, using Fubini's theorem, derive that for almost all g, the set

$$\left\{\omega: \sum_{n=1}^{\infty} 1_A(g^{-1} X_n(\omega)) < \infty\right\}$$

is P_e-negligible.

(3) Prove the difficult part of Theorem 3.9. For that purpose one may use the stopping time of Exercise 3.20.

Exercise 3.22. Show that if μ is spread out, bounded harmonic functions are continuous (solution in ch. 5 §1). The result can be extended to random walks on homogeneous spaces.

Exercise 3.23. If X is irreducible with respect to Haar measure, the random walk induced on a homogeneous space is irreducible with respect to quasi-invariant measures.

Exercise 3.24. Suppose that μ is an adapted probability measure on a discrete group $\boldsymbol{G}$. If $\boldsymbol{H}$ is a subgroup, either the random walk of law μ on $\boldsymbol{G}/\boldsymbol{H}$ (ch. 1 §4.4) is irreducible recurrent or all its states are transient. Using the group of positive affine motions of the real line prove that this result does not carry over to non-discrete groups. (See however ch. 4 §4.12.)

4. Recurrence criteria for random walks and applications

The computation of the potential kernels of the preceding section is not, as a rule, an easy matter. We need therefore a more tractable criterion to decide whether a particular random walk is recurrent or not. A related problem is to find the groups for which there exist recurrent random walks. For abelian groups, the harmonic analysis of transition functions provides satisfactory answers to both questions.

In this section, $\boldsymbol{G}$ will be a locally compact *abelian* and metrizable group. We denote by Γ the dual group of $\boldsymbol{G}$ and 0 the identity in Γ. If $\gamma \in \Gamma$, we denote by $\gamma(g)$ the value of the character γ on the point $g \in \boldsymbol{G}$. The Haar measure of Γ will be denoted by m_Γ or $\mathrm{d}\gamma$, and will be adjusted so that the inversion formula holds. We proceed to study a random walk of law μ on $\boldsymbol{G}$, whose Fourier transform will be written

$$\varphi(\gamma) = \int_{\boldsymbol{G}} \gamma(g)\, \mathrm{d}\mu(g).$$

Proposition 4.1. *The law μ is adapted if and only if 0 is the only point $\gamma \in \Gamma$ such that $\varphi(\gamma) = 1$.*

Proof. As 1 is an extremal point in the unit disc, if $\varphi(\gamma) = 1$ then $\gamma(g)$ must be equal to one μ-a.s., and therefore if $\boldsymbol{G}_\mu = \boldsymbol{G}$, $\gamma(g)$ equals 1 identically. The converse is clear.

In the following, t will stand for a positive real number less than one.

It is convenient to remark once for all that if N is a symmetric compact neighbourhood of 0 in Γ, then

$$\int_N \operatorname{Im}\left(\frac{1}{1 - t\,\varphi(\gamma)}\right) \mathrm{d}\gamma = 0.$$

Theorem 4.2. *The random walk X is transient if and only if there exists a neighbourhood N of 0 in Γ such that*

$$\varlimsup_{t \uparrow 1} \int_N \operatorname{Re}\left(\frac{1}{1 - t\,\varphi(\gamma)}\right) \mathrm{d}\gamma < +\infty.$$

Proof. Let N_1 be a symmetric compact neighbourhood of 0 in Γ. Making use of the convolution in Γ, we get

$$\left(\int_{N_1} \gamma(g)\,\mathrm{d}\gamma\right)^2 = \int_\Gamma \gamma(g)\left(\int_\Gamma 1_{N_1}(\gamma')\,1_{N_1}(\gamma - \gamma')\,\mathrm{d}\gamma'\right)\mathrm{d}\gamma$$
$$= \int_\Gamma \gamma(g)\, m_\Gamma\,(N_1 \cap \gamma N_1)\,\mathrm{d}\gamma;$$

taking the integrals of both members with respect to μ yields

$$\int_\Gamma \varphi(\gamma)\, m_\Gamma\,(N_1 \cap \gamma N_1)\,\mathrm{d}\gamma = \int_G \mathrm{d}\mu(g)\left(\int_{N_1} \gamma(g)\,\mathrm{d}\gamma\right)^2.$$

The analogous formula being true for all powers of μ, we get, by summing up,

$$\int_\Gamma \frac{m_\Gamma(N_1 \cap \gamma N_1)}{1 - t\,\varphi(\gamma)}\,\mathrm{d}\gamma = \sum_{n=0}^{\infty} t^n \int_G \mu^n(\mathrm{d}g)\left(\int_{N_1} \gamma(g)\,\mathrm{d}\gamma\right)^2.$$

Let N be a symmetric compact neighbourhood of 0 in Γ and choose N_1 such that $N_1^2 \subset N$. The function $\gamma \to m_\Gamma(N_1 \cap \gamma N_1)$ is zero outside N, so the left-hand side of the preceding equality is a real number less than

$$m_\Gamma(N_1)\int_N \operatorname{Re}\left(\frac{1}{1 - t\,\varphi(\gamma)}\right)\mathrm{d}\gamma.$$

In the right-hand side, the function $g \to (\int_{N_1} \gamma(g)\,\mathrm{d}\gamma)^2$ is continuous and takes the value $(m_\Gamma(N_1))^2$ at e; it is therefore greater than $2^{-1}(m_\Gamma(N_1))^2$ on a neighbourhood V of e in G. We thus get

$$2\int_N \mathrm{Re}\left(\frac{1}{1-t\varphi(\gamma)}\right)\mathrm{d}\gamma \geqslant m_\Gamma(N_1)\left(\sum_{n=0}^{\infty} t^n P_e[X_n \in V]\right),$$

and if X is recurrent, then

$$\lim_{t\uparrow 1}\int_N \mathrm{Re}\left(\frac{1}{1-t\varphi(\gamma)}\right)\mathrm{d}\gamma = +\infty.$$

Conversely, assume X to be transient, and let V_1 be a symmetric compact neighbourhood of e in $\boldsymbol{G}$. The function

$$(1_{V_1} * 1_{V_1} * \mu)\,(h) = \int_{\boldsymbol{G}} m_{\boldsymbol{G}}(hV_1 \cap gV_1)\,\mu(\mathrm{d}g)$$

is continuous, positive definite, and in $L^1(\boldsymbol{G})$, so that it is the inverse Fourier transform of $\varphi(\gamma)\,(\int_{V_1}\gamma(g)\,\mathrm{d}g)^2$. We thus get

$$\int_\Gamma \varphi(\gamma)\left(\int_{V_1}\gamma(g)\,\mathrm{d}g\right)^2 \mathrm{d}\gamma = (1_{V_1} * 1_{V_1} * \mu)\,(e)$$

$$= \int_{\boldsymbol{G}} m_{\boldsymbol{G}}(V_1 \cap gV_1)\,\mu(\mathrm{d}g) \leqslant m_{\boldsymbol{G}}(V_1)\,\mu(V_1^2).$$

The similar inequality is also true for μ^n, and we get, by summing up,

$$\int_\Gamma \left(\int_{V_1}\gamma(g)\,\mathrm{d}g\right)^2 \frac{1}{1-t\varphi(\gamma)}\,\mathrm{d}\gamma \leqslant m_{\boldsymbol{G}}(V_1)\left(\sum_{n=0}^{\infty} t^n P_e[X_n \in V_1^2]\right).$$

Let us choose a symmetric compact neighbourhood N of 0 in Γ and the neighbourhood V_1 such that $G(e, V_1^2) < \infty$ and $\mathrm{Re}\,\gamma(g) \geqslant \frac{1}{2}$ for all $(g \times \gamma) \in V_1 \times N$. We then have

$$m_{\boldsymbol{G}}(V_1)\int_N \mathrm{Re}\left(\frac{1}{1-t\varphi(\gamma)}\right)\mathrm{d}\gamma \leqslant 4\sum_{n=0}^{\infty} t^n P_e[X_n \in V_1^2],$$

and letting t converge to one, we get

$$\overline{\lim_{t\uparrow 1}}\int_N \mathrm{Re}\left(\frac{1}{1-t\varphi(\gamma)}\right)\mathrm{d}\gamma < +\infty.$$

Corollary 4.3. *If X is transient, for every compact neighbourhood N of 0 in Γ, $\mathrm{Re}(1-\varphi(\gamma))^{-1}$ is integrable on N.*

Proof. For all t and γ, $\operatorname{Re}(1 - t\varphi(\gamma))^{-1}$ is $\geqslant 0$ and converges to $\operatorname{Re}(1 - \varphi(\gamma))^{-1}$. One can therefore apply Fatou's lemma to the effect that, with the same N as in the above proof,

$$\int_N \operatorname{Re}\left(\frac{1}{1-\varphi(\gamma)}\right) \mathrm{d}\gamma \leqslant \varliminf_{t\uparrow 1} \int_N \operatorname{Re}\left(\frac{1}{1-t\varphi(\gamma)}\right) \mathrm{d}\gamma < \infty.$$

Since $\operatorname{Re}((1 - \varphi(\gamma))^{-1})$ is continuous and non-zero outside 0, the result is true for all N.

Remarks 4.4. (1) This criterion is in fact necessary and sufficient for X to be transient, but the proof of the sufficiency relies on a profound study of potential theory for recurrent random walks, which will be dealt with in ch. 9.

(2) If G is compact, Γ is discrete, so that $m_\Gamma(\{0\}) > 0$, and it follows from Corollary 4.3 that all random walks on G are recurrent. This is actually straightforward even if G is non-abelian. (See Exercise 3.11.)

The remaining of the section is devoted to applications of the preceding criterion. We first consider the groups $\mathbf{R}^{d_1} \oplus \mathbf{Z}^{d_2}$. We recall that their dual group is isomorphic to $\mathbf{R}^{d_1} \oplus \mathbf{T}^{d_2}$. We shall denote by y the generic element of this group, $y(x)$ its scalar product with $x \in G$, and $\varphi(y) = \int e^{iy(x)} \,\mathrm{d}\mu(x)$, the Fourier transform of μ.

Proposition 4.5. *The random walk of law μ is recurrent in the following two cases:*

(i) $G = \mathbf{R}$ *or* $\mathbf{Z}$, $\int |x| \,\mathrm{d}\mu(x) < \infty$ *and* $\int x \,\mathrm{d}\mu(x) = 0$.

(ii) $G = \mathbf{R}^{d_1} \oplus \mathbf{Z}^{d_2}$ *with*

$$d_1 + d_2 = 2, \qquad \int |x|^2 \,\mathrm{d}\mu(x) < \infty, \qquad \int x \,\mathrm{d}\mu(x) = 0.$$

Proof. (i) It is true that

$$\operatorname{Re}\left(\frac{1}{1-t\varphi}\right) \geqslant \frac{1-t}{(\operatorname{Re}(1-t\varphi))^2 + t^2(\operatorname{Im}\varphi)^2},$$

and

$$(\operatorname{Re}(1-t\varphi))^2 = [(1-t) + \operatorname{Re} t(1-\varphi)]^2 \leqslant 2(1-t)^2 + 2t^2(\operatorname{Re}(1-\varphi))^2.$$

The hypothesis implies that φ is continuously differentiable and that its derivative is zero in 0. Take $\varepsilon > 0$; by Taylor's formula there exists $\alpha > 0$

such that for $|y| < \alpha$ we have

$$|\operatorname{Im} \varphi(y)| \leqslant \varepsilon |y|, \qquad \operatorname{Re}(1 - \varphi(y)) \leqslant \varepsilon |y|.$$

It follows that

$$\int_{-\alpha}^{+\alpha} \operatorname{Re}\left(\frac{1}{1 - t(y)}\right) dy \geqslant (1 - t) \int_{-\alpha}^{+\alpha} \frac{dy}{2(1 - t)^2 + 3t^2\varepsilon^2 y^2}$$

$$\geqslant \frac{1}{3} \int_{-\alpha(1-t)^{-1}}^{\alpha(1-t)^{-1}} \frac{dy}{1 + \varepsilon^2 y^2},$$

and thus

$$\lim_{t \uparrow 1} \int_{-\alpha}^{+\alpha} \operatorname{Re}\left(\frac{1}{1 - t\,\varphi(y)}\right) dy \geqslant \frac{\pi}{3\varepsilon}.$$

If X were transient, there would exist $\alpha_0 > 0$ such that $\operatorname{Re}((1 - t\,\varphi(y))^{-1})$ was positive on $[-\alpha_0, \alpha_0]$ and

$$\lim_{t \uparrow 1} \int_{-\alpha_0}^{\alpha_0} \operatorname{Re}\left(\frac{1}{1 - t\,\varphi(y)}\right) dy = M < \infty.$$

For $\alpha < \alpha_0$, we still would have

$$\lim_{t \uparrow 1} \int_{-\alpha}^{+\alpha} \operatorname{Re}\left(\frac{1}{1 - t\,\varphi(y)}\right) dy \leqslant M < \infty,$$

but this contradicts the fact that ε is arbitrary.

The proof of (ii) is almost identical using polar coordinates and the open ball of radius α and is left to the reader as an exercise.

We now show that, conversely, if $d_1 + d_2 > 2$, all the random walks are transient. For that purpose we need the following

Proposition 4.6. *If $G = \mathbf{R}^{d_1} \oplus \mathbf{Z}^{d_2}$, there exists a constant c such that*

$$\operatorname{Re}[1 - \varphi(y)] \geqslant c|y|^2$$

on a neighbourhood of 0 *in* Γ.

Proof. Since G is generated by the support of μ, one can find $d = d_1 + d_2$ elements $x_1, \ldots, x_d$ linearly independent and belonging to the support of μ. We set $L = \max|x_i|$; then the quadratic form

$$Q(y) = \int_{\{|x| \leqslant L\}} y(x)^2 \, d\mu(x)$$

is positive-definite. Indeed, if $y \neq 0$, we have $y(x_i) \neq 0$ for at least one x_i and therefore $y(x_i)^2 > 0$ on a neighbourhood of x_i; as a result $Q(y)$ is strictly positive. The smallest, proper value λ of Q is then strictly positive and $Q(y) \geqslant \lambda |y|^2$.

On the other hand,

$$\operatorname{Re}[1 - \varphi(y)] = \int_G (1 - \cos y(x)) \, d\mu(x)$$

$$\geqslant 2 \int_{\{|x| \leqslant L\}} \sin^2 \tfrac{1}{2} y(x) \, d\mu(x).$$

But $|\sin(\frac{1}{2}y(x))| \geqslant \pi^{-1}|y(x)|$ if $|y(x)| \leqslant \pi$ and therefore

$$\operatorname{Re}[1 - \varphi(y)] \geqslant 2\pi^{-2} \int_S |y(x)|^2 \, d\mu(x),$$

where $S = \{x : |x| \leqslant L, |y(x)| \leqslant \pi\}$. If $|y| \leqslant \pi L^{-1}$ and $|x| \leqslant L$, then $|y(x)| \leqslant \pi$; thus on the neighbourhood $\{y \colon |y| \leqslant \pi L^{-1}\}$ of 0 in Γ, we have

$$\operatorname{Re}[1 - \varphi(y)] \geqslant 2\pi^{-2} \int_{\{|x| \leqslant L\}} |y(x)|^2 \, d\mu(x) \geqslant 2\lambda\pi^{-2}|y|^2,$$

which yields the result with $c = 2\lambda\pi^{-2}$.

Theorem 4.7. *If $d_1 + d_2 > 2$, all random walks on $\mathbf{R}^{d_1} \oplus \mathbf{Z}^{d_2}$ are transient.*

Proof. Let P be a neighbourhood of 0 in Γ on which $\operatorname{Re} \varphi$ is $\geqslant 0$ and the inequality of Proposition 4.6 holds. Then, for $t < 1$,

$$\operatorname{Re}[(1 - t\,\varphi(y))^{-1}] \leqslant (\operatorname{Re}[1 - \varphi(y)])^{-1},$$

and therefore if $d_1 + d_2 > 2$,

$$\int_P \operatorname{Re}\left[\frac{1}{1 - t\,\varphi(y)}\right] dy \leqslant c^{-1} \int_P \frac{dy}{|y|^2} = M < \infty,$$

which implies that all random walks are transient.

We now deal with the characterization of the groups for which there exist recurrent random walks.

Definition 4.8. A group G (not necessarily abelian) will be called *recurrent* if there exists a probability measure μ on the Borel sets of G such that the random walk of law μ is recurrent.

Compact groups, groups isomorphic to $\mathbf{R}^{d_1} \oplus \mathbf{Z}^{d_2}$ with $d_1 + d_2 \leqslant 2$ are recurrent, but the most general recurrent group is not yet known. We deal below with abelian groups, but the first lemma is true for all groups.

Lemma 4.9. *An open sub-group of a recurrent group is recurrent. In particular every sub-group of a discrete recurrent group is recurrent.*

Proof. Left to the reader as an exercise.

Lemma 4.10. *Let H be a compact sub-group of an abelian group G and σ the canonical continuous and open mapping from G to G/H. Let μ be a probability on G, and $\bar{\mu}$ its image by σ; then, if the random walk of law $\bar{\mu}$ is recurrent, the random walk of law μ is recurrent.*

Proof. Let V be a compact neighbourhood of the identity in G. The sets VH and $\sigma(VH)$ are then also compact neighbourhoods of the identity in G and G/H. Since, as is easily seen,

$$\sum_0^\infty \mu^n(VH) = \sum_0^\infty \bar{\mu}^n(\sigma(VH)),$$

it follows that the random walk of law μ is recurrent if the random walk of law $\bar{\mu}$ is recurrent.

Theorem 4.11. *A denumerable abelian group is recurrent if and only if its rank (maximal number of linearly independent elements) is at most 2.*

Proof. If the rank of G is greater than 2, there exists a sub-group isomorphic to $\mathbf{Z}^3$ which is not a recurrent group; by Lemma 4.9 the group G cannot be recurrent.

Conversely, let us assume that the rank of G is less than or equal to 2. If G is of rank 1, let a_1 be any element of infinite order; if G is of rank 2, let a_1, a_2 be two linearly independent elements and in all cases (rank 0, 1 or 2) let $\{a_n\}_{n\in\mathbf{N}}$ be a sequence generating G and such that a_{m+1} is not in the sub-group G_m generated by $a_1, \ldots, a_m$. The fundamental theorem on finitely generated abelian groups asserts that, for every m, G_m is of the form $\mathbf{Z}^2 \oplus K$ or $\mathbf{Z} \oplus K$ or K, where K is a finite group. It then follows from Proposition 4.5 and Lemma 4.10 that all the random walks we are about to define on $\boldsymbol{G}_m$ are recurrent.

We define inductively a sequence $\mu^{(m)}$ of probability measures carried by $\boldsymbol{G}_m$ by setting $\mu^{(1)}(a_1) = \mu^{(1)}(-a_1) = \frac{1}{2}$, and for $m \geqslant 2$,

$$\mu^{(m)}(g) = (1 - q_m)\,\mu^{(m-1)}(g)$$

if $g \in \boldsymbol{G}_{m-1}$, and

$$\mu^{(m)}(a_m) = \mu^{(m)}(-a_m) = \tfrac{1}{2}q_m.$$

Every point $g \in \boldsymbol{G}$ belongs to at least one $\boldsymbol{G}_m$; we can thus define

$$\mu(g) = \prod_{i=m+1}^{\infty} (1 - q_i)\,\mu^{(m)}(g),$$

and it is easily checked that this number does not depend on m. Now if the infinite product $\prod_{i=2}^{\infty} (1 - q_i)$ is convergent, we have

$$\sum_{g\in\boldsymbol{G}} \mu(g) = \lim_m \sum_{g\in\boldsymbol{G}_m} \mu(g) = \lim_m \prod_{i=m+1}^{\infty} (1 - q_i) = 1,$$

and μ defines a probability measure on $\boldsymbol{G}$.

Plainly $\boldsymbol{G}$ is generated by the support of μ and we shall show that the q_i may be choosen in such a way that μ will be recurrent. Call X^m and X the random walks of law $\mu^{(m)}$ and μ, P_ν^m and P_ν the corresponding probability measures. We choose once for all a sequence of numbers $r_n \in\,]0, 1[$ such that $\prod_1^\infty (1 - r_n)$ converges. We choose a number N_1 and a sequence of numbers $A_j^1 \in\,]0, 1[$ such that

$$\prod_{j=1}^{\infty} (1 - A_j^1)^{N_1} \sum_{k=1}^{N_1} P_e^1[X_k^1 = e] > 1,$$

and we set $q_2 = \min(r_1, A_1^1)$. The random walk X^2 is then known and we choose N_2 and A_j^2 such that

$$\prod_{j=1}^{\infty} (1 - A_j^2)^{N_2} \sum_{k=N_1+1}^{N_2} P_e^2[X_k^2 = e] > 1,$$

and we set $q_3 = \min(r_2, A_2^1, A_2^2)$, which allows us to define X^3. We follow this process inductively. Given N_{n-1} and q_{n-1}, we choose N_n and A_j^n such that

$$\prod_{j=1}^{\infty} (1 - A_j^n)^{N_n} \sum_{k=N_{n-1}+1}^{N_n} P_e^{n-1}[X_k^{n-1} = e] > 1,$$

and define

$$q_n = \min(r_{n-1}, A^1_{n-1}, A^2_{n-1}, \ldots, A^{n-1}_{n-1}).$$

Such choices are always possible since the random walks X^n are recurrent.

Plainly, the infinite product $\prod_1^\infty (1 - q_i)$ converges, and we have, $\bar{G}$ being the potential measure of X,

$$\bar{G}(e) \geqslant \sum_{k=1}^{\infty} P_e[X_k = e] = \sum_{n=1}^{\infty} \sum_{k=N_{n-1}+1}^{N_n} P_e[X_k = e]$$

$$\geqslant \sum_{n=1}^{\infty} \sum_{k=N_{n-1}+1}^{N_n} \prod_{i=n+1}^{\infty} (1 - q_i)^k P_e^{n-1}[X_k^{n-1} = e]$$

since

$$\prod_{i=n+1}^{\infty} (1 - q_i)^k P_e^{n-1}[X_k^{n-1} = e]$$

is the probability for X to be in e at time k without having gone outside G_{n-1}. We further have

$$\bar{G}(e) \geqslant \sum_{n=1}^{\infty} \sum_{k=N_{n-1}+1}^{N_n} \prod_{i=n+1}^{\infty} (1 - q_i)^{N_n} P_e^{n-1}[X_k^n = e]$$

$$\geqslant \sum_{n=1}^{\infty} \sum_{k=N_{n-1}+1}^{N_n} \prod_{i=1}^{\infty} (1 - A_i^n)^{N_n} P_e^{n-1}[X_k^n = e] = +\infty.$$

Thus X is recurrent and the proof is completed.

4.12. The above result leads to a characterization of abelian locally compact metrizable recurrent groups. By the structure theorems, such a group G is isomorphic to $\mathbf{R}^n \times G_0$, where G_0 is a locally compact abelian group containing a compact open sub-group K. From Lemma **4.10** the group G is recurrent if and only if $G/\{e\} \times K$ is recurrent. But this latter group is isomorphic to $\mathbf{R}^n \times G_1$, where G_1 is countable and of rank r. The group G is recurrent if and only if $n + r \leqslant 2$. Indeed, if $n + r \leqslant 2$, it contains the recurrent and dense sub-group $\mathbf{Q}^n \times G_1$.

For instance, the groups $\mathbf{Q}^2 \times K$, where K is compact, and the groups of p-adic numbers, are recurrent.

Exercise 4.13. Show that $\gamma(X_n)/\varphi(\gamma)^n$ is a complex martingale for P_e.

Exercise 4.14. If G is abelian, X is aperiodic if and only if $|\varphi(\gamma)| < 1$ for every $\gamma \neq 0$.

Exercise 4.15. Prove Exercise 3.12 by means of Theorem 5.2.

Exercise 4.16. If G is discrete, Γ is compact. Show that one can then replace N by Γ in Theorems 4.2 and 4.3.

[*Hint*: Express $\mu^n(e)$ by mean of inverse Fourier transform.]

Show that, if $G = \mathbf{Z}^d$, the inequality 4.6 is true on the whole group Γ.

Exercise 4.17. Show that, if the law μ is *symmetric* ($\mu = \hat{\mu}$), the criterion 4.3 is necessary and sufficient for X to be transient.

[*Hint*: Notice that φ is real.]

Exercise 4.18. If G is non-discrete, thus Γ non-compact, and if μ is spread out, then

$$\overline{\lim_{\gamma\to\infty}} |\varphi(\gamma)| < 1.$$

This allows one to show that some non-discrete singular laws are not spread out. For example, the probability measure on $\mathbf{R}$ whose characteristic function equals $\prod_{n=1}^{\infty} \cos(t/n!)$ is singular and is not spread out.

The following exercises deal with recurrent, not necessarily abelian groups.

Exercise 4.19. With the notation of Lemma 4.10 but H not necessarily abelian, show that if μ is recurrent, then $\bar{\mu}$ is recurrent.

Exercise 4.20. Every discrete group such that every finite subset generates a finite sub-group is recurrent.

Exercise 4.21. Let G be the free group with two generators a and b.

(1) Define a law μ on G by $\mu(a) = \mu(b) = \mu(a^{-1}) = \mu(b^{-1}) = \frac{1}{4}$. Show that the random walk of law μ is transient.

[*Hint*: If $|g|$ is the minimal number of symbols a, b, a^{-1}, b^{-1}, necessary to write $g \in G$, the function $g \to 3^{-|g|}$ is superharmonic.]

(2) More generally show that G is not recurrent.

[*Hint*: Every denumerable group is isomorphic to a sub-group of a quotient group of the free group with two generators.]

Exercise 4.22. A group G is said to have the *fixed point property* if, whenever G acts continuously on a compact convex subset Q of a locally convex linear topological space by affine transformations, then G has a fixed point in Q.

(1) Show that a group $\boldsymbol{G}$ has the fixed point property if and only if for every compact $\boldsymbol{G}$-space M there exists a probability measure on M invariant by $\boldsymbol{G}$.

[*Hint*: For the "if" part, use the barycenter of a probability measure on Q invariant by $\boldsymbol{G}$.]

(2) Show that a recurrent group has the fixed point property and is therefore amenable.

[*Hint*: See ch. 2, Exercise **4.18** and ch. 5, Exercise **1.10**.]

Deduce from this fact that free groups with more than one generator and semi-simple Lie groups are not recurrent.

CHAPTER 4

POINTWISE ERGODIC THEORY

This chapter may be seen as the most important in the whole book. It culminates with the celebrated theorems of Birkhoff and Chacon–Ornstein for which we will give or sketch three different proofs; they have many applications in Markov chain theory as well as in other fields which are thus related to one another. We will also touch on subadditive ergodic theory which has proved extremely useful in many situations.

1. Preliminaries

We have already noticed that with a transition probability P we may associate two positive contractions on the Banach spaces $b\mathscr{E}$ and $b\mathscr{M}(\mathscr{E})$; these two contractions are dual in the duality between bounded functions and bounded measures, as is clear from the formula

$$\langle mP, f\rangle = \langle m, Pf\rangle.$$

In this chapter we shall deal mainly with the contraction on $b\mathscr{M}(\mathscr{E})$. Among the closed subspaces of $b\mathscr{M}(\mathscr{E})$, we find the space of bounded measures absolutely continuous with respect to a σ-finite positive measure μ; this space is isomorphic to $L^1(E, \mathscr{E}, \mu)$. We shall be interested in the subspaces of this kind which are invariant by P.

Proposition 1.1. *The space of bounded measures absolutely continuous with respect to the σ-finite positive measure μ is invariant by P if and only if $\mu P \ll \mu$.*

Proof. If $\mu P \ll \mu$ and if $\mu(A) = 0$, then $P(\cdot, A) = 0$ μ-a.e., and consequently for every f in $L^1(\mu)$ we have $\langle f\mu, P(\cdot, A)\rangle = 0$.

Conversely, since μ is σ-finite, there is a sequence $\{f_n\}$ of elements of $L^1(\mu)$ increasing to 1; if $\mu(A) = 0$, the hypothesis implies that, for every n, we have $\langle f_n \mu, P(\cdot, A)\rangle = 0$, and passing to the limit yields $\mu P(A) = 0$.

Remark. The condition $\mu P \ll \mu$ is satisfied if μ is excessive. It is also satisfied if $\mu = \sum_0^\infty 2^{-n} \nu P_n$, where ν is any bounded measure. As a result, for any sequence of bounded measures, there is a measure μ such that $\mu P \ll \mu$ and stronger than any measure in the given sequence.

Definition 1.2. If $\mu P \ll \mu$ the linear operator which maps $f \in L^1(\mu)$ onto the Radon–Nykodim derivative of $(f\,\mu)\,P$ with respect to μ is called *the operator induced* by P on $L^1(\mu)$.

Proposition 1.3. *If $\mu P \ll \mu$, the operator induced by P is a positive contraction.*

Proof. This follows at once from the fact that the map $f \to (f\,\mu)$ from $L^1(\mu)$ into $b\mathscr{M}(\mathscr{E})$ is an isometry, and that P is a contraction on the latter space. The positivity is obvious.

The adjoint operator of an operator induced by a T.P. on $L^1(\mu)$ is an operator on $L^\infty(\mu)$ which is described in the following result.

Proposition 1.4. *If $\mu P \ll \mu$, then P maps functions equal μ-a.e. to functions in $b\mathscr{E}$ into functions of the same kind, and thus defines on $L^\infty(\mu)$ a positive contraction which is the adjoint operator of the operator induced by P on $L^1(\mu)$.*

Proof. If $f \in \mathscr{E}$ is bounded above μ-a.e. by a number M, the set $A = \{f > M\}$ has μP-measure zero; in other words, $P(\cdot, A) = 0$ μ-a.e. and consequently $Pf \leqslant M$ μ-a.e. In the same way it is seen that if $f = g$ μ-a.e. then $Pf = Pg$ μ-a.e. The proof is now easily completed.

Rather than study the operators just described, we shall study an arbitrary positive contraction T on a space $L^1(E, \mathscr{E}, \mu)$. Its adjoint operator which is a positive contraction on $L^\infty(E, \mathscr{E}, \mu)$ will be denoted T^*. We shall need the following lemma, which is obvious when T is induced by a T.P. For f in $L^1(\mu)$ and g in $L^\infty(\mu)$, we shall set

$$\langle f, g \rangle_\mu = \int_E fg \, d\mu,$$

and often drop the subscript μ in the left-hand side when there is no risk of ambiguity.

Lemma 1.5. *Let $\{h_n\}$ be a sequence of elements of $L^\infty(\mu)$ increasing μ-a.e. to an element h of $L^\infty(\mu)$; then $\{T^*h_n\}$ increases μ-a.e. to T^*h.*

Proof. The sequence $\{T^*h_n\}$ increases to a function $g \leqslant T^*h$. If $\mu\{g < T^*h\} > 0$, there is a non-zero element f in L^1_+ such that $\langle f, g\rangle < \langle f, T^*h\rangle$; but for every f in L^1_+ we have

$$\langle f, g\rangle = \lim_n \langle f, T^*h_n\rangle = \lim_n \langle Tf, h_n\rangle$$

$$= \langle Tf, h\rangle = \langle f, T^*h\rangle,$$

so that we get a contradiction.

In what follows, the equalities between sets and functions must always be understood to hold up to μ-equivalence, and we shall indulge in the usual confusion between functions and their equivalence classes.

The powers of T and T^* will be denoted T^n and T^{*n}. We shall use with respect to T^* the notations and definitions introduced for transition probabilities. For instance we say that $\varphi \in L^\infty_+(\mu)$ is superharmonic if $T^*\varphi \leqslant \varphi$ μ-a.e., in particular $\varphi = 1$ is superharmonic, and we set

$$P^*_A = \sum_{n=0}^{\infty} (I_{A^c}T^*)^n I_A.$$

In ch. 2, the proof of the properties of P_A used only the linearity and positivity of the relevant operators. We thus have in the same way that $P^*_A 1$ is the limit of the sequence defined inductively by $g_0 = 1_A$, $g_n = 1_A \vee T^*g_{n-1}$, hence that $P^*_A 1 \leqslant 1$ and that $\lim_n (I_{A^c}T^*)^n P^*_A 1 = 0$. Moreover, if $B \supset A$, then $P^*_B P^*_A = P^*_A$.

Most relationships between the notions for T^* and the same notions for P, when T is induced by P, are of the same kind as that described in the following

Proposition 1.6. *If T is induced by P, there is a one-to-one correspondence between the space of T^*-harmonic functions and the space of equivalence classes of bounded P-harmonic functions. More precisely, if $T^*h = h$, there is a bounded function h' belonging to the equivalence class h such that $Ph' = h'$.*

Proof. We prove only the second sentence. Pick a bounded function g in the class h; by Proposition 1.4, we have $Pg = g$ μ-a.e. Define $g_0 = g$ and g_{n+1}

$= g_n \wedge Pg_n$; this sequence decreases to a function g_∞ equal to g μ-a.e. and such that $Pg_\infty \geqslant g_\infty$. The function $h' = \lim_n P_n g_\infty$ answers our question.

We will finally need the following

Proposition 1.7. *The operator T may be uniquely extended to an operator, still denoted by T, on $\mathscr{E}_+$, in such a way that for every increasing sequence $\{f_n\}$ of functions in $\mathscr{E}_+$, $T(\lim_n f_n) = \lim_n Tf_n$.*

Proof. Let h be a function in $\mathscr{E}_+$ and define Th as the limit of the sequence $\{Th_n\}$ where $\{h_n\}$ is a sequence in L^1_+ increasing to h. If $\{g_p\}$ is another sequence increasing to h, we have

$$\lim_n Th_n \geqslant \lim_n T(h_n \wedge g_p) = Tg_p,$$

since $\{h_n \wedge g_p\}$ converges to g_p in L^1. It is now clear that Th does not depend on the particular sequence $\{h_n\}$ and the proof is easily completed.

The following proposition is, to some extent, a converse to the preceding discussion. It shows that positive contractions of a space L^1 are not much more general than those induced by transition probabilities. We recall that a compact class $\mathscr{K}$ is a family of subsets of E such that, for every sequence $\{C_n, n \geqslant 1\}$ of sets in $\mathscr{K}$ with intersection $\bigcap C_n$ empty, there exists an integer N such that $\bigcap_{n \leqslant N} C_n$ is empty. It is always possible to assume that such a class is closed under finite unions. The family of compact subsets of a topological space provides an example of compact class.

Proposition 1.8. *Suppose $(E, \mathscr{E})$ is separable, and there is a compact class $\mathscr{K}$ such that μ has the approximation property with respect to $\mathscr{K}$, that is, for every F in $\mathscr{E}$,*

$$\mu(F) = \sup\{\mu(K), K \in \mathscr{K}, K \subset F\}.$$

Then every positive contraction T of $L^1(E, \mathscr{E}, \mu)$ is induced by at least one transition probability.

Proof. Let us call B_n, $n \geqslant 1$, the elements of a countable boolean algebra $\mathscr{B}$ generating $\mathscr{E}$. Since $\mathscr{K}$ may be taken closed under finite unions, there exists for every n an increasing sequence $\{C_n^k\}_{k \geqslant 1}$ of sets in $\mathscr{K}$ such that $\mu(B_n) = \lim_k \mu(C_n^k)$. The boolean algebra $\mathscr{D}$ generated by $\mathscr{B}$ and by the sets C_n^k is countable and generates $\mathscr{E}$.

For $F \in \mathscr{E}$, we pick a function $\bar{P}(\cdot, F)$ within the equivalence class of $T^* 1_F$. These functions possess the following properties:

(i) for every $D \in \mathscr{D}$, $0 \leqslant \bar{P}(\cdot, D) \leqslant 1$ μ-a.e.;

(ii) for every pair D, D' of disjoint sets in $\mathscr{D}$

$$\bar{P}(\cdot, D + D') = \bar{P}(\cdot, D) + \bar{P}(\cdot, D') \quad \mu\text{-a.e.};$$

(iii) for every n, the sequence $\{\bar{P}(\cdot, C_n^k)\}_{k \geqslant 1}$ increases μ-a.e. to $\bar{P}(\cdot, B_n)$; this follows by applying Lemma 1.5 to the sequence $1_{C_n^k}$ which increases μ-a.e. to 1_{B_n}.

Since $\mathscr{D}$ is countable, there is a negligible set N such that, for $x \in N^c$, the set function $\bar{P}(x, \cdot)$ satisfies the above properties, hence is a positive and additive set function on $\mathscr{B}$. We claim that it is a countably additive set function.

Let $\{B_j\}$ be a subsequence of $\{B_n\}$, decreasing to $\emptyset$. For every j, every $\varepsilon > 0$ and $x \in N^c$, we may find a set $C_j \in \mathscr{K} \cap \mathscr{D}$ such that $C_j \subset B_j$ and

$$\bar{P}(x, B_j) \leqslant \bar{P}(x, C_j) + 2^{-j}\varepsilon.$$

Since $\bigcap C_j \subset \bigcap B_j = \emptyset$, there is an integer J such that $\bigcap_{j \leqslant J} C_j = \emptyset$. From the formula

$$B_J = \bigcap_{j \leqslant J} B_j \subset \bigcup_{j \leqslant J} (B_j - C_j)$$

we get

$$\bar{P}(x, B_J) \leqslant \sum_{j \leqslant J} \bar{P}(B_j - C_j) \leqslant \varepsilon.$$

Since ε is arbitrary it follows that $\bar{P}(x, \bigcap B_j) = 0$ and therefore $\bar{P}(x, \cdot)$ has a unique extension to a countably additive set function $P(x, \cdot)$ on $\mathscr{E}$. For x in N, we set $P(x, \cdot) = \nu(\cdot)$, where ν is any probability measure equivalent to μ.

By construction, the function $P(\cdot, B)$ is measurable for $B \in \mathscr{B}$, and by the monotone class theorem it is measurable for $B \in \mathscr{E}$. It is now plain that P is a transition probability on $(E, \mathscr{E})$ such that $\mu P \ll \mu$ and that T is induced by P.

Remarks. (1) If E is an LCCB space, or more generally a Polish space, and $\mathscr{E}$ is the σ-algebra of Borel sets, and if μ has the approximation property with respect to compact sets, the conditions of the preceding proposition are satisfied.

(2) If μ is a probability measure on $\mathscr{E}$ and $\mathscr{G}$ is a sub-σ-algebra of $\mathscr{E}$, the map $f \to E[f \mid \mathscr{G}]$ is a positive contraction of $L^1(E, \mathscr{E}, \mu)$ to which we may

apply the above result. In fact we may even choose the functions $\bar{P}(\cdot\,, F)$ in $\mathscr{G}$ and the set N will be in $\mathscr{G}$. It is then easily seen that the T.P. thus constructed has the additional property that $P(\cdot\,, F)$ is $\mathscr{G}$-measurable for every $F \in \mathscr{E}$. Such a T.P. is called a *conditional probability distribution* because for every $f \in \mathscr{E}_+$,

$$E[f \mid \mathscr{G}] = \int_E P(\cdot\,, dy)\, f(y) \quad \mu\text{-a.s.}$$

Exercise 1.9. If μ is excessive for P and T is the contraction induced by P on $L^1(\mu)$, then $T1 \leqslant 1$. (For μ unbounded, $T1$ is defined in Proposition 1.7.)

Exercise 1.10. Let T be a bounded linear operator on the space $L^1_{\mathbf{C}}$ of equivalence classes of complex integrable functions. Prove that there exists a positive (that is, mapping positive real functions into positive real functions) operator τ on $L^1_{\mathbf{C}}$ such that

(i) $|Tf| \leqslant \tau|f|$ for every $f \in L^1_{\mathbf{C}}$,
(ii) $\|\tau\| \leqslant \|T\|$.

For $f \in L^1_+$, τ is given by

$$\tau f = \sup_{|g| \leqslant f} |Tg|.$$

The operator τ is called the *linear modulus* of T. Show that $\tau^n|f| \geqslant |T^n f|$ for every f in $L^1_{\mathbf{C}}$.

Exercise 1.11. Prove for T^* the result similar to Proposition 1.7.

Exercise 1.12. Retain the situation of Remark (2) below Proposition 1.8, but assume in addition that $\mathscr{G}$ is generated by a countable algebra $\mathscr{A}$. Prove that for any x in E the atom A_x containing x, that is the set $\bigcap \{B: B \in \mathscr{G}, x \in B\}$ is equal to $\bigcap \{A : A \in \mathscr{A}, x \in A\}$ and is therefore a set of $\mathscr{G}$. Prove then that the conditional probability distribution may be chosen with the additional feature that $P(x, A_x) = 1$; it is then said to be *regular*.

2. Maximal ergodic lemma. Hopf's decomposition

Theorem 2.1 (Maximal ergodic lemma). *If $f \in L^1(\mu)$ and*

$$E_f = \bigcup_{n \geqslant 0} \left\{ \sum_{k=0}^{n} T^k f > 0 \right\},$$

then

$$\int_{E_f} \varphi f \, \mathrm{d}\mu \geqslant 0$$

for every superharmonic function φ *in* $L_+^\infty(\mu)$.

Proof. Let

$$h_N = \sup(0, f, f + Tf, \ldots, f + Tf + \cdots + T^N f);$$

the sequence $\{h_N\}$ is increasing and the sets $E_N = \{h_N > 0\}$ increase to E_f. Since T is a positive operator, we have

$$Th_N \geqslant \sup(0, Tf, Tf + T^2 f, \ldots, Tf + T^2 f + \cdots + T^{N+1} f)$$

and consequently

$$f + Th_N \geqslant \sup(f, f + Tf, \ldots, f + Tf + \cdots + T^{N+1} f).$$

The right term of this inequality being equal to h_{N+1} on E_{N+1}, we have

$$\begin{aligned}
\int_{E_{N+1}} \varphi f \, \mathrm{d}\mu &\geqslant \int \varphi \, h_{N+1} \, \mathrm{d}\mu - \int_{E_{N+1}} \varphi \, Th_N \, \mathrm{d}\mu \\
&\geqslant \int \varphi \, h_{N+1} \, \mathrm{d}\mu - \int \varphi \, Th_N \, \mathrm{d}\mu \\
&= \int \varphi \, h_{N+1} \, \mathrm{d}\mu - \int T^*\varphi \, h_N \, \mathrm{d}\mu \\
&\geqslant \int \varphi \, h_{N+1} \, \mathrm{d}\mu - \int \varphi \, h_N \, \mathrm{d}\mu \geqslant 0.
\end{aligned}$$

By letting N tend to infinity we thus get the desired result.

In the sequel, for every pair (f, g) of elements of $L^1(\mu)$ we set

$$D_n(f, g) = \sum_{m=0}^{n} T^m f \Big/ \sum_{m=0}^{n} T^m g$$

on the set where the denominator does not vanish. Our goal is to describe the asymptotic behaviour of these ratios when n tends to infinity.

Lemma 2.2. *If* f *and* g *are in* L_+^1, *the function* g *vanishes almost everywhere on the set* $F = \{\sup_n D_n(f, g) = \infty\}$.

Proof. By applying the maximal ergodic lemma with $\varphi = 1$ to the function $f - cg$, where c is a strictly positive real, we find

$$\int_{E_{f-cg}} g \, d\mu \leqslant c^{-1}\|f\|_1.$$

The result follows by letting c tend to infinity, since $F \subset E_{f-cg}$ for every c.

Theorem 2.3 (Hopf's decomposition theorem). *There exists a set* $\boldsymbol{C}$ *in* $\mathscr{E}$, *unique up to equivalence, such that for every* f *in* L^1_+,

$$\sum_{n=0}^{\infty} T^n f = 0 \quad \text{or} + \infty \quad \text{on } \boldsymbol{C},$$

$$\sum_{n=0}^{\infty} T^n f < +\infty \qquad \text{on } \boldsymbol{D} = \boldsymbol{C}^c.$$

Proof. Let g be a strictly positive element in L^1. For every $f \in L^1_+$, Lemma 2.2 implies that $\sum_{n=0}^{\infty} T^n f$ is finite on $\{\sum_{n=0}^{\infty} T^n g < \infty\}$. If f is also strictly positive, it follows that

$$\left\{\sum_{n=0}^{\infty} T^n g < \infty\right\} = \left\{\sum_{n=0}^{\infty} T^n f < \infty\right\} \quad \mu\text{-a.e.}$$

The set $\boldsymbol{C} = \{\sum_{n=0}^{\infty} T^n g = \infty\}$ thus does not depend on the choice of g, provided that g is chosen strictly positive.

Let now $h \in L^1_+$ be such that $\sum_{n=0}^{\infty} T^n h < +\infty$ on a subset B of $\boldsymbol{C}$; then for every k, $\sum_{n=0}^{\infty} T^n(T^k h) < +\infty$ on B, hence $\sup_n D_n(g, T^k h) = +\infty$ on B, and by Lemma 2.2, $T^k h = 0$ μ-a.e. on B. It follows that $\sum_{k=0}^{\infty} T^k h = 0$ μ-a.e. on B and the proof is thus complete.

Definitions 2.4. The set $\boldsymbol{C}$ is called the *conservative part* and the set $\boldsymbol{D}$ the *dissipative part* of E with respect to T. The decomposition of E into its conservative and dissipative part is called *Hopf's decomposition* of E. Finally, if $\boldsymbol{D}$ is empty, T is said to be a *conservative* contraction.

We shall, for a while, restrict ourselves to the study of conservative contractions.

Proposition 2.5. *Assume* T *to be conservative. The class* $\mathscr{C}$ *of sets which are in the equivalence classes*

$$C_f = \left\{\sum_{n=0}^{\infty} T^n f = \infty\right\},$$

where f runs through L^1_+, is a sub-σ-algebra of $\mathscr{E}$. For a function $h \in L^\infty_+$, the following properties are equivalent:

(i) *$T^*h \leqslant h$ μ-a.e.;*

(ii) *$T^*h = h$ μ-a.e.;*

(iii) *h is $\mathscr{C}$-measurable.*

In particular, a set B is in $\mathscr{C}$ if $T^*1_B = 1_B$ on B and only if $T^*1_B = 1_B$ everywhere.

Proof. Let $h \in L^\infty_+$ be such that $T^*h \leqslant h$ and pick a function f in L^1 and strictly positive; for every integer N,

$$\left\langle \sum_{n=0}^{N-1} T^n f, h - T^*h \right\rangle = \langle f, h - T^{*N}h\rangle \leqslant \|f\|_1 \|h\|_\infty < \infty.$$

Since T is conservative, this is impossible unless $h = T^*h$ μ-a.e., which shows the equivalence between (i) and (ii). In particular $T^*1 = 1$.

Call $\mathscr{H}$ the sub-space of L^∞ of functions h such that $T^*h = h$. It contains the function 1 and is closed under increasing limits. Moreover, if h and h' are in $\mathscr{H}$, then $h \wedge h'$ is in $\mathscr{H}$; indeed if a is a constant such that $a + h \geqslant 0$ and $a + h' \geqslant 0$ then

$$a + T^*(h \wedge h') = T^*(a + h \wedge h') \leqslant (a + h) \wedge (a + h') = a + h \wedge h',$$

which gives our claim by the equivalence of (i) and (ii). Consequently, there exists a sub-σ-algebra $\mathscr{C}$ of $\mathscr{E}$ such that $\mathscr{H} = b\mathscr{C}$. If $B \in \mathscr{C}$ we have therefore $T^*1_B = 1_B$. On the other hand, if $T^*1_B = 1_B$ on B, then $T^*1_{B^c} = 0$ on B; hence $T^*1_{B^c} \leqslant 1_{B^c}$ so that, by definition of $\mathscr{C}$, B^c, and hence B, is in $\mathscr{C}$.

Next, we get from Theorem 2.3 that

$$0 = \left\langle 1_{C_f^c}, \sum_{n=1}^{\infty} T^n f \right\rangle = \left\langle T^*1_{C_f^c}, \sum_{n=0}^{\infty} T^n f \right\rangle,$$

which proves that $T^*1_{C_f^c} = 0$ on C_f and consequently that C_f is in $\mathscr{C}$.

Finally, let B be any set in $\mathscr{C}$ and choose f in L^1_+ such that $B = \{f > 0\}$; we have $\sum_{n=0}^{\infty} T^n f = +\infty$ on B, since the first summand is strictly positive on B. On the other hand, for every $n \geqslant 0$, we have

$$\langle 1_{B^c}, T^n f\rangle = \langle T^{*n}1_{B^c}, f\rangle = \langle 1_{B^c}, f\rangle = 0$$

which implies that $\sum_{n=0}^{\infty} T^n f = 0$ on B^c, hence $B = C_f$.

Definition 2.6. The sets of $\mathscr{C}$ are called the *invariant sets*. The $\mathscr{C}$-measurable functions are called *invariant*. If $\mathscr{C}$ is trivial μ-a.e., the conservative contraction T is said to be *ergodic*.

When T is induced by a transition probability P the superharmonic function for P are invariant. Finally we observe that all $\mathscr{E}$-measurable μ-negligible sets are in $\mathscr{C}$.

In the sequel we shall need the following property of invariant functions.

Lemma 2.7. *If h is invariant and bounded, $T(hf) = h(Tf)$ for every f in L^1.*

Proof. We shall first show that $T^*(hg) = h(T^*g)$ for every g in L^∞. It suffices to prove this property when $h = 1_B$ for $B \in \mathscr{C}$ and $g = 1_A$. We have

$$T^*(1_A 1_B) \leqslant \inf(T^*1_A, T^*1_B) = \inf(T^*1_A, 1_B) = 1_B(T^*1_A),$$

and in the same way $T^*(1_{A^c} 1_B) \leqslant 1_B(T^*1_{A^c})$; adding these two inequalities yields that $T^*1_B \leqslant 1_B$. Since by hypothesis $T^*1_B = 1_B$, this shows that the above inequalities are in fact equalities; thus $T^*(1_A 1_B) = 1_B(T^*1_A)$.

The lemma is then a consequence of the following equalities, where g runs through L^∞:

$$\langle h(Tf), g\rangle = \langle Tf, hg\rangle = \langle f, T^*(hg)\rangle = \langle f, h(T^*g)\rangle = \langle T(hf), g\rangle.$$

The dissipative part may be viewed as a "transient" part, while the conservative part may be viewed as a "recurrent" part and the invariant sets as "recurrent classes". The following result as well as Exercise 2.10 support this somewhat heuristic assertion.

Proposition 2.8. *We have $T^*1_{\mathbf{D}} = 0$ μ-a.e. on $\mathbf{C}$. Moreover $\mathbf{D} = \operatorname{ess\,sup} \mathscr{B}$ where*

$$\mathscr{B} = \left\{B : \sum_0^\infty T^{*n} 1_B \in L^\infty\right\}.$$

Proof. Pick a strictly positive function f in L^1. If $Tf = 0$ on a set A, it is easily seen that $T^*1_A = 0$ μ-a.e. and it is enough to prove that $T^*1_{\mathbf{D}\setminus A} = 0$ μ-a.e. on $\mathbf{C}$. Equivalently we may suppose $Tf > 0$ on $\mathbf{D}$. Choose then a sequence $\{D_k\}$ of sets of finite measure increasing to $\mathbf{D}$ and for $c > 0$ set

$$h_c = \inf\left\{1, c\Big/\sum_{n=1}^{\infty} T^n f\right\}.$$

Plainly, h_c increases to 1_D as c tends to infinity. On the other hand

$$\left\langle T^*(1_{D_k} h_c), \sum_{n=0}^{\infty} T^n f \right\rangle = \int_{D_k} h_c \left(\sum_{n=1}^{\infty} T^n f\right) d\mu \leqslant c\,\mu(D_k) < \infty;$$

since on $\boldsymbol{C}$, the sum $\sum_{n=0}^{\infty} T^n f$ is infinite, it follows that $T^*(h_c 1_{D_k}) = 0$ on $\boldsymbol{C}$. By letting c, then k, tend to infinity, we get the first part of the statement.

Set now $F_n = \{h_n > \frac{1}{2}\}$; since $h_n > 2\, 1_{F_n}$, it follows from the above inequalities that $\sum_{l=1}^{\infty} T^{*l}(1_{D_k} 1_{F_n})$ is a.e.finite. By using the complete maximum principle for T^* as in ch. 2, Proposition 1.14, it follows readily that $\boldsymbol{D}$ is the union of a sequence of sets of $\mathscr{B}$.

On the other hand, if $B \in \mathscr{B}$ and f is strictly positive and in L^1 then

$$\left\langle f, \sum_{0}^{\infty} T^{*n} 1_B \right\rangle = \left\langle \sum_{0}^{\infty} T^n f, 1_B \right\rangle < \infty,$$

which is impossible unless $B \subset \boldsymbol{D}$ a.e. The proof of the second part of the statement is thus complete.

If we refer to the case in which T is induced by a T.P., the above results say that the chain does not pass from $\boldsymbol{C}$ to $\boldsymbol{D}$. We also have that if f is a positive function vanishing on $\boldsymbol{D}$, then Tf vanishes on $\boldsymbol{D}$; indeed

$$\int_{\boldsymbol{D}} Tf\, d\mu = \int f\, T^* 1_{\boldsymbol{D}}\, d\mu = 0.$$

We can therefore study the restriction of T to $\boldsymbol{C}$, that is, to the functions vanishing on $\boldsymbol{D}$, and this restriction is clearly a conservative contraction. In the sequel, by invariant functions we shall mean functions vanishing on $\boldsymbol{D}$ and invariant for the restriction of T to $\boldsymbol{C}$, and the σ algebra $\mathscr{C}$ will be a σ-algebra of subsets of $\boldsymbol{C}$. In this setting we prove

Lemma 2.9. *For every measurable subset A of $\boldsymbol{C}$ there exists a smallest (up to equivalence) invariant set $\bar{A}$ containing A, and*

$$\bar{A} = \left\{\sum_{n=0}^{\infty} T^n 1_A = \infty\right\} = \left\{\sum_{n=0}^{\infty} T^n 1_A > 0\right\} = \{E[1_A \mid \mathscr{C}] > 0\},$$

where the conditional expectation is taken with respect to any probability measure equivalent to μ. Moreover, $1_{\bar{A}} = 1_{\boldsymbol{C}} P_A^* 1$.

Proof. Since $\mathscr{C}$ is complete with respect to μ, it is a classical exercise to show that the set $\bar{A} = \{E[1_A \mid \mathscr{C}] > 0\}$ does not depend on the particular probability measure used in the conditional expectation and that it is the smallest set in $\mathscr{C}$ containing A. Moreover even if A has infinite measure it follows easily from Propositions 1.7 and 2.5 that

$$\left\{\sum_0^\infty T^n 1_A > 0\right\} = \left\{\sum_0^\infty T^n 1_A = \infty\right\}$$

and that this set is in $\mathscr{C}$ and contains A, hence $\bar{A}$. On the other hand, for every n,

$$\langle 1_{\bar{A}^c}, T^n 1_A\rangle = \langle T^{*n} 1_{\bar{A}^c}, 1_A\rangle = \langle 1_{\bar{A}^c}, 1_A\rangle = 0,$$

which implies that $\sum_0^\infty T^n 1_A = 0$ on $\bar{A}^c$ and therefore that $\{\sum_{n=0}^\infty T^n 1_A = \infty\} = \bar{A}$.

In the same way as in ch. 2, it is seen that $P_A^* 1$ is the smallest superharmonic function greater than 1_A; it is therefore equal to $1_{\bar{A}}$ on C.

Exercise 2.10. Use the results of this section to give another proof of some results in ch. 3 §1, namely that there is a partition $D, C_1, C_2, \ldots$ of the countable space E such that:

(i) $\sum_{n=0}^\infty P_n(x, y) < \infty$ for any x in E if $y \in D$;

(ii) $\sum_{n=0}^\infty P_n(x, y) = \infty$ or 0 if $x \in C_i$, according as y is in C_i or not.

Exercise 2.11. (1) Let P be a T.P. possessing a bounded invariant measure μ. Show that the contraction induced by P on $L^1(\mu)$ is conservative and moreover that it is ergodic if and only if μ cannot be written as the sum of two different (non-proportional) invariant measures. If in addition E is countable all points with positive μ-measure are recurrent.

(2) (Poincaré's recurrence theorem). Let θ be a measure-preserving point transformation of the probability measure space $(E, \mathscr{E}, m)$. Prove that for every set A with $m(A) > 0$ and for almost every $x \in A$ there exists an infinite sequence of numbers n_i for which $\theta^{n_i}(x) \in A$.

(3) (Kac's recurrence theorem). The situation being that of (2), set

$$\nu_A(x) = \inf\{n > 0\colon \theta^n(x) \in A\}$$

if this set is non-empty, $\nu_A(x) = \infty$ otherwise. Prove that

$$m(\{x\colon \nu_A(x) < \infty\}) = \int_A \nu_A \, dm.$$

[*Hint*: Show that $\{\nu_A < \infty\} = \bar{A}$, then use the formula $mH_A \leqslant m$ of ch. 2 §4, which in the present case is an equality.]

Exercise 2.12. Let T be a conservative contraction of $L^1(E, \mathscr{E}, \mu)$.

(1) Prove that a function f in L^1 belongs to the closure of the image of $I - T$ if and only if it is orthogonal to $L^\infty(E, \mathscr{C}, \mu)$.

(2) Given that f and g are two functions in $L^1(E, \mathscr{E}, \mu)$, prove that $n^{-1}\sum_{k=0}^{n-1} T^k(f - g)$ converges to zero in the L^1 sense as $n \to \infty$ if and only if, for every $C \in \mathscr{C}$,

$$\int_C f \, d\mu = \int_C g \, d\mu.$$

Exercise 2.13. Continuation of Exercise 1.10. Assume that τ is conservative and let $\mathscr{C}$ be the relevant σ-algebra of invariant sets. Prove that there is a function s in $L_{\mathscr{C}}^\infty$ and a set $\Gamma \in \mathscr{C}$ such that

(i) $|s| = 1$ μ-a.e. on Γ and $Tf = \bar{s}\tau(sf)$ for every $f \in L^1(\Gamma)$ (we denote by $L^1(A)$ the set of functions of L^1 vanishing outside A);

(ii) if $\Delta = E \backslash \Gamma$, $(I - T)\, L^1(\Delta)$ is dense in $L^1(\Delta)$;

(iii) this partition of E is unique up to equivalence;

(iv) a function r satisfies the properties stated for s if and only if there is a function $l \in L_{\mathscr{C}}^\infty(\mu)$ such that $|l| = 1$ μ-a.e. on Γ, $\tau^* l = l$ μ-a.e. on Γ, and $r = sl$.

Exercise 2.14. Prove the following converse to Proposition 2.5: if $T^*f \leqslant f$, $f \in L_+^\infty$, implies $Tf = f$, then T is conservative.

Exercise 2.15. (1) Let P be a T.P. and μ be such that $\mu P \ll \mu$. Let θ be the shift operator of the associated canonical Markov chain. Prove that $Z \to Z \circ \theta$ is a positive contraction on $L^1(P_\mu)$ and call S its adjoint.

(2) If $\boldsymbol{C} \cup \boldsymbol{D}$ is a Hopf decomposition for the contraction T induced by P on $L^1(\mu)$, prove that $\{\omega: X_0(\omega) \in \boldsymbol{C}\} \cup \{\omega: X_0(\omega) \in \boldsymbol{D}\}$ is a Hopf decomposition for S. Show that S is ergodic if and only if T is ergodic.

Exercise 2.16. If P satisfies the equivalent conditions of ch. 2, Exercise 3.13 for every x in E, and if the contraction it induces on $L^1(E, \mathscr{E}, \nu)$ is conservative, the σ-algebra $\mathscr{C}$ is atomic.

Exercise 2.17. If T is conservative and induced by P, the σ-algebra $\mathscr{C}$ is equal to the class of sets which may be written $A \cup N$ where A is absorbing for P and $\mu(N) = 0$.

[*Hint*: Use Proposition 1.6.]

Exercise 2.18 (Another proof of Hopf's decomposition theorem). We assume that $\mu(E) < \infty$.

(1) For $c > 1$ and $f \in L_+^\infty$ with $\|f\|_\infty \leqslant 1$, set

$$h_c = \left(cf - \sum_{m=1}^{\infty} T^{*m}f\right)^+.$$

Show that on $\{h_c > 0\}$, $\sum_{n=0}^{N} T^* h_c \leqslant c^2$ for every integer N. Prove that the same inequality holds on $\{h_c = 0\}$, hence everywhere.

(2) When c tends to infinity, the sets $B_c = \{h_c \geqslant c^{-1}\}$ increase to

$$\{f > 0\} \cap \left\{\sum_{n\geqslant 1} T^{*n}f < \infty\right\};$$

prove that B belongs to the class $\mathscr{B}$ of Proposition 2.8.

(3) Show that $\mathscr{B}$ is closed under finite unions. Set $\boldsymbol{D} = \operatorname{ess\,sup} \mathscr{B}$ and $\boldsymbol{C} = \boldsymbol{D}^c$ and prove that for $h \in L_+^\infty$ one has $\sum_0^\infty T^{*n}h = 0$ or $+\infty$ on $\boldsymbol{C}$. Prove that $T^* 1_{\boldsymbol{D}} = 0$ on $\boldsymbol{C}$.

(4) Prove that the partition of E into $\boldsymbol{C}$ and $\boldsymbol{D}$ is a Hopf decomposition of E for T.

Exercise 2.19. For $f \in L^1$ define inductively the sequence $\{f_n\}$ by

$$f_0 = f, \ldots, f_{n+1} = Tf_n^+ - f_n^-.$$

Notice that if T is induced by P, this is nothing else that the filling scheme applied to the pair $(f^+\mu, f^-\mu)$, which makes some of the results below intuitively obvious.

(1) Prove by induction that

$$\sum_{k<n} T^k f^+ - \sum_{k<n} f_k^+ = \sum_{k<n} T^k f^- - \sum_{k<n} T^{n-1-k} f_k^- \geqslant 0.$$

(2) Set $f_\infty^- = \lim f_n^-$ and notice that $\int f \,\mathrm{d}\mu \geqslant - \int f_\infty^- \,\mathrm{d}\mu$. Furthermore deduce from (1) that $f_\infty^- = 0$ on the set E_f and prove that this has Theorem 2.1 with $\varphi = 1$ as an immediate consequence.

Exercise 2.20. This exercise is designed to show that a property stronger than the maximal ergodic lemma may be derived from the properties of the filling scheme. Take T induced by P and for $f \in L^1(\mu)$ set $\lambda = f^+\mu$, $\nu = f^-\mu$.

(1) Prove that for every n,

$$\sum_0^n (f\mu) P_k \leqslant \sum_0^n \lambda_k.$$

[*Hint*: Suppose the result to be true up to the order $n - 1$ and apply this induction hypothesis to $f' = Tf^+ - f^-$.]

(2) Using the fact that $\nu - \nu_\infty \prec \lambda$ prove that for any $A \subset E_f$, one has $\nu I_A \prec \lambda$ and that this implies Theorem 2.1.

3. The Chacon–Ornstein theorem for conservative contractions

In this section we study a conservative contraction and prove the Chacon–Ornstein theorem; we will get Birkhoff's theorem as a corollary. The study of general contractions is postponed to §5. We begin with the technical

Lemma 3.1. *Let g be a strictly positive element in $L^1(\mu)$; then for every f in $L^1(\mu)$ and integer k,*

$$\lim_n \left(T^{n+k}f \Big/ \sum_{n=0}^{\infty} T^m g \right) = 0 \quad \mu\text{-a.e.}$$

Proof. We can obviously make $k = 0$ in the proof. Denote by ν the bounded measure $g\mu$, and for $\varepsilon > 0$, set

$$f_0 = f - \varepsilon g, \ldots, f_n = T^n f - \varepsilon \left(\sum_{m=0}^{n} T^m g \right), \qquad A_n = \{f_n > 0\}.$$

We have $f_n = Tf_{n-1} - \varepsilon g$, and consequently

$$f_n^+ = f_n 1_{A_n} \leqslant Tf_{n-1}^+ - \varepsilon 1_{A_n} g,$$

which implies

$$\varepsilon \nu(A_n) \leqslant \langle Tf_{n-1}^+, 1\rangle - \langle f_n^+, 1\rangle \leqslant \langle f_{n-1}^+, 1\rangle - \langle f_n^+, 1\rangle,$$

and summing up on n, we obtain

$$\varepsilon \sum_1^{\infty} \nu(A_n) \leqslant \int f^+ \, d\mu < \infty.$$

By the Borel–Cantelli lemma we conclude that $\nu(\overline{\lim}\, A_n) = 0$, which completes the proof.

Theorem 3.2 (Chacon–Ornstein theorem). *If T is conservative and g is a strictly positive element in $L^1(\mu)$, then for every element f in $L^1(\mu)$ the ratios $D_n(f, g)$ converge almost everywhere to $E[f \mid \mathscr{C}]/E[g \mid \mathscr{C}]$ as n tends to infinity.*

The meaning of the ratios of conditional expectations when μ is not bounded will be explained in the proof.

Proof. We first assume that μ is bounded. The principle of the proof is to show the result for larger and larger classes of functions.

(a) If $f = hg$ with h invariant, then by Lemma 2.7 we have $T^n f = h(T^n g)$; hence $D_n(f, g) = h$. The result follows at once since

$$E[hg \mid \mathscr{C}] = hE[g \mid \mathscr{C}].$$

(b) If $f = (I - T)\, k$ with $k \in L^1$, then $E[f \mid \mathscr{C}] = 0$; indeed

$$\langle h, f\rangle = \langle h, (I - T)\, k\rangle = 0$$

for every invariant h. Since $\sum_{m=0}^{n} T^m f = k - T^{n+1}k$, it suffices to show that $T^{n+1}k/\sum_0^n T^m g$ converges to zero as n tends to infinity, but this follows from Lemma 3.1.

(c) The subspace L_g of L^1 of functions of the form $hg + (I - T)\, k$ is dense in L^1. Indeed, let l be an element of L^∞ orthogonal to L_g. Then $\langle l, (I - T)\, k\rangle = 0$ for every $k \in L^1$, which proves that $T^*l = l$, that is that l is invariant. Thus lg is in L_g, and therefore we have also $\langle l, lg\rangle = \int l^2 g \, d\mu = 0$, which shows that $l = 0$, since g is strictly positive.

It remains thus to show the result for the functions in the closure $\bar{L}_g$ of L_g. Let $\{f_p\}$ be a sequence of elements in L^1 such that $\sum_p \|f_p\|_1 < +\infty$; we claim that the sequence $\sup_n |D_n(f_p, g)|$ converges to zero a.s. Pick $\varepsilon > 0$; by the maximal ergodic lemma applied to $|f_p| - \varepsilon g$ it follows that

$$\varepsilon \sum_p \int_{E_{|f_p| - \varepsilon g}} g \, d\mu \leqslant \sum_p \int |f_p| \, d\mu < \infty.$$

By the Borel–Cantelli lemma,

$$(g\mu)\left(\overline{\lim_p}\, E_{|f_p| - \varepsilon g}\right) = 0,$$

which proves that $\sup_n D_n(f_p, g)$ converges to zero μ-a.e. as p tends to infinity.

Let now f be in $\bar{L}_g$; we may find a sequence of functions f_p in L_g converging to f in the L^1-sense and such that $\sum_p \|f - f_p\|_1 < \infty$ (this may be achieved by taking a sub-sequence); we have therefore

$$\lim_p \sup_n D_n(f - f_p, g) = 0 \quad \text{a.s.}$$

But

$$|D_n(f, g) - E[f \mid \mathscr{C}]/E[g \mid \mathscr{C}]| \leqslant |D_n(f_p, g) - E[f_p \mid \mathscr{C}]/E[g \mid \mathscr{C}]| \\ + \sup_n |D_n(f - f_p, g)| + |E[f - f_p \mid \mathscr{C}]/E[g \mid \mathscr{C}]|.$$

If we let n and then p tend to infinity, we see that on the right-hand side of the inequality the first term converges to zero because f_p is in L_g, the second one on account of the above remark and the third one because the conditional expectation being a contraction for the L^1-norm, we have $\sum_p \|E[f - f_p \mid \mathscr{C}]\|_1 < \infty$ hence the series $\sum_p |E[f - f_p \mid \mathscr{C}]|$ is a.s. finite and the conditional expectations themselves converge to zero a.s. The proof is thus complete for the case in which μ is a bounded measure.

When μ is unbounded, the conditional expectations are no longer meaningful. For instance, if $\mathscr{C}$ contains an atom B such that $\mu(B) = \infty$, the conditional expectation of a function $f \in L^1_+$ should be constant on B, which is impossible, since we should then have $\int_B E[f \mid \mathscr{C}]\, d\mu = 0$ or $+\infty$, whereas $0 < \int_B f\, d\mu < \infty$. This is due to the fact that the restriction of μ to $\mathscr{C}$ is not bound to be σ-finite, and therefore that the Radon–Nykodim theorem does not apply. This difficulty will be dealt with in the following way.

Suppose now that μ is σ-finite, and let h be a strictly positive function in $L^1(\mu)$; call μ' the bounded measure $h\mu$ which is equivalent to μ. A function f is in $L^1(\mu')$ if and only if hf is in $L^1(\mu)$. For $f \in L^1(\mu')$, we define $T'f = h^{-1}T(fh)$; T' is a positive and conservative contraction of $L^1(\mu')$, as is easily seen. Moreover, the spaces $L^\infty(\mu)$ and $L^\infty(\mu')$ are identical, and the adjoint contraction of T' is still T^*, since

$$\langle T'f, g\rangle_{\mu'} = \langle T(fh), g\rangle_\mu = \langle f, T^*g\rangle_{\mu'}.$$

The σ-algebra of invariant sets is thus the same for T and T', and for $f, g \in L^1(\mu')$ such that $\{g > 0\} = E$,

$$\lim_n D_n'(f, g) = E_{\mu'}[f \mid \mathscr{C}]/E_{\mu'}[g \mid \mathscr{C}] \quad \mu\text{-a.e.}$$

Let f, g be now in $L^1(\mu)$ and such that $\{g > 0\} = E$; by applying the preceding result to f/h and g/h, we find that

$$\lim_n D_n(f, g) = E_{\mu'}[f/h \mid \mathscr{C}]/E_{\mu'}[g/h \mid \mathscr{C}] \quad \mu\text{-a.e.}$$

The ratio on the right-hand side thus does not depend on the particular function h – this can be seen directly – and we say that it is equal to $E_\mu[f \mid \mathscr{C}]/E_\mu[g \mid \mathscr{C}]$, thus defining the ratio of two conditional expectations even if conditional expectations themselves are meaningless. The proof of the Chacon–Ornstein theorem is now complete.

The hypothesis of the Chacon–Ornstein theorem may be weakened in several ways that we shall indicate in the sequel and in the exercises. We prove below that it is not necessary to take g strictly positive.

Indeed if we call $A = \{g > 0\}$, the set $\bar{A}$ being in $\mathscr{C}$, by the same argument as below Proposition 2.8, we may restrict T to $\bar{A}$ and then apply the Chacon–Ornstein theorem to the functions g and $E[g \mid \mathscr{C}]$, since the latter is strictly positive on $\bar{A}$. We thus find that $D_n(g, E[g \mid \mathscr{C}])$ tends to 1 on $\bar{A}$. If f is now any function in L^1, by considering the product

$$D_n(f, E[g \mid \mathscr{C}])\, D_n(E[g \mid \mathscr{C}], g)$$

we obtain, on account of Lemma 2.9, the following.

Theorem 3.3. *If T is conservative and g positive, for every function f in L^1, the ratio $D_n(f, g)$ converges a.e. to $E[f \mid \mathscr{C}]/E[g \mid \mathscr{C}]$ on the set*

$$\{E[g \mid \mathscr{C}] > 0\} = \left\{\sum_0^\infty T^n g = \infty\right\} = \left\{\sum_0^\infty T^n g > 0\right\}.$$

If we think of the case in which T is induced by a T.P., we see that we have obtained a theorem about T, but that it is T^* which has probabilistic significance. It is therefore advisable to prove an ergodic theorem for T^*. To this end we shall need an additional hypothesis.

Proposition 3.4. *If h is a finite and strictly positive function (not necessarily in L^1) such that $Th \leqslant h$ μ-a.e., the space $\mathscr{H} = \{f \in L^1 : f/h \in L^\infty\}$ is dense in L^1 and the map $f \to hT^*(f/h)$ from $\mathscr{H}$ to L^1 extends to a unique positive contraction S of L^1. The Hopf decompositions are the same for S and T and the σ-algebras of invariant sets are equal.*

Proof. Notice that f/h is in L^∞ if and only if there is a constant M such that $|f| \leqslant Mh$ a.e.; if g is in L^1_+, the functions $g_n = g \wedge nh$ are in $\mathscr{H}$ and increase to g, hence converge to g in the L^1-sense. It follows that $\bar{\mathscr{H}} = L^1$.

Next, for $f \in \mathscr{H}$, set $Sf = hT^*(g/h)$; for $f \geqslant 0$ we have

$$\|Sf\|_1 = \int h\, T^*(f/h)\, \mathrm{d}\mu = \int Th(f/h)\, \mathrm{d}\mu \leqslant \int f\, \mathrm{d}\mu,$$

and consequently the operator S may be uniquely extended into a positive contraction of $L^1(\mu)$. Using the result in Exercise 1.11, it is easily seen that for the extended operator S we still have $Sf = hT^*(f/h)$.

We now observe that S and T play completely symmetric roles. If $f \in L^\infty_+$ and $g \in \mathscr{H}$, we have

$$\begin{aligned}\langle S^*f, g\rangle &= \langle f, Sg\rangle = \langle f, hT^*(g/h)\rangle \\ &= \langle fh, T^*(g/h)\rangle = \langle T(fh), g/h\rangle,\end{aligned}$$

which proves that for f in $\mathscr{H}$ we have $Tf = hS^*(f/h)$. (Note that $Sh \leqslant h$.)

Let now C_T (resp. C_S) be the conservative part for T (resp. S) and f be a strictly positive element in L^1. Set $D_S = \{\sum_0^\infty S^n f < \infty\}$ and let g be a bounded positive function such that $D_S = \{g > 0\}$ and $\langle g, \sum_0^\infty S^n f\rangle < \infty$. By duality, this implies that

$$\left\langle h^{-1} \sum_0^\infty T^n(gh), f \right\rangle = \left\langle g, h \sum_0^\infty T^{*n}(f/h) \right\rangle < \infty;$$

the series in the left-hand side diverges a.e. on the set $D_S \cap C_T$ which perforce must be empty. It follows that $C_T \subset C_S$ and by symmetry we get $C_S = C_T$.

Finally the invariant sets are the same for S and T. Indeed, assume S and T to be conservative and let g be a bounded invariant function for T; on account of Lemma 2.7 it is easily checked that $T(gh) = g(Th)$, hence that $S^*g = g(Th/h) \leqslant g$; the function g is thus invariant for S and we conclude once again from the symmetric roles played by S and T.

Remark. If T is induced by a T.P. P and if μ is P-excessive, the hypothesis of the preceding proposition is satisfied with $h = 1$ (see Exercise 1.9); in that case, $S = T^* = P$. If moreover there exists a T.P. $\hat{P}$ in duality with P relative to μ the contraction S is the contraction induced by $\hat{P}$.

We now turn to the ergodic theorem for T^*.

Theorem 3.5. *If T is conservative, and if there exists a finite and strictly positive function h (not necessarily in L^1), and such that $Th \leqslant h$ μ-a.e., then for every pair f, g of functions in $L^1_+(h\mu)$, the ratios*

$$\sum_0^n T^{*m} f \Big/ \sum_0^n T^{*m} g$$

converge μ-a.e. to $E[fh \mid \mathscr{C}]/E[gh \mid \mathscr{C}]$ on the set $\{E[gh \mid \mathscr{C}] > 0\}$.

Proof. By the above discussion, all we have to do is to apply Theorem **3.3** to the operator S and to the functions fh and gh.

We will end this section by proving as a corollary to the Chacon–Ornstein theorem the fundamental theorem of Birkhoff which will be generalized in §6.

Theorem 3.6. *Let θ be a measure preserving point transformation of the probability measure space $(E, \mathscr{E}, m)$ and $\mathscr{I}$ the σ-algebra of sets A such that $A = \theta^{-1}(A)$ m-a.s. Then for every function f in $L^1(m)$,*

$$\lim_n \, (n+1)^{-1} \sum_0^n f \circ \theta^k = E[f \mid \mathscr{I}]$$

m-a.s. and in the L^1-sense.

Proof. Because of the invariance of m under θ, the map $T: f \to f \circ \theta$ is a positive contraction of $L^1(m)$; it is conservative since $\sum_0^\infty f \circ \theta^k$ is obviously infinite for the function $f = 1$ which is in $L^1(m)$. Furthermore $\mathscr{I}$ is the σ-algebra of invariant sets for T; indeed the sets $\{\sum_0^\infty f \circ \theta^k = \infty\}$, $f \in L^1_+(m)$, are plainly invariant by θ and on the other hand if $A = \theta^{-1}(A)$ m-a.e., then $A = \{\sum_0^\infty 1_A \circ \theta^k = \infty\}$ m-a.e.

The almost-sure convergence in the statement follows now by applying the Chacon–Ornstein theorem to T and to the functions f and 1. If f is bounded the convergence holds also in the L^1-sense by Lebesgue's theorem. The usual density argument then gives the L^1-convergence for any $f \in L^1(m)$.

Exercise 3.7. (1) Prove that in Lemma 2.7 one may drop the assumption that h is bounded, and take h merely $\mathscr{C}$-measurable.

(2) Prove that the Chacon–Ornstein theorem is still true if f, g are not supposed integrable but merely such that $E[f \mid \mathscr{C}] < \infty$ and $E[g \mid \mathscr{C}] < \infty$.

[*Hint*: $f/E[f \mid \mathscr{C}]$ is integrable.]

Exercise 3.8. (1) Let f, g be in L^1_+ and set $h = \sup_n D_n(f, g)$. Show that, for $p > 1$, one gets

$$\|h\|_p \leqslant (p/(p-1)) \, \|f/g\|_p$$

where the norm is that of $L^p(E, \mathscr{E}, g\mu)$.

[*Hint*: Prove that

$$\int_{\{h>a\}} f \, d\mu \geqslant a \int_{\{h>a\}} g \, d\mu$$

and integrate this inequality with respect to the measure $pa^{p-2} \, da$ on $[0, \infty[$.]

(2) Given that T is conservative and g strictly positive, prove that

$$\lim_n D_n(f, g) = E[f \mid \mathscr{C}]/E[g \mid \mathscr{C}]$$

in the sense of $L^p(E, \mathscr{E}, g\mu)$, provided that $f/g \in L^p(E, \mathscr{E}, g\mu)$.

(3) By using the restriction of the Lebesgue measure to $[1, \infty[$ and the inequality $a \log b \leqslant a \log^+ a + be^{-1}$, solve the same problem for $p = 1$ and prove that the convergence holds in the sense of $L^1(g\mu)$ provided that

$$\int fg^{-1} \log^+(fg^{-1}) \, g \, d\mu < \infty.$$

Exercise 3.9 (continuation of Exercise 2.13). Let g be a real, strictly positive function in L^1 and f any element in L^1_C. Prove that the ratio $\sum_1^n T^k f / \sum_1^n \tau^k g$ converges μ-a.s. to a $\mathscr{C}$-measurable limit $l(f, g)$. Moreover, for every $C \in \mathscr{C}$, show that

$$\lim_n \int_C s \, T^n f \, d\mu = \int_C s \, l(f, g) \, g \, d\mu.$$

Exercise 3.10. Prove that if $Th \leqslant h$ and if T is conservative, then $Th = h$, and h is invariant.

Exercise 3.11. For $A \in \mathscr{E}$, define $H_A = I_A \sum_0^\infty (TI_{A^c})^n$. In the setting of Theorem 3.2 call L'_g the space generated by the functions $H_A g$ where A runs through $\mathscr{E}$. Prove that L'_g is dense in L^1 and could be used in lieu of L_g in the proof of Theorem 3.2.

Exercise 3.12. If μ is excessive for P and $C \cup D$ is the Hopf decomposition of E for the contraction $f \to Pf$ of $L^1(\mu)$ then $P1_D \leqslant 1_D$ μ-a.e. and $P1_C \leqslant 1_C$ μ-a.e.

Exercise 3.13. Retain the situation and notation of Theorem 3.6.

(1) (*Ergodicity*). In accordance with §2 we say that θ is ergodic if $\mathscr{I}$ is m-a.s.trivial. Prove that the following three conditions are equivalent:

(i) θ is ergodic;

(ii) if $\theta^{-1}(A) \subset A$ m-a.s. then $m(A) = 0$ or $m(A) = 1$;

(iii) for every pair (A, B) of sets in

$$\lim_n n^{-1} \sum_0^{n-1} m(A \cap \theta^{-k}(B)) = m(A)\, m(B).$$

If θ is invertible and θ^{-1} is a measurable mapping, m is also invariant with respect to θ^{-1} and θ^{-1} is ergodic if and only if θ is.

(2) If m_1 and m_2 are two probability measures invariant by the same point transformation θ and if θ is ergodic with respect to both, then either $m_1 = m_2$ or the two measures are mutually singular.

(3) If θ is the shift operator of a canonical Markov chain, under what condition on ν does the space $(\Omega, \mathscr{F}, P_\nu)$ satisfy the conditions of (1)?

(4) Use (1) and the space $(W, \mathscr{A}, P)$ of ch. 1, Exercise 4.12 to give a proof of the law of large numbers.

Exercise 3.14. Let P be a Feller T.P. on a compact metrizable space E.

(1) Prove that there is at least one invariant measure for P.

[*Hint*: Use the same device as in ch. 2 §4.18.]

(2) The following three conditions are equivalent:

(i) P has only one invariant probability measure;

(ii) for every continuous function f on E, the sequence $n^{-1} \sum_0^{n-1} P_k f$ converges to a constant uniformly on E;

(iii) for every continuous function f on E, there is a sequence n_j of integers such that the sequence $n_j^{-1} \sum_0^{n_j-1} P_k f$ converges to a constant pointwise.

(3) The set of invariant probability measures for P is convex and compact for the topology of vague convergence. Characterize the extremal points of this set.

[*Hint*: See (2) in the preceding exercise.]

4. Applications to Harris chains

In this section we will apply the results of the previous three to the case of Harris chains. We will moreover study the relationship between conservativity and the recurrence in the sense of Harris. Throughout the section the σ-algebra $\mathscr{E}$ is assumed separable although some of the results extend to the general case. We begin with

Proposition 4.1. *If X is Harris with invariant measure m, the map $f \to Pf$ is a conservative and ergodic contraction of $L^1(m)$.*

Proof. By the remark following Proposition 3.4 this map is the operator S of the preceding section. It is obvious that for $f \in L^1_+(m)$, the sets $C_f = \{\sum_0^\infty P_n f = \infty\}$ are m-a.e. equal to E or to the empty set according as $m(f) > 0$ or $m(f) = 0$, and this proves the proposition.

As an immediate corollary we get

Theorem 4.2. *If X is a Harris chain with invariant measure m and if f, g are in $L^1_+(m)$ with $m(g) > 0$, then*

$$\lim_n E.\left[\sum_{k=0}^{n} f(X_k)\right] \Big/ E.\left[\sum_{k=0}^{n} g(X_k)\right] = m(f)/m(g) \quad m\text{-a.e.}$$

Proof. Since $P_n f(x) = E_x[f(X_n)]$ the result follows at once from Theorem 3.3 and from the preceding result.

If X is an irreducible recurrent discrete chain, the empty set is the only set of measure zero, so that the above convergence holds everywhere. If f and g are characteristic functions of sets in $\mathscr{E}$, the preceding theorem expresses that the ratio of the mean times spent by the chain in these two sets tends eventually to the ratio of their measures. This result will be sharpened in ch. 6. We proceed with another application to Harris chains.

Let X be a canonical chain recurrent in the sense of Harris, with invariant measure m. The map $T: Z \to T \circ \theta$ is a positive contraction of $L^1(\Omega, \mathscr{F}, P_m)$, since by the Markov property, we have

$$\int_\Omega Z \circ \theta \, dP_m = \int_E m(dx)\, E_x[Z \circ \theta] = \int_E m\, P(dx)\, E_x[Z] = \int_\Omega Z \, dP_m.$$

We claim that this contraction is conservative and ergodic. If $m(A) > 0$ and if $Z = 1_A(X_0)$, then by ch. 3 §2,

$$\sum_{n=0}^{\infty} T^n Z = \sum_{n=0}^{\infty} 1_A(X_n) = \infty \quad P_m\text{-a.e.}$$

On the other hand the sets

$$\left\{\sum_{n=0}^{\infty} T^n Z = \infty\right\} = \left\{\sum_{n=0}^{\infty} Z \circ \theta_n = \infty\right\}$$

are invariant by θ, hence equal P_m-a.e. to Ω or to $\emptyset$.

Theorem 4.3. *Let f, g be in $L^1_+(E, \mathscr{E}, m)$ with $m(g) > 0$, then the ratios $\sum_0^n f(X_k)/\sum_0^n g(X_k)$ converge P_ν-a.s. to $m(f)/m(g)$ for any probability measure ν.*

Proof. Let Λ be the set of ω's for which the above ratios converge to the assigned limit, that is

$$\Lambda = \left\{\omega \in \Omega : \lim \sum_0^n f(X_k(\omega)) \Big/ \sum_0^n g(X_k(\omega)) = m(f)/m(g)\right\}.$$

By the above discussion and the Chacon–Ornstein theorem we have $P_m[\Lambda^c] = 0$. On the other hand it is easily seen that the equivalence class of the set Λ is invariant by θ, so that, by the results in ch. 3 §2, the function $P.[\Lambda]$ is identically zero or one. The proof is now easily completed.

Remark. If X is positive and $m(E) = 1$, we may take $g = 1$ and it turns out that $n^{-1} \sum_0^{n-1} f(X_k)$ converges a.s. to $m(f)$. This is the so-called "*Law of large numbers for functionals of a Markov chain*".

We now want to investigate the converse to Proposition 4.1, namely the conditions under which a chain which induces a conservative and ergodic contraction on a L^1-space is a Harris chain. We will need the following

Definition 4.4. Let ν be a probability measure on $\mathscr{E}$. The chain X is said to be *ν-essentially irreducible* if for every $f \in \mathscr{E}_+$ such that $\nu(f) > 0$ we have $U_c(f) > 0$ ν-a.s. for one (hence for all) $c \in]0, 1[$.

This condition is obviously weaker than irreducibility. It is easily checked that a chain which induces a conservative and ergodic contraction on the space $L^1(\nu)$ of a probability measure ν is ν-essentially irreducible; this is in particular the case for Harris chains if ν is taken equivalent to m. We will now work in the converse direction and see how ν-essential irreducibility relates to the ν-irreducibility of ch. 3 §2.

Theorem 4.5. *For a ν-essentially irreducible chain X such that $\nu P \ll \nu$, there are only two possibilities:*

(i) *there is an absorbing set E_1 such that $\nu(E_1^c) = 0$ and the restriction of X to E_1 is ν-irreducible;*

(ii) *there is an absorbing set* E_2 *such that* $\nu(E_2^c) = 0$ *and for all* x *in* E_2, *the measures* ν *and* $U_c(x,\cdot)$ *are mutually singular for one, hence for all*, $c \in]0, 1[$.

Proof. Pick $c \in]0, 1[$ and set

$$A = \{x \in E : \nu \ll U_c(x, \cdot)\};$$

we claim that A is in $\mathscr{E}$. Let $\mathscr{G}$ be a denumerable algebra of sets generating $\mathscr{E}$ and put

$$\mathscr{G}_n = \{G \in \mathscr{G} : \nu(G) \geqslant 1/n\};$$

then ν is not absolutely continuous with respect to $U_c(x, \cdot)$ if and only if there is an integer $n > 0$ such that, for all $\delta > 0$, there is a $G \in \mathscr{G}$ such that $U_c(x, G) \leqslant \delta$ and $\nu(G) \geqslant 1/n$. Consequently,

$$A^c = \bigcup_n \left\{x : \inf_{G \in \mathscr{G}_n} U_c(x, G) = 0\right\},$$

and since each $\mathscr{G}_n$ is countable the set A is in $\mathscr{E}$.

Plainly A does not depend on the choice of c within $]0, 1[$. We proceed to show that $A^c = \{x : \nu \perp U_c(x, \cdot)\}$. Let x be a point in E such that $U_c(x, \cdot)$ is not singular with respect to ν; if $\nu(F) > 0$ the resolvent equation and the ν-essential irreducibility imply that, for $d < c$,

$$U_d(x, F) \geqslant (c - d) \int U_c(x, \mathrm{d}y)\, U_d(y, F) > 0.$$

Thus $\nu \ll U_d(x,\cdot)$, and x is in A.

Let x be a point in E such that $U_d(x, A) > 0$. For every F in $\mathscr{E}$ such that $\nu(F) > 0$, we have $U_c(\cdot, F) > 0$ on A, and consequently

$$U_d(x, F) \geqslant (c - d) \int_A U_0(x, \mathrm{d}y)\, U_c(y, F) > 0;$$

it follows that $\nu \ll U_d(x,\cdot)$, hence that x is in A. Thus $U_d(\cdot\,, A) = 0$ on A^c and A^c is an absorbing set. Moreover, we have either $\nu(A) = 0$ or $\nu(A^c) = 0$; indeed, if $\nu(A) > 0$, by hypothesis $U_c(\cdot\,, A) > 0$ ν-a.s., and since $U_c(\cdot\,, A) = 0$ on A^c, we get $\nu(A^c) = 0$.

Thus either $\nu(A) = 0$ and (ii) is verified for $E_2 = A^c$, or $\nu(A^c) = 0$ and we shall show that (i) is verified for $E_1 = \{x : U_c(x, A^c) = 0\}$. Since $\nu P \ll \nu$ we have $\nu U_c \ll \nu$; hence $\nu U_c(A^c) = 0$, which proves that $\nu(E_1^c) = 0$. Moreover one easily sees that E_1 is absorbing. Finally, since A^c is absorbing, we have $U_c(\cdot\,, A^c) > 0$ on A^c; hence $E_1 \subset A$ and therefore $\nu \ll U_c(x,\cdot)$ for any x in E_1.

The second case in the above theorem is usually what happens for ergodic point transformations, the first one leads to the characterization of Harris chains which we are looking for. Roughly speaking a conservative and ergodic chain is Harris provided there is some absolute continuity of the T.P. with respect to the given measure.

Theorem 4.6. *If P induces a conservative and ergodic contraction on $L^1(\nu)$, set $A_c = \{x\colon U_c(x,\cdot\,) \perp \nu,\ c \in\,]0,1[\}$. Then P is Harris, after possible deletion of a ν-null set, if and only if $\nu(A_c^{\mathrm{c}}) > 0$ for some, hence for any, c in $]0,1[$.*

Proof. The "only if" part is given by Proposition 4.1 and the results in ch. 3 §2. To prove the "if" part we observe that we are in the setting of the preceding result, but that the hypothesis $\nu(A_c^{\mathrm{c}}) > 0$ rules out the possibility (ii). There is therefore an absorbing set E_1 which carries ν and the restriction of X to E_1 is ν-irreducible. By Theorems 2.3 and 2.5 in ch. 3 the chain is consequently either Harris (after possibly deleting a ν-null set) or the potential kernel is proper. In the latter case we would be able to find a set F such that $\nu(F) > 0$ and $G(\cdot\,, F)$ be bounded and this would contradict the conservativity of P. The proof is complete.

We will finally use the foregoing results to study duality for Harris chains. In ch. 3, we saw that an irreducible recurrent discrete chain always has a dual chain which is also irreducible recurrent. Under suitable regularity assumptions we will prove the analogous property for Harris chains. Another proof yielding more, but relying on more elaborate notions, will be given in ch. 8.

We begin with a preliminary lemma. Let m be a positive σ-finite measure and N a positive kernel on $(E, \mathscr{E})$. We define a σ-finite measure Π_N on $\mathscr{E} \otimes \mathscr{E}$ by setting

$$\Pi_N(A \times B) = \int_A m(\mathrm{d}x)\, N(x, B).$$

Lemma 4.7. *The density $\bar{\omega}$ of Π_N with respect to $m \otimes m$ is equal $m \otimes m$-a.e. to the density n of N with respect to m.*

Proof. Clearly $\Pi_N \geqslant n(m \otimes m)$, hence $\bar{\omega} \geqslant n$ $m \otimes m$-a.e. If this inequality were not an equality there would exist a positive non-negligible function φ on $E \times E$ such that

$$\Pi_{N^1} \geqslant \varphi(m \otimes m),$$

where N^1 is the singular part of N (ch. 1, §5); hence for all A

$$N^1(\cdot\,, A) \geqslant \int \varphi(\cdot\,, y)\, 1_A(y)\, m(\mathrm{d}y) \quad m\text{-a.e.}$$

Letting A run through a denumerable algebra of sets generating $\mathscr{E}$, we obtain that for almost all x,

$$N^1(x, \mathrm{d}y) \geqslant \varphi(x, y)\, m(\mathrm{d}y)$$

which contradicts the definition of N^1.

Theorem 4.8. *If P and $\hat{P}$ are two T.P.'s in duality with respect to the measure m, and if P is Harris, then there is modification of $\hat{P}$ which is also Harris.*

Proof. Since P is Harris, the measure m is the invariant measure for P. Let p_c be a strictly positive density for U_c with respect to m and $\hat{p}_c$ a version of the density of $\hat{U}_c$. The duality formula of ch. 2 §4 between U_c and $\hat{U}_c$ implies immediately that $\Pi_{U_c} = \Pi_{\hat{U}_c}$, hence that $\hat{p}_c = p_c\; m \otimes m$-a.e.

On the other hand, by the discussion in the preceding section, the contraction induced by $\hat{P}$ is conservative and ergodic since that induced by P is. By Theorem 4.6, $\hat{P}$ is Harris up to a m-null set N, which can be stated by saying that there is a modification of $\hat{P}$ which is Harris. Indeed we get the desired modification by setting $\hat{P}'(x,\cdot\,) = \hat{P}(x,\cdot\,)$ for $x \notin N$ and $\hat{P}'(x,\cdot\,) = 1 \otimes fm$ for $x \in N$, where $f \in \mathscr{E}_+$ and $m(f) = 1$.

Finally, the existence of a dual T.P. is settled by the following theorem which applies in particular to Harris chains.

Theorem 4.9. *If the measure m is excessive for P and has the approximation property with respect to a compact class, there is at least one T.P. which is in duality with P with respect to m.*

Proof. The operator $S : f \to Pf$ is a positive contraction of $L^1(m)$. By Proposition 1.8, there exists therefore a T.P. denoted by $\hat{P}$ such that $m\hat{P} \ll m$ and $Sf = \mathrm{d}(fm)\,\hat{P}/\mathrm{d}m$. If f and g are in $b\mathscr{E} \cap L^1_+(m)$ we have, confusing functions and equivalence classes of functions,

$$\langle Pf, g\rangle = \langle Sf, g\rangle = \int_E g\,\frac{\mathrm{d}(fm)\,\hat{P}}{\mathrm{d}m}\,\mathrm{d}m = \langle f, \hat{P}g\rangle,$$

and therefore $\hat{P}$ is in duality with P with respect to m.

As a result of the foregoing discussions, every Harris chain on a Polish space such that m has the approximation property with respect to compact sets has at least one dual chain which is also Harris. It may have several which are modifications of one another; in most usual cases however, for instance random walks, one of them is naturally given and has a canonical character; it will often be referred to as *the* dual chain.

Exercise 4.10 (Doeblin's ratio limit theorem). Let X be discrete and irreducible recurrent; then for any four states x, y, z, t show that

$$\lim_n \left(\sum_0^n P_k(x, y) \Big/ \sum_0^n P_k(z, t) \right) = m(y)/m(t);$$

[*Hint*: Use the dual chain.]

Exercise 4.11. Let X be a Harris chain and, with the notations of ch. 3 §2.14, call S_n the positive contraction of $L^1(m)$ induced by P^1_n. Prove that for every $f \in L^1(m)$,

$$\sum_0^\infty S_n f < +\infty \quad m\text{-a.e.}$$

Exercise 4.12. Let G be a LCCB group, H a closed subgroup and μ a probability measure on G adapted and spread-out. Set $M = G/H$ and call Y the Markov chain on M with transition probability $P(x, \cdot\,) = \mu * \varepsilon_x$.

(1) Prove that if A is a Borel subset of M such that $P1_A = 1_A$ m-a.e., where m is a quasi-invariant measure on M, then either $m(A) = 0$ or $m(A^c) = 0$.

[*Hint*: One may translate the hypothesis on G and use Proposition 1.6 together with the fact (ch. 5 §1) that bounded harmonic functions are continuous.]

(2) If $M = \bigcup_{k=0}^\infty M_k$, where $h_{M_k} = 0$ for every k, every compact set in M has a bounded potential.

(3) When $T\mu = G$, prove the following dichotomy property for Y: either Y is recurrent in the sense of Harris with respect to m or the potential of any compact set is bounded.

(4) Prove the same result with the following set of hypothesis: the measure m is relatively invariant that is there exists a continuous function χ on G such that $\varepsilon_g * m = \chi(g)\, m$ for every $g \in G$, and $C = \int \chi(g)\, d\mu(g) \leqslant 1$. If $C < 1$ we are in the latter case and if $C = 1$ and we are in the former case then m is the invariant measure for Y.

[*Hint*: Notice that m is excessive and use Exercise 3.12.]

Exercise 4.13. Let μ be a probability measure on a group G; the random walk of law μ ($\hat{\mu}$) is supposed to be recurrent but not necessarily Harris. Recall that the convolution by μ ($\hat{\mu}$) is a positive contraction of $L^1(m_G)$.

(1) Prove that these contractions are conservative. Derive that $G(\cdot, A) = \infty$ m_G-a.e. if and only if $m_G(A) > 0$; otherwise it is zero m_G-a.e.

(2) The corresponding σ-algebras $\mathscr{C}$ are trivial, in other words, if $f \in b\mathscr{E}$ and $f = f * \mu$ m_G-a.e., then f is constant m_G-a.e.

(3) Prove in this setting a result similar to Theorem 4.1.

(4) Derive from (1) and (2) that if $\nu \ll m_G$ and $A \in \mathscr{E}$, then $P_\nu[R(A)] = 0$ or 1 according as $m_G(A)$ is positive or zero. (See ch. 2, Exercise 3.13.)

Exercise 4.14. Let X be a positive Harris chain and f be a bounded measurable function on $(E^k, \mathscr{E}^{\otimes k})$. Prove that

$$n^{-1} \sum_{l=0}^{n-1} f(X_l, X_{l+1}, \ldots, X_{l+k-1})$$

converges a.s. and compute the limit.

5. Brunel's lemma and the general Chacon–Ornstein theorem

In this section, we give another method of proof, which will enable us to state the Chacon–Ornstein theorem in a slightly more general fashion. This method also sheds more light on the relationship with the potential theory of ch. 2. Finally, it is easily extended to resolvents, as will be seen in the exercises at the end of the section.

We could write this section independently of the results in §§ 2 and 3, but this would involve repeating some definitions and proofs already given, and therefore we shall not do so. As before, T is a positive contraction of $L^1(E, \mathscr{E}, \mu)$.

Theorem 5.1 (Brunel's maximal ergodic lemma). *For $f \in L^1(\mu)$ and any set*

$$A \subset E'_f = \overline{\lim_n} \left\{ \sum_{k=0}^{k=n} T^k f > 0 \right\},$$

we have

$$\int f P_A^* 1 \, d\mu \geq 0.$$

Proof. Let h be a strictly positive element of L^1, and assume that for $\varepsilon > 0$, we have

$$\sum_0^n T^k f \geqslant \varepsilon \sum_0^n T^k h$$

on A for infinitely many n. If $g_m = \sum_0^m (TI_{A^c})^k f$, we have

$$f = 1_A g_m + (TI_{A^c})^{m+1} f + 1_{A^c} g_m - (TI_{A^c})\, g_m,$$

and consequently

$$\sum_0^n T^k f = \sum_0^n T^k (1_A g_m + (TI_{A^c})^{m+1} f) + 1_{A^c} g_m - T^{n+1}(1_{A^c} g_m).$$

Set $\psi = 1_A g_m + (TI_{A^c})^{m+1} f$; this function is in $L^1(\mu)$. Dividing the above equality by $\sum_{k=0}^n T^k h$, we easily see that, on the set A,

$$\sum_{k=0}^n T^k \psi \Big/ \sum_{k=0}^n T^k h \geqslant \varepsilon + T^{n+1}(1_{A^c} g_m) \Big/ \sum_{k=0}^n T^k h$$

for infinitely many n. It follows from Lemma 3.1 that $A \subset E_\psi$; also since $P_A^* 1$ is superharmonic, the maximal ergodic Lemma 2.1 yields

$$\int_{E_\psi} (1_A g_m + (TI_{A^c})^{m+1} f)\, P_A^* 1 \,\mathrm{d}\mu \geqslant 0.$$

It follows that

$$\int_E f \left(\sum_0^m (I_{A^c} T^*)^k 1_A \right) \mathrm{d}\mu = \int_E 1_A g_m \,\mathrm{d}\mu$$

$$= \int_{E_\psi} 1_A g_m P_A^* 1 \,\mathrm{d}\mu \geqslant - \int_{E_\psi} P_A^* 1 (TI_{A^c})^{m+1} |f| \,\mathrm{d}\mu$$

$$\geqslant - \int_E |f| (I_{A^c} T^*)^{m+1} P_A^* 1 \,\mathrm{d}\mu.$$

When m tends to infinity, the last term tends to zero, and the first one to $\int_E f P_A^* 1 \,\mathrm{d}\mu$, which gives, in that case, the desired conclusion. Now the function $f + \varepsilon h$ and any set $A \subset E'_f$ always satisfy the requirements of the beginning of the proof; thus

$$\int (f + \varepsilon h)\, P_A^* 1 \,\mathrm{d}\mu \geqslant 0,$$

and since ε is arbitrary the proof is complete.

Our next step is to use the above result to show the existence of a limit for the ratios D_n.

Proposition 5.2. *If $f \in L^1(\mu)$ and $g \in L^1_+(\mu)$, the sequence $\{D_n(f, g)\}$ has μ-a.e. a finite limit on the set $\{\sum_0^\infty T^n g > 0\}$.*

Proof. Set

$$A = \{T^n g > 0\} \cap \{\overline{\lim}\, D_n(f, g) = +\infty\}.$$

For every real number $c > 0$, one has $A \subset E'_{f-cg}$; hence by Theorem 5.1,

$$\int f\, P_A^* 1 \,\mathrm{d}\mu \geqslant c \int g\, P_A^* 1 \,\mathrm{d}\mu.$$

Since $T^* P_A^* 1 \leqslant P_A^* 1$, one has furthermore

$$\int f\, P_A^* 1 \,\mathrm{d}\mu \geqslant c \int g\, T^{*n}\, P_A^* 1 \,\mathrm{d}\mu = c \int T^n g\, P_A^* 1 \,\mathrm{d}\mu.$$

Since $T^n g$ is strictly positive on A and $P_A^* 1 \geqslant 1_A$, by taking c arbitrarily large we see that $\mu(A) = 0$. As a result, $\overline{\lim}\, D_n(f, g)$ is finite μ-a.e. on

$$\bigcup_0^\infty \{T^n g > 0\} = \left\{\sum_0^\infty T^n g > 0\right\}.$$

Let a, b be two rational numbers such that $a < b$. The set

$$A = \{T^n g > 0\} \cap \{\underline{\lim}\, D_n(f, g) < a,\ \overline{\lim}\, D_n(f, g) > b\}$$

is contained in $E'_{ag-f} \cap E'_{f-bg}$ so that, by Theorem 5.1,

$$\int (ag - f)\, P_A^* 1 \,\mathrm{d}\mu \geqslant 0 \quad \text{and} \int (f - bg)\, P_A^* 1 \,\mathrm{d}\mu \geqslant 0;$$

summing up these two inequalities gives

$$(a - b) \int g\, P_A^* 1 \,\mathrm{d}\mu \geqslant 0,$$

which is impossible unless the integral is zero. But

$$\int g\, P_A^* 1 \, \mathrm{d}\mu \geqslant \int g\, T^{*n} P_A^* 1 \, \mathrm{d}\mu = \int T^n g \, P_A^* 1 \, \mathrm{d}\mu$$

which is > 0 unless A is negligible. The proof is thus complete.

We then derive Hopf's decomposition theorem.

Theorem 5.3. *There exists a set $\boldsymbol{C}$, unique up to equivalence, such that for every f in L^1_+,*

$$\sum_0^\infty T^n f = 0 \quad \text{or } +\infty \quad \text{on } \boldsymbol{C},$$

$$\sum_0^\infty T^n f < +\infty \qquad \text{on } \boldsymbol{D} = \boldsymbol{C}^c.$$

Proof. Let h be a strictly positive element of L^1, and set $\boldsymbol{C} = \{\sum_0^\infty T^n h = \infty\}$. If $f \in L^1_+$, the preceding proposition applied to $D_n(h, f)$ implies that $\sum_0^\infty T^n f = +\infty$ on $\boldsymbol{C} \cap \{\sum_0^\infty T^n f > 0\}$; applied to $D_n(f, h)$, it implies that $\sum_0^\infty T^n f < +\infty$ on $\boldsymbol{D}$. The proof is complete, since it is now plain that $\boldsymbol{C}$ is independent of the choice of h.

As in §§ 2 and 3, one may restrict T to $\boldsymbol{C}$ and define the σ-algebra $\mathscr{C}$ of invariant sets. We are now going to identify the limit, the existence of which was shown in Proposition 5.2.

Proposition 5.4. *If T is conservative, the limit in Proposition 5.2 is equal to $E[f \mid \mathscr{C}]/E[g \mid \mathscr{C}]$.*

Proof. We assume that μ is bounded; the extension to σ-finite μ, as well as the definition for that case of the ratios of conditional expectations, is dealt with as in Theorem 3.2.

We may also assume f positive. Given two real numbers a, b such that $0 \leqslant a < b < \infty$, we will prove that

$$a \leqslant E[f \mid \mathscr{C}]/E[g \mid \mathscr{C}] \leqslant b \quad \mu\text{-a.e.}$$

on the set $A = \{a < \lim D_n(f, g) < b\}$; this will imply the desired conclusion, because simple functions which approximate $\lim D_n(f, g)$ will also approximate μ-a.s. the ratio of conditional expectations.

Since this ratio is $\mathscr{C}$-measurable we must prove that it lies between a and b on the set $\bar{A}$, hence that for $B \in \mathscr{C}$, $B \subset \bar{A}$ we have

$$\int_B (E[f \mid \mathscr{C}] - a\, E[g \mid \mathscr{C}])\, d\mu \geqslant 0 \quad \text{and} \int_B (b\, E[g \mid \mathscr{C}] - E[f \mid \mathscr{C}])\, d\mu \geqslant 0,$$

which is equivalent to

$$\int_B (f - ag)\, d\mu \geqslant 0 \quad \text{and} \int_B (bg - f)\, d\mu \geqslant 0.$$

Clearly, $B \supset \overline{B \cap A}$; on the other hand, the set $B \setminus \overline{B \cap A}$ is invariant, contained in $\bar{A}$ and disjoint from A; by the definition of $\bar{A}$ it is thus empty and $B = \overline{B \cap A}$. By Lemma 2.9 we can then replace 1_B by $P^*_{A \cap B}1$, and the above integrals may be written

$$\int P^*_{A \cap B}1\, (f - ag)\, d\mu \quad \text{and} \int P^*_{A \cap B}1\, (bg - f)\, d\mu;$$

it follows from Theorem 5.1 and the definition of A that they are positive, thus completing the proof.

Now let T be such that the conservative and dissipative parts are non-empty. For $f, g \in L^1_+$, it is plain that $D_n(f, g)$ converges on $\boldsymbol{D} \cap \{\sum_0^\infty T^n g > 0\}$ to $\sum_0^\infty T^n f / \sum_0^\infty T^n g$. If f, g vanish on $\boldsymbol{D}$, then the preceding result gives the limit on $\boldsymbol{C}$. In order to find this limit when these functions do not vanish on $\boldsymbol{D}$, we introduce the operator $H_C = I_C \sum_0^\infty (TI_D)^n$, which is adjoint to P^*_C and similar to the operator $\hat{H}_C$ of ch. 2. For $g \in L^1_+$, the functions $H_C g$ vanish on $\boldsymbol{D}$, and if g vanishes on $\boldsymbol{D}$, then $H_C g = g$. Furthermore

Proposition 5.5. *The ratio $D_n(H_C g, g)$ converges to 1 on $\boldsymbol{C} \cap \{\sum_0^\infty T^n g > 0\}$ whenever n tends to infinity.*

Proof. On $\boldsymbol{C}$, the ratios $D_n(H_C g, g)$ and $D_n(T^k H_C g, T^k g)$ have the same limit. For $a > 1$, the set $\{\lim D_n(H_C g, g) > a\}$ is contained in $E'_{H_C g - ag}$. If

$$A \subset \{T^k g > 0\} \cap \{\lim D_n(H_C g, g) > a\} \cap \boldsymbol{C},$$

Theorem 5.1 implies that

$$\int P^*_A 1\, (T^k H_C g - a T^k g)\, d\mu \geqslant 0.$$

But on the one hand $P_A^* 1$ is harmonic for T^*, since $P_A^* 1 = 1_{\bar{A}}$ on C and $T^* P_A^* 1 = P_A^* 1$ on D, as is easily checked; on the other hand, since $C \supset A$, we have $P_C^* P_A^* 1 = P_A^* 1$. Thus we may write

$$\begin{aligned}\langle P_A^* 1, T^k H_C g\rangle &= \langle T^{*k} P_A^* 1, H_C g\rangle = \langle P_A^* 1, H_C g\rangle \\ &= \langle P_C^* P_A^* 1, g\rangle = \langle P_A^* 1, g\rangle, \\ &= \langle T^{*k} P_A^* 1, g\rangle = \langle P_A^* 1, T^k g\rangle,\end{aligned}$$

and consequently the above inequality involving integrals reduces to

$$(1-a)\int P_A^* 1\, T^k g \, d\mu \geqslant 0,$$

which is impossible unless $P_A^* 1$, and hence A, is negligible. That the limit cannot be less than one is shown in the same way.

We are now ready to summarize all these results and state the general form of the Chacon–Ornstein theorem.

Theorem 5.6. *If f, g are in $L_+^1(\mu)$, the sequence of ratios $D_n(f, g)$ converges μ-a.e. as n tends to infinity to*

$$\sum_0^\infty T^n f \Big/ \sum_0^\infty T^n g \quad \text{on } D \cap \left\{\sum_0^\infty T^n g > 0\right\},$$

$$E[H_C f \mid \mathscr{C}]/E[H_C g \mid \mathscr{C}] \quad \text{on } C \cap \left\{\sum_0^\infty T^n g > 0\right\}.$$

Proof. By the preceding proposition,

$$\lim D_n(f, g) = \lim D_n(H_C f, H_C g),$$

and since $H_C f$ and $H_C g$ vanish outside C, the result follows from Proposition 5.4.

Exercise 5.7 (Ergodic theory for Abel sums). (1) Let T be a positive contraction of L^1 and set, for $\lambda \in [0, 1]$,

$$G_\lambda = \sum_0^\infty \lambda^n T^n.$$

For $f \in L^1$, set $E_f'' = \{\overline{\lim}_{\lambda \to 1} G_\lambda f > 0\}$. Prove that E_f'' is measurable, that $E_f'' \subset E_f'$, and derive that if $A \subset E_f''$, then

$$\int P_A^* 1 \, f \, \mathrm{d}\mu \geqslant 0.$$

(2) If T is conservative, then for $f, g \in L_+^1$ the ratio $D_\lambda(f, g) = G_\lambda f / G_\lambda g$ converges μ-a.e., when λ tends to 1, to $E[f \mid \mathscr{C}]/E[g \mid \mathscr{C}]$ on $\{G_1 g > 0\}$. For the case in which T is not conservative, prove a result similar to Theorem 4.6.

(3) Let $\{V_\alpha\}_{\alpha > 0}$ be a Harris resolvent with invariant measure m; prove that if f, g are in $L_+^1(m)$, with $m(g) > 0$, then

$$\lim_{\alpha \to 0} (V_\alpha f / V_\alpha g) = m(f)/m(g) \quad m\text{-a.e.}$$

Exercise 5.8. (1) In the situation of Exercise 2.20, Brunel's lemma may be stated:

$$(f^- m) \, P_A \prec (f^+ m)$$

for any $A \subset E_f'$.

(2) Prove that $\lim D_n(f, g) \leqslant 1$ on A, if and only if $(fm) \, P_A \prec (gm)$. In particular this limit is $\leqslant 1$ μ-a.e., if and only if $(fm) \prec (gm)$; this result is obvious for a dissipative T.

Exercise 5.9 (continuation of Exercise 2.19). This exercise describes another proof of the Chacon–Ornstein theorem for conservative contractions.

(1) Suppose that T is conservative and that $A = \{h_\infty^- > 0\}$ is not empty. Prove that $\sum_0^\infty h_n^+ < \infty$ on $\bar{A}$.

[*Hint*: Prove that $\int h_n^+ (I_{A^c} T^*)^p 1_A \, \mathrm{d}\mu \leqslant \int h^+ (I_{A^c} T^*)^{p+n} 1_A \, \mathrm{d}\mu$.]

(2) For f, g in L_+^1 and $b > 0$, set $B = \{bE[g \mid \mathscr{C}] > E[f \mid \mathscr{C}]\}$. By applying (1) and Exercise 2.19(1) to the function $h = (f - bg)$, prove that $\overline{\lim}\, D_n(f, g) \leqslant b$ on the set B. Derive therefrom the Chacon–Ornstein theorem for conservative T.

Exercise 5.10. If T is conservative, Brunel's lemma follows directly from Theorem 2.1 and the fact that the bounded superharmonic function $P_A^* 1$ vanishes outside E_f.

6. Subadditive ergodic theory

Throughout this section we will use the setting and notation of §3.6; we will often write Tf instead of $f \circ \theta$. We are going to prove that the convergence in Theorem **3.6** still holds under a weaker hypothesis. The result thus obtained has proved very useful in widespread situations and in particular in the study of Markov chains.

Definition 6.1. A sequence $\{s_n, n > 0\}$ of functions in L^1 is *subadditive* (with respect to θ) if, for every pair (n, k) of integers > 0,

$$s_{n+k} \leqslant s_n + s_k \circ \theta_n.$$

It is *additive* if the inequality is replaced by an equality.

Of course an additive sequence is nothing else than the sequence of partial sums $\sum_{k=0}^{n-1} T^k s_1$. In all cases, it will be useful to set $s_0 = 0$. If E has only one point, subadditive sequences boil down to subadditive sequences $\{a_n\}$ of real numbers. For such sequences we have the following well-known

Lemma 6.2. *If* $\gamma = \inf_n\{n^{-1}a_n\} > -\infty$, *then*

$$\lim_n n^{-1}a_n = \gamma.$$

Proof. Pick $\varepsilon > 0$; one can find n_0 such that $n_0^{-1}a_{n_0} \leqslant \gamma + \varepsilon$. Any integer n may be uniquely written $n = kn_0 + r$ with $r < n_0$. Thanks to the subadditivity of $\{a_n\}$, we have

$$n^{-1}a_n \leqslant n_0^{-1}a_{n_0} + n^{-1}a_r;$$

passing to the limit in this inequality yields

$$\gamma \leqslant \varliminf_n n^{-1}a_n \leqslant \varlimsup_n n^{-1}a_n \leqslant n_0^{-1}a_{n_0} \leqslant \gamma + \varepsilon,$$

and since ε is arbitrary the proof is complete.

The theorem of Kingman which is the main goal of this section subsumes both Birkhoff's theorem and the above lemma.

Definition 6.3. A subadditive sequence $\{s_n\}$ is *integrable* if $\gamma = \inf_n\{n^{-1}m(s_n)\} > -\infty$. The number γ is then called the *time-constant* of the sequence s_n.

An additive sequence is always integrable and its time-constant is $m(s_1)$. In what follows we study a given subadditive and integrable sequence $\{s_n\}$; we will set

$$\varphi_k = k^{-1} \sum_{i=1}^{k} (s_i - Ts_{i-1}), \quad k > 0.$$

Lemma 6.4. (i) $\lim_n n^{-1}m(s_n) = \gamma$;
(ii) $m(\varphi_k) = k^{-1}m(s_k) \geqslant \gamma$ *for every* $k > 0$.

Proof. Using the invariance of m with respect to θ, it is easily seen that the sequence $\{m(s_n)\}$ is subadditive, so that (i) follows from Lemma 6.2. The proof of (ii) is straightforward.

We now turn to a very important result.

Definition 6.5. An *exact minorant* of $\{s_n\}$ is an additive sequence $\{t_n\} = \{\sum_0^{n-1} T^k f\}$ such that $\{s_n - t_n\}$ is positive and has time-constant zero. The function $f = t_1$ itself will also be referred to as an exact minorant.

Proposition 6.6. *For a subadditive and integrable sequence, any limit point of* $\{\varphi_m\}$ *in the* $\sigma(L^1, L^\infty)$*-topology is an exact minorant.*

Proof. For an integer $n < k$, we have

$$\begin{aligned} J = k \sum_{i=0}^{n-1} T^i \varphi_k &= (I - T^n) \sum_{i=1}^{k-1} s_i + \sum_{j=0}^{n-1} T^j s_k \\ &= \sum_{i=1}^{n} s_i + \sum_{i=1}^{k-n-1} (s_{n+i} - T^n s_i) - \sum_{i=k-n}^{k-1} T^n s_i + \sum_{j=0}^{n-1} T^j s_k; \end{aligned}$$

since $s_{n+i} - T^n s_i \leqslant s_n$ for every i, we get

$$J \leqslant \sum_{i=1}^{n} s_i + (k - n - 1)\, s_n + \sum_{j=0}^{n-1} T^j (s_k - T^{n-j} s_{k-n+j}).$$

Using once more the subadditivity of $\{s_n\}$ we obtain

$$k \sum_{i=0}^{n-1} T^i \varphi_k \leqslant \sum_{i=l}^{n} s_i + (k - n - 1)\, s_n + \sum_{j=0}^{n-1} T^j s_{n-j}. \tag{6.1}$$

Suppose that there exists a limit point f in $\sigma(L^1, L^\infty)$ for the sequence $\{\varphi_k\}$. Then by Lemma 6.4 we have $m(f) = \gamma$ and passing to the limit in the inequality (6.1) yields, since T is $\sigma(L^1, L^\infty)$-continuous,

$$\sum_{j=0}^{n-1} T^j f \leqslant s_n.$$

The existence of exact minorants could thus be proved by showing that there does exist limit points as in the preceding result; unfortunately this cannot be yet done in a completely elementary way and we are going to perform the proof in a slightly different fashion. However, in practical situations the above proposition allows one to find explicit exact minorants.

Theorem 6.7 (decomposition theorem). *Every subadditive and integrable sequence has at least one exact minorant.*

Proof. By subtracting if necessary the additive sequence $\{\sum_0^{n-1} T^i s_1\}$ we may suppose that $\{s_n\}$ is negative. By the criterion for relative $\sigma(L^1, L^\infty)$-compactness of sets in $L^1(m)$, we may extract from $\{\varphi_m\}$ a subsequence $\{\varphi_{m_k}\}$ such that for every integers i and p, the sequence $\{T^i\varphi_{m_k} \vee (-p)\}$ converges in $\sigma(L^1, L^\infty)$ to a function λ_{ip}.

The sequence $\{\lambda_{ip}\}$ is obviously decreasing in p for every i and we may therefore set $\lambda_i = \lim_p \lambda_{ip}$. It follows from eq. (6.1) that

$$\sum_{i=0}^{n-1} \lambda_i \leqslant s_n$$

for every $n \geqslant 1$.

Now, because T is $\sigma(L^1, L^\infty)$-continuous and positive,

$$T\lambda_{ip} = T\,(\lim_k\,(T^i\varphi_{m_k} \vee (-p))) = \lim_k\,(T(T^i\varphi_{m_k} \vee (-p)))$$

$$\geqslant \lim_k\,(T^{i+1}\varphi_{m_k} \vee (-p)) = \lambda_{(i+1)p}.$$

Consequently $T\lambda_i \geqslant \lambda_{i+1}$, and we can write

$$\lambda_i = (\lambda_i - T\lambda_{i-1}) + T(\lambda_{i-1} - T\lambda_{i-2}) + \cdots + T^i\lambda_0,$$

where all the summands are negative. Let us put

$$f = \lambda_0 + \sum_{i=0}^{\infty} (\lambda_{i+1} - T\lambda_i);$$

the function f is negative and $m(f) = m(\lambda_0) \geqslant \gamma$ by Lemma 6.4. The function f is thus in $L^1(m)$ and moreover

$$\sum_{i=0}^{n-1} T^i f \leqslant \sum_{i=0}^{n-1} \lambda_i \leqslant s_n,$$

which proves that f is an exact minorant.

Remark. The above decomposition is not unique as may be surmised from the fact that any limit point of $\{\varphi_k\}$ in the $\sigma(L^1, L^\infty)$-topology will do. Actually even when $\{\varphi_k\}$ is convergent, there still may exist several exact minorants, as will be seen in Exercise 6.11.

The preceding theorem allows us to prove the convergence theorem which is the main result of this section.

Theorem 6.8 (Kingman's theorem). *The sequence $\{n^{-1}s_n\}$ has a limit ζ m-a.s. and in L^1. The limit ζ is $\mathscr{I}$-measurable and for any $A \in \mathscr{I}$*

$$\int_A \zeta \, dm = \lim_n n^{-1} \int_A s_n \, dm.$$

Proof. By Birkhoff's theorem and the decomposition theorem above it is enough to prove that for a positive subadditive sequence $\{s_n\}$ with time constant zero, $\{s_n/n\}$ converges to zero m-a.s. and in L^1 and for this it is enough, since $s_n/n \leqslant s_1$, to prove that $\overline{\lim}_n\, (s_n/n) = 0$ m-a.s.

Pick an integer k; any integer n may be uniquely written $n = pk + r$ with $r < k$. Using the subadditivity we have

$$n^{-1}s_n \leqslant n^{-1} \sum_{i=0}^{p-1} T^{ki} s_k + n^{-1} T^{pk} s_r.$$

As n goes to infinity, the last term on the right of this inequality goes to zero; indeed, for any $\varepsilon > 0$, because of the invariance of m by θ we have

$$\sum_{n=1}^{\infty} m(n^{-1} T^{pk} s_r \geqslant \varepsilon) \leqslant \sum_{n=1}^{\infty} m\left(\sup_{0 \leqslant r < k} s_r \geqslant \varepsilon n \right)$$
$$\leqslant \varepsilon^{-1} \int \left(\sup_{0 \leqslant r < k} s_r \right) dm < \infty,$$

and the result follows from the Borel–Cantelli lemma. Passing to the limit in the above inequality thus yields

$$\overline{\lim_n}\,(s_n/n) \leqslant E[(T^k s_k/k) \mid \mathscr{I}]$$

and consequently

$$\int \overline{\lim_n}\,(s_n/n)\,dm \leqslant k^{-1}m(s_k).$$

Since k is arbitrary the integral on the left is less than or equal to the time constant which is zero; this completes the proof.

Remark. If θ is ergodic the limit in the preceding theorem is equal to γ. It is more generally equal to $E[f \mid \mathscr{I}]$ where f is an exact minorant of s_n. It is important to observe that in some applications a limit point of $\{\varphi_k\}$ in the $\sigma(L^1, L^\infty)$-topology may be computed and thus provides the value of the limit; this is in particular the case when $\{\varphi_k\}$ or, even better, $\{s_n - s_{n-1} \circ \theta\}$ has a limit.

We must finally record that some applications of the foregoing results may be obtained in the following setting.

Corollary 6.9. *Let $\{V_{k,l},\ 1 \leqslant k < l\}$ be a family of doubly integer indexed random variables such that*

(i) $V_{k,p} \leqslant V_{k,l} + V_{l,p}$ *whenever* $k < l < p$;

(ii) *the joint distributions of the sequence* $\{V_{k+1,l+1}\}$ *are the same as those of the sequence* $\{V_{k,l}\}$;

(iii) *the expectations of the variables* $V_{1,n}$ *exist and* $\inf_n n^{-1}E[V_{1,n}] > -\infty$. *Then* $\lim_n n^{-1}V_{1,n}$ *exists a.s. and in* L^1.

Proof. Let $(W, \mathscr{A}, Q)$ be the probability space on which the variables $V_{k,l}$ are defined. Let E be the space of all real valued functions x defined on the set of pairs of integers (k,l) with $0<k<l$ and such that $x(k,p) \leqslant x(k,l)+x(l,p)$ if $k < l < p$. Endow the set E with the σ-field $\mathscr{E}$ which is the coarsest for which all coordinates mappings are measurable. Then the mapping J from W to E defined by $J(w)\,(k,l) = V_{k,l}(w)$ is measurable. Set $m = J(Q)$.

Finally let θ be the shift operator on E defined by $\theta(x)\,(k,l) = x(k+1, l+1)$. Thanks to (ii) the probability measure m is invariant by θ; moreover by (i) and (iii) we may apply Theorem 6.8 to the random variables $x(1, n)$ which completes the proof.

This corollary is actually equivalent to Theorem 6.6 as is seen in

Exercise 6.10. Derive Theorem 6.8 from Corollary 6.9.

Exercise 6.11. In the setting of ch. 1 §4.12 suppose that $G = \mathbf{R}$ and that $\mu = \frac{1}{2}(\varepsilon_1 + \varepsilon_{-1})$. Set $Y_n = \sum_1^n Z_k$ and $X_n = Y_n^+$.

(1) The sequence $\{X_n\}$ is subadditive relative to the shift θ on W and its time constant is zero. Prove that for any $a \in [0, 1]$ the sequence $\{aY_n\}$ is an exact minorant of X_n.

(2) Prove that the sequence $\{X_n - X_{n-1} \circ \theta\}$, hence the sequence $\{\varphi_k\}$ associated with $\{X_n\}$, converge in $\sigma(L^1, L^\infty)$ to $Y_1/2$ but do not converge in the L^1-sense. This shows that the convergence in $\sigma(L^1, L^\infty)$ of the sequence $\{\varphi_k\}$ is not a sufficient condition to ensure uniqueness in the decomposition theorem.

Exercise 6.12. If $E_s = \bigcup_{n \geqslant 1} \{s_n > 0\}$, then

$$\int_{E_s} s_1 \, dm \geqslant 0.$$

Exercise 6.13. Let $\{Y_n\}$ be a sequence of independent, equally distributed strictly positive, random matrices and call x_n^{ij} the $(i, j)^{\text{th}}$ entry of the matrix $X_n = Y_1 Y_2 \cdots Y_n$. State an integrability condition under which $\lim_n n^{-1} \log x_n^{ij}$ exists a.s. and is independent of (i, j).

Exercise 6.14 (law of large numbers for random walks on groups). Let G be a group generated by a compact neighbourhood V of e. Set

$$\delta(g) = \inf\{n > 0 \colon g \in V^n\}.$$

If μ is a probability measure on G such that $\int \delta \, d\mu < \infty$, prove that

$$\lim_n n^{-1} \delta(X_n) = \lim_n n^{-1} \int \delta \, d\mu^n \quad P_e\text{-a.s.},$$

where X is the right random walk of law μ.

Exercise 6.15. Let $A \in \mathscr{E}$ and m be excessive for the Markov chain X. Prove that the function g on $\mathbf{N}$ defined by $g(n) = P_m[S_A \leqslant n]$ is subadditive. Suppose further that $m(A) < \infty$ and prove that $n^{-1} g(n)$ has a finite limit L as n tends to infinity.

(2) If X is in duality with $\hat{X}$ with respect to m and if $\hat{P}1 = 1$ then $L = C(A)$ where $C(A)$ is the capacity of A defined in ch. 2, Exercise 4.13.

[*Hint*: Compute $g(n) - g(n-1) - \hat{C}(A)$ and prove that this quantity goes to zero as n tends to infinity.]

(3) The chain X is henceforth a right random walk on a group $\mathbf{G}$ and m a right Haar measure. Put

$$V_n(\omega) = m\left(\bigcup_{k=1}^{k=n} \{x: xX_k(\omega) \in A\}\right),$$

and prove that $E_y[V_n] = g(n)$ for any $y \in \mathbf{G}$. The reader will visualize the above set in relation to the set "swept out" by the random walk when started at e. If $\mathbf{G}$ is discrete and $A = \{e\}$, $V_n(\omega)$ is the cardinality of the *range* of X, that is the set $\{X_1(\omega), X_2(\omega), \ldots, X_n(\omega)\}$.

(4) Prove that $n^{-1}V_n$ converges a.s. to $C(A)$. If X is Harris this limit is zero; explain this fact.

(5) Prove that the sequence $\{V_n - V_{n-1} \circ \theta\}$ has a limit pointwise and find an exact minorant of $\{V_n\}$. Use this to give another proof of the above results.

Exercise 6.16. (1) For an additive sequence $\{s_n\}$ put $M_n = \sup_{0<k\leqslant n} s_k$, $I_n = \inf_{0<k\leqslant n} s_k$ and $O_n = M_n - I_n$. Prove that $n^{-1}O_n$ converges a.s. and in L^1 to $|E[s_1 \mid \mathscr{I}]|$.

(2) Set $M = \lim M_n$ and $I = \lim I_n$ and prove that $(M \circ \theta)^- + (I \circ \theta)^+$ is an exact minorant for O_n. Prove that $|E[s_1 \mid \mathscr{I}]| = E[M^- + I^+ \mid \mathscr{I}]$.

(3) Let X be a random walk on $\mathbf{R}$ with mean $\lambda > 0$ (see ch. 3 §3.12) and set $I = \inf_n(X_n)$. Prove that

$$\lambda = \int_\Omega I^+ \mathrm{d}P_e.$$

CHAPTER 5

TRANSIENT RANDOM WALKS. RENEWAL THEORY

This chapter is devoted to the study of renewal theory for transient random walks on abelian groups. For non-abelian groups the theory is now almost as complete as in the abelian case but it lies beyond the scope of this book. In the first section we show the fundamental theorem of Choquet and Deny. The renewal theorem will be shown in §§3 and 4, and will be applied to the study of hitting probabilities and recurrent sets in the last section.

1. The theorem of Choquet and Deny

For a random walk on an abelian group, the theorem of Choquet and Deny asserts that all bounded continuous harmonic functions are constant. We shall need the following

Lemma 1.1. *Let T_1 and T_2 be two commuting contractions of a Banach space B. If α_1 and α_2 are two positive real numbers such that $\alpha_1 + \alpha_2 = 1$ and $(\alpha_1 T_1 + \alpha_2 T_2)\, x = x$ for some $x \in B$, then $T_1 x = T_2 x = x$.*

Proof. If T is a contraction of B, then

$$\lim_n \|(I - T) \exp[- n(I - T)\| = 0.$$

Indeed,

$$\exp[- n(I - T)] = \sum_{j \geqslant 0} \alpha_{j,n} T^j,$$

where $\alpha_{j,n} = n^j / e^n j!$. For fixed n, the sequence $\alpha_{j,n}$ increases up to $j = n$ and then decreases. Therefore

$$\|(I - T) \exp[- n(I - T)]\| \leqslant \sum_j |\alpha_{j,n} - \alpha_{j-1,n}| \leqslant 2\alpha_{n,n},$$

and the Stirling formula shows that $\alpha_{n,n}$ tends to zero as n tends to infinity.

Let us now set $S = \alpha_1 T_1 + \alpha_2 T_2$; as T_1 and T_2 commute, we have

$$\exp[-\alpha_1^{-1}]\exp[\alpha_1^{-1}S] = \exp[-(I - T_1)]\exp[1 - \alpha_1^{-1}]\exp[(\alpha_1^{-1} - 1)T_2],$$

and taking the nth power,

$$\exp[-n\alpha_1^{-1}]\exp[n\alpha_1^{-1}S] = \exp[-n(I - T_1)]\,U^n,$$

where U is a contraction. From $Sx = x$ we get

$$x = \exp[-n(I - T_1)]\,U^n x,$$

$$(I - T_1)x = (I - T_1)\exp[-n(I - T_1)]\,U^n x,$$

and letting n tend to infinity yields $(I - T_1)\,x = 0$.

We shall now prove the theorem of Choquet and Deny in a slightly more general setting. Let $\boldsymbol{S}$ be a locally compact separable semi-group, and μ a probability measure on $\boldsymbol{S}$. As usual, a Borel function h on $\boldsymbol{S}$ is called μ-harmonic if

$$h(x) = \int_{\boldsymbol{S}} h(xy)\,\mu(\mathrm{d}y).$$

Theorem 1.2. *If $\boldsymbol{S}$ is abelian and if g belongs to the support of μ, then $h(xg) = h(x)$ for all $x \in \boldsymbol{S}$ and all bounded continuous μ-harmonic functions.*

Proof. Let g be a point in the support of μ and V_n, $n \in \boldsymbol{N}$, a basis of neighbourhoods of g. For each n, the restriction μ_n of μ to V_n is non-zero; hence if h is a bounded, continuous harmonic function, we can write

$$h(x) = \|\mu_n\| \int_{\boldsymbol{S}} h(xy)\,\|\mu_n\|^{-1}\,\mu_n(\mathrm{d}y)$$
$$+ \|\mu - \mu_n\| \int_{\boldsymbol{S}} h(xy)\,\|\mu - \mu_n\|^{-1}\,(\mu - \mu_n)\,(\mathrm{d}y).$$

For every probability measure ν on $\boldsymbol{S}$, the mapping $h \to Th$, where

$$Th(x) = \int h(xy)\,\nu(\mathrm{d}y)$$

is a positive contraction on the Banach space of bounded continuous functions. The semi-group $\boldsymbol{S}$ being abelian, we may apply the preceding lemma and find that

$$h(x) = \int_S h(xy) \, ||\mu_n||^{-1} \, \mu_n(dy).$$

But as n tends to infinity, the sequence $||\mu_n||^{-1} \mu_n$ converges weakly to ε_g, so that $h(x) = h(xg)$ for all $x \in \boldsymbol{S}$, and the proof is complete.

From now on we study a random walk of law μ on an abelian separable locally compact group $\boldsymbol{G}$. We shall denote by e or $\mathbf{0}$ the identity of $\boldsymbol{G}$, and by m a Haar measure of $\boldsymbol{G}$. We suppose that, as usual, μ is adapted. In this context, the preceding result reads as follows.

Theorem 1.3 (Choquet–Deny theorem). *The bounded continuous harmonic functions are constant.*

Proof. It is easily checked that the set of points y in $\boldsymbol{G}$ such that $h(x + y) = h(x)$ for all x in $\boldsymbol{G}$ and every bounded continuous harmonic function h is a closed sub-group of $\boldsymbol{G}$. The result thus follows immediately from Theorem 1.2.

This theorem implies the following corollaries.

Corollary 1.4. *If ν is an invariant measure for X (i.e. $\mu * \nu = \nu$), and if the set $\{\varepsilon_x * \nu;\, x \in \boldsymbol{G}\}$ is vaguely relatively compact, then ν is a Haar measure.*

Proof. For any $\varphi \in C_K^+$, the function f defined by

$$f(x) = \int \varphi(x + y) \, \nu(dy) = \langle \varepsilon_x * \nu, \varphi \rangle$$

is bounded and continuous. Furthermore, f is μ-harmonic, since

$$\int f(x + y) \, \mu(dy) = \iint \varphi(x + y + z) \, \mu(dy) \, \nu(dz) = \int \varphi(x + y) \, \nu(dy) = f(x).$$

The function f is therefore constant and it follows that ν is a Haar measure.

Corollary 1.5. *The bounded harmonic functions are constant m-a.e.*

Proof. If h is bounded harmonic, the measure $\nu = hm$ satisfies the conditions of Corollary 1.4 for the dual random walk. Indeed

$$\hat{\mu} * (hm) = (\hat{\mu} * h) \, m = hm,$$

and it is easily checked that $\{\varepsilon_x * \nu, x \in G\}$ is vaguely relatively compact. Therefore ν is a Haar measure and h is constant m-a.e.

Remark. These results may be carried over to some classes of non-abelian groups but are not true for every group, as is shown in Exercise 1.10.

When the law μ is spread out, matters become even nicer: all harmonic bounded functions are constant, so that the σ-algebra of invariant events is trivial (ch. 2 §3). This follows from the following important proposition, which does not require G to be abelian.

Proposition 1.6. *If μ is spread out, every bounded harmonic function is continuous.*

Proof. Choose n such that $\mu^n = \alpha + \beta$, where α is absolutely continuous with respect to m and β singular with $\|\beta\| < 1$. Call α_p and β_p the absolutely continuous and singular parts of μ^{np}; it is easily seen that $\|\beta_p\| < \|\beta\|^p$.

If h is bounded and harmonic, for every p we have

$$\|h - h * \hat{\alpha}_p\| = \|h * \hat{\mu}^{np} - h * \hat{\alpha}_p\| \leqslant \|h\|_\infty \|\beta\|^p.$$

The function $h * \hat{\alpha}_p$ is continuous, since it is the convolution of a bounded function and an integrable one. The function h is thus the uniform limit of continuous functions, and is therefore continuous.

In the subsequent section the Choquet–Deny theorem will be used as a tool in renewal theory. Exercise 1.11 and ch. 6, Exercise 1.14 give an insight into the relationship between the Choquet–Deny theorem and the limit theorems for powers of transition functions. This relationship is necessary to enlarge the scope of validity of the Choquet–Deny theorem to certain classes of non-abelian groups, but it will not be used in the remainder of the book.

Exercise 1.7 (another proof of the Choquet–Deny theorem).

(1) Let φ be a μ-harmonic bounded function. Prove that for every $x \in G$ the sequence of random variables $\varphi(x + Z_1 + \cdots + Z_n)$ defined on the probability space $(W, \mathscr{A}, P)$ of ch. 1, Exercise 4.12 is a martingale which converges a.s. and in the mean to a random variable H_x which is symmetric and thus constant a.s. The reader will notice that the fact that G is abelian is necessary.

(2) Derive from (1) that $\varphi(x + Z_1) = \varphi(x)$ a.s.; hence $\varphi(x + x') = \varphi(x)$ for φ-almost every x'.

(3) Prove the Choquet–Deny theorem.

Exercise 1.8. (1) Let X be a random walk on an abelian group G. Prove that $\mathscr{T}$ is P_ν-trivial whenever ν is absolutely continuous with respect to m.

(2) Derive that for a recurrent random walk (not necessarily Harris) the function $P.[R(A)] = 1$ m-a.e. or 0 m-a.e. according as $m(A) > 0$ or $m(A) = 0$.

(3) If $\mathscr{I}$ is a.s. trivial, then μ is spread out. (See also ch. 7, Theorem 1.13.)

Exercise 1.9. A continuous homomorphism of an abelian group G into the multiplicative group of positive real numbers is called an *exponential*. Show that extremal (cf. ch. 7 §3) positive harmonic functions are exponentials.

Exercise 1.10. The theorem of Choquet–Deny is not true for all groups. Every element in the group $G = \mathrm{SL}(2, \mathbf{R})$ of real, unimodular matrices of order 2 can be written uniquely as the product of an element of the group K of rotations by an element of the group S of upper-triangular unimodular matrices (a special case of the Iwasawa decomposition of semi-simple Lie groups with finite centre). The circle S^1 may be identified with G/S, and is thus a compact homogeneous space of G.

(1) Prove that there does not exist any measure on S^1 invariant by G.

(2) Let μ be a spread-out probability measure on G. Prove that there exists at least one probability measure ν on S^1 such that $\mu * \nu = \nu$, and derive the existence of non-constant μ-harmonic functions.

[*Hint*: See ch. 2, Exercise 4.18.]

Exercise 1.11. If μ is a probability measure on a group G such that $\lim_n \|\mu^n * \nu\| = 0$ for all $\nu \ll m$ and such that $\nu(E) = 0$, then the Choquet–Deny theorem is true for the left random walk of law μ on G. What can be said if the assumption $\nu \ll m$ is dropped?

Exercise 1.12. If μ is spread-out deduce Theorem 3.3 of ch. 3 from the Choquet–Deny theorem.

Exercise 1.13. For any $g \in S^1$ the map $x \to g + x$ from S^1 into itself preserves the Lebesgue measure on S^1. Prove that this map is ergodic (ch. 4 §3.13) if and only if g is irrational. Generalize to the n-dimensional torus.

2. General lemmas

From now on, the random walk X is supposed to be transient. According to ch. 3 § 4, the potential measure $\bar{G}$ is then a Radon measure, the asymptotic behaviour of which we are going to study. Throughout the sequel, f stands for an arbitrary function in C_K^+.

Proposition 2.1. (i) *The set of measures* $\{G(x,\cdot), x \in \boldsymbol{G}\}$ *is vaguely relatively compact.*

(ii) *The function* Gf *is uniformly continuous.*

Proof. The statement (i) was proved in ch. 3, Corollary 3.6. For every compact K, there exists therefore a constant M_K such that $G(x, K) \leqslant M_K$ for all $x \in \boldsymbol{G}$.

Let H be the compact support of f. Since f is uniformly continuous, for every $\varepsilon > 0$, we can pick a compact symmetric neighbourhood J of e, such that $|f(x+y) - f(x)| < \varepsilon$ for $y \in J$ and all $x \in \boldsymbol{G}$. For $y \in J$, the function $g(\cdot) = f(\cdot + y) - f(\cdot)$ vanishes outside the compact $K = H + J$, and it is easily checked that $Gg(x) = Gf(x+y) - Gf(x)$. We thus get, from (i),

$$|Gf(x+y) - Gf(x)| \leqslant \varepsilon M_K$$

for all $x \in \boldsymbol{G}$ and $y \in J$, which yields (ii).

We now aim at finding the closure of the set $\{G(x,\cdot), x \in \boldsymbol{G}\}$ in the vague topology. Since the mapping $x \to G(x,\cdot)$ is continuous, as follows from Proposition 2.1, this is equivalent to the problem of finding the limit points of the set $\{G(x,\cdot), x \in \boldsymbol{G}\}$ when x tends to Δ, the point at infinity in the Alexandroff compactification of $\boldsymbol{G}$.

Theorem 2.2. *The limit points of the set* $\{G(x,\cdot), x \in \boldsymbol{G}\}$ *when* x *tends to* Δ *are Haar measures. Moreover they always include the zero measure.*

Proof. Let $\{x_n\}$ be a sequence of points in $\boldsymbol{G}$, converging to Δ and such that the measures $G(x_n,\cdot)$ converge to the measure ν in the vague topology. Now the measures $\varepsilon_{x_n} * \bar{G} * \mu$ converge vaguely to $\nu * \mu$; indeed

$$\langle \varepsilon_{x_n} * \bar{G} * \mu, f \rangle = \int Gf(x_n + y)\, \mu(\mathrm{d}y)$$

and since $|Gf(x_n + y)| \leqslant ||Gf||$ and $Gf(x_n + y)$ converges to $\langle \nu * \varepsilon_y, f \rangle$ for every y, our claim follows from Lebesgue's theorem. Passing to the limit in the equality

$$\varepsilon_{x_n} * \bar{G} = \varepsilon_{x_n} + \varepsilon_{x_n} * \bar{G} * \mu$$

yields $\nu = \nu * \mu$. It is now clear from Proposition 2.1 that ν satisfies the hypothesis of Corollary 1.4 and is therefore a Haar measure.

To prove the second sentence take f non-zero and observe that by ch. 2 §1.8, we have $\lim_n Gf(X_n) = 0$ a.s. Since every compact set is transient and P is markovian, we may consequently find a path ω such that $\lim_n Gf(X_n(\omega)) = 0$, $\lim_n X_n(\omega) = \Delta$ and $X_n(\omega) \neq \Delta$ for every n. By the first sentence, the measures $G(X_n(\omega), \cdot\,)$ converge to a Haar measure which gives the value zero to f and is therefore the zero measure. The proof is complete.

Remark. Using a countable dense subset of C_K^+, it is easy to see that the second sentence is still true for non-abelian groups even though the first sentence is not. The second sentence is actually true for any markovian T.P., in a suitable analytical set-up. It is no longer true if the T.P. is not markovian as is shown by important examples in ch. 9.

We will now work to prove that there is at most one limit point beside the zero measure.

Lemma 2.3. *For every compact K,*

$$\lim_{x \to \Delta}(Gf(x + y) - Gf(x)) = 0$$

uniformly in $y \in K$.

Proof. After Propositions 2.1–2.2 every sequence converging to Δ contains a subsequence $\{x_n\}$ such that the measures $G(x_n, \cdot\,)$ converge in the vague sense towards a Haar measure αm. Then

$$\lim_{x_n \to \Delta}(Gf(x_n + y) - Gf(x_n)) = \alpha(\langle \varepsilon_y * m, f \rangle - \langle m, f \rangle) = 0,$$

which shows that the limit in the statement exists. For every $\varepsilon > 0$, there exists a compact K_y such that $|Gf(x + y) - Gf(x)| < \frac{1}{2}\varepsilon$ for $x \notin K_y$.

Moreover, Gf being uniformly continuous, for every $\varepsilon > 0$, every $y \in \boldsymbol{G}$ possesses a neighbourhood V_y such that $z \in V_y$ implies

$$|Gf(x + z) - Gf(x + y)| < \tfrac{1}{2}\varepsilon$$

uniformly in $x \in \boldsymbol{G}$. Then, for $z \in V_y$ and $x \notin K_y$,

$$|Gf(x + z) - Gf(x)| < \varepsilon,$$

and the lemma follows by applying the Heine–Borel property.

The following proposition is a key for further results. It can be extended to large classes of non-abelian groups. (See Exercise 2.9.)

Proposition 2.4. *For every* $f \in C_K^+$,

$$\lim_{x \to \Delta} Gf(x)\, Gf(-x) = 0.$$

Proof. Take $\varepsilon > 0$; by Lemma 2.3, there is a compact set K_ε such that for $x \notin K_\varepsilon$ and every y in the compact support K of f we have

$$Gf(x) \leqslant Gf(x + y) + \varepsilon,$$

and a fortiori,

$$Gf(x)\, Gf(y)/\|Gf\| \leqslant Gf(x + y) + \varepsilon. \tag{2.1}$$

Applying the complete maximum principle to $Gf(y)$ yields that the inequality (2.1) holds for every $x \notin K_\varepsilon$ and every $y \in \boldsymbol{G}$.

Now by Theorem 2.2, we may choose a point y_ε such that $Gf(y_\varepsilon) \leqslant \varepsilon$, and putting $-x + y_\varepsilon$ in place of y in eq. (2.1) yields

$$Gf(x)\, Gf(-x + y_\varepsilon) \leqslant 2\|Gf\|\, \varepsilon.$$

The proposition then follows from the fact that $Gf(-x) - Gf(-x + y_\varepsilon)$ tends to zero when x tends to Δ.

Theorem 2.5. *The set* $\{G(x, \cdot),\ x \in \boldsymbol{G}\}$ *has at most one non-zero limit point when* x *tends to* Δ.

Proof. Let us suppose that there exist two non-zero limit points. Then there is a function $f \in C_K^+$, of support K, such that the set $\{Gf(x),\ x \in \boldsymbol{G}\}$ has two strictly positive limit points when x tends to Δ. Let us call l the smaller one.

We take $\varepsilon > 0$ and define inductively a sequence $\{x_n\}$ of points in $\boldsymbol{G}$ in the following way: we set $x_0 = e$; then by Lemma 2.3 and Proposition 2.4 we may choose a point $x_1 \in \boldsymbol{G}$ such that

$$Gf(x_1) < l + 2^{-2}\varepsilon, \qquad Gf(-x_1) < 2^{-2}\varepsilon,$$

$$|Gf(x_1) - Gf(x_1 + x)| < 2^{-2}\varepsilon \quad \text{for } x \in K,$$

$$|Gf(-x_1) - Gf(-x_1 + x)| < 2^{-2}\varepsilon \quad \text{for } x \in K;$$

the sets $K \cup \{x_1 + K\}$ and $K \cup \{-x_1 + K\}$ being also compact, we may choose x_2 such that

$$Gf(x_2) < l + 2^{-3}\varepsilon, \qquad Gf(-x_2) < 2^{-3}\varepsilon,$$

$$|Gf(x_2) - Gf(x_2 + x)| < 2^{-3}\varepsilon \quad \text{for } x \in K \cup \{-x_1 + K\},$$

$$|Gf(-x_2) - Gf(-x_2 + x)| < 2^{-3}\varepsilon \quad \text{for } x \in K \cup \{x_1 + K\};$$

in the same way, for every n, we choose x_n such that

$$Gf(x_n) < l + 2^{-n-1}\varepsilon, \qquad Gf(-x_n) < 2^{-n-1}\varepsilon,$$

$$|Gf(x_n) - Gf(x_n + x)| < 2^{-n-1}\varepsilon \quad \text{for } x \in \bigcup_{i=0}^{n-1} \{-x_i + K\},$$

$$|Gf(-x_n) - Gf(-x_n + x)| < 2^{-n-1}\varepsilon \quad \text{for } x \in \bigcup_{i=0}^{n-1} \{x_i + K\}.$$

The support of the function

$$f(\cdot) + f(\cdot + x_1) + \cdots + f(\cdot + x_n)$$

is contained in $\bigcup_{i=0}^{n} \{-x_i + K\}$. Let $x \in \{-x_p + K\}$, $1 \leqslant p \leqslant n$; we have

(i) $Gf(x + x_p) \leqslant \|Gf\|$;

(ii) for $i = 1, 2, \ldots, n - p$, since $x \in \{-x_p + K\}$,

$$Gf(x + x_{p+i}) = Gf(x_{p+i}) + Gf(x + x_{p+i}) - Gf(x_{p+i})$$
$$\leqslant l + 2^{-(p+i)-1}\varepsilon + 2^{-(p+i)-1}\varepsilon;$$

(iii) for $i = 1, 2, \ldots, p$,

$$Gf(x + x_{p-i}) = Gf(z - x_p),$$

where $z = x_p + x + x_{p-i} \in \{x_{p-i} + K\}$, and this can be written

$$Gf(-x_p) + Gf(z - x_p) - Gf(-x_p) \leqslant 2^{-p-1}\varepsilon + 2^{-p-1}\varepsilon.$$

Summing up all these inequalities yields

$$G[f(\cdot) + f(\cdot + x_1) + \cdots + f(\cdot + x_n)]\,(x) \leqslant \|Gf\| + nl + \varepsilon,$$

for every $x \in \bigcup_{i=0}^{n} \{-x_i + K\}$. The complete maximum principle implies that this inequality holds everywhere. By Lemma 2.3, we have

$$\overline{\lim_{x \to \Delta}} G[f(\cdot) + f(\cdot + x_1) + \cdots + f(\cdot + x_n)](x) = (n+1) \overline{\lim_{x \to \Delta}} Gf(x);$$

hence, for every n,

$$\overline{\lim_{x \to \Delta}} Gf(x) \leqslant (nl/(n+1)) + (\|Gf\| + \varepsilon)/n.$$

Since l is the smallest of the positive limits, we have a contradiction.

This leads us to lay down the following

Definition 2.6. A transient random walk X will be said to be of *type I* if the measures $G(x, \cdot)$ converge vaguely to zero as x tends to Δ. It will be said to be of *type II* if there exist a non-zero limit point.

The preceding theorem says that for type II random walks there exists only one non-zero limit point and we know that this limit point is then a Haar measure. In the following sections we shall classify all random walks according to these two types and compute the actual value of the non-zero limit in the case of type II walks.

Exercise 2.7. Any symmetric random walk, that is such that $\mu = \hat{\mu}$, on an abelian group is type I.

Exercise 2.8. A random walk X on a general non-abelian group $\boldsymbol{G}$ has the property (P) if the set of measures $\{\varepsilon_x * \bar{G} * \varepsilon_y\}$ is vaguely relatively compact.

(1) Prove that (P) holds if and only if the set $\{\varepsilon_x * \bar{G} * \varepsilon_{x^{-1}}\}$ is vaguely relatively compact.

(2) Prove that (P) holds whenever $\boldsymbol{G}$ is unimodular and μ is spread out. [*Hint*: Use Theorem 5.1.]

(3) Assume that (P) holds and that the Choquet–Deny theorem is true for X, and show that Theorem 2.2 and Lemma 2.3 are still valid. Prove that Proposition 2.4 is true for x converging to Δ in the center of $\boldsymbol{G}$.

(4) Using the setting of Exercise 4.11 in ch. 1 with $\boldsymbol{H}$ the multiplicative group $\mathbf{R}_+$ and $\boldsymbol{K} = \mathbf{R}$ ($\boldsymbol{G}$ is then the group of positive affine motions of the real line) prove that (P) no longer holds for a non-unimodular group and that Theorem 2.2 cannot be extended to this case.

Exercise 2.9. Show that Proposition 2.4 is true for all random walks on a (not necessarily abelian) discrete group; more precisely

$$\lim_{x \to 0} G(x, e)\, G(xy^{-1}, e) = 0$$

for y arbitrary but fixed in G.

[*Hint*: Setting $N(x, e) = P_{S_{\{e\}}}(x, e)$, show that

$$G(x, e) = I(x, e) + N(x, e)\, G(e, e);$$

then for $x \neq e$ and $x \neq y^{-1}$

$$E_x\left[\sum_{n > S_{\{e\}}} 1_{xy}(X_n)\right] = G(x, e)\, G(e, xy)/G(e, e).$$

For every p, the first member of this equality is less than

$$\sum_{n=0}^{p} P_x[S_{\{e\}} < n] + \sum_{n=p+1}^{\infty} P_x[X_n = xy],$$

and therefore tends to zero as x tends to Δ.]

3. The renewal theorem for the groups R and Z

In this section we study the cases of the group **R** of real numbers and of the group **Z** of integers. In the latter case the proofs will be written down only if they are really different from those in the former case. The letter m will stand for the Lebesgue measure (counting measure for **Z**) and we shall often abbreviate $m(\mathrm{d}x)$ to $\mathrm{d}x$. The notations $\lim_{x \to +\infty}$ and $\lim_{x \to -\infty}$ have their usual meaning.

Theorem 3.1. *When x tends to $+\infty$ (resp. $-\infty$), the measures $G(x, \cdot\,)$ converge vaguely towards a multiple $c_+ m$ (resp. $c_- m$) of the Lebesgue measure. At least one of the two constants c_+ and c_- is zero, the other one being positive or zero.*

Proof. First consider the case of **Z**. Suppose that when x tends to $+\infty$ there exist two limit points, which by Theorem 2.5 are 0 and cm with $c > 0$. Then, for any function $f \in C_K^+$ and any $\varepsilon > 0$, and for x sufficiently large, $Gf(x)$ is either greater than $cm(f) - \varepsilon$ or less than ε. One can therefore find a sequence $\{x_n\}$ converging to $+\infty$ and such that for every n

$$Gf(x_n) < \varepsilon, \qquad Gf(x_n + 1) > cm(f) - \varepsilon.$$

By Lemma 2.3, this is impossible, and we have a contradiction. Let us now consider the group $\mathbf{R}$. As the function Gf is continuous, if there exist two limit points 0 and $cm(f)$, every point in the interval $[0, cm(f)]$ would also be a limit point, and by Theorem 2.5 this is impossible.

In the following, we want to identify the constants c_+ and c_-. For this purpose we need some prerequisites.

Let f be a measurable function on $\mathbf{R}$. For $h > 0$ we call $\underline{U}_n$ and $\bar{U}_n$ the minimum and maximum of f in the interval $[(n-1)h, nh]$, $n \in \mathbf{Z}$, and we set

$$\underline{\sigma} = h \sum_{-\infty}^{+\infty} \underline{U}_n, \qquad \bar{\sigma} = h \sum_{-\infty}^{+\infty} \bar{U}_n.$$

Definition 3.2. The function f will be said to be directly Riemann integrable (abbreviated R-integrable) if the series $\underline{\sigma}$ and $\bar{\sigma}$ converge, and if for every $\varepsilon > 0$, $\bar{\sigma} - \underline{\sigma} < \varepsilon$, for h sufficiently small.

The functions in C_K are R-integrable. Let f be a function vanishing on $]-\infty, 0[$, decreasing on $[0, \infty[$ and such that $f(\infty) = 0$; we have $\sum_0^\infty (\bar{U}_n - \underline{U}_n) \leqslant f(0)$, so that the two series $\underline{\sigma}$ and $\bar{\sigma}$ either both diverge or both converge. The function f is then R-integrable if and only if f is integrable in the ordinary Lebesgue sense. This example will be useful below in connection with the following

Proposition 3.3. *If f is R-integrable, the function Gf is bounded and converges to $c_+m(f)$ (resp. $c_-m(f)$) as x tends to $+$ (resp. $-\infty$).*

Proof. From the vague convergence it is easy to derive the convergence of functions Gf, where f is the characteristic function of an interval.

Fix now $h > 0$ and set $f_n = 1_{[(n-1)h, nh[}$. By the maximum principle, there is a constant M such that $Gf_n \leqslant M$ for every n. Let $\{a_n\}_{n\in\mathbf{Z}}$ be a sequence of positive real numbers such that $\sum_{-\infty}^{+\infty} a_n < +\infty$, and set $f = \sum_{-\infty}^{+\infty} a_n f_n$. For every n

$$\sum_{-n}^{+n} a_k Gf_k(x) \leqslant Gf(x) \leqslant \sum_{-n}^{+n} a_k Gf_k(x) + M \sum_{|k|>n} a_k,$$

which shows that Gf is bounded. Then letting x tend to $\pm\infty$ shows that the proposition is true for a function f of this kind.

Let now f be a positive R-integrable function; with the previous notation, we set

$$\underline{f} = \sum_{-\infty}^{+\infty} U_n \underline{f}_n, \qquad \bar{f} = \sum_{-\infty}^{+\infty} \bar{U}_n f_n.$$

Obviously $G\underline{f} \leqslant Gf \leqslant G\bar{f}$, and as $\underline{f}$ and $\bar{f}$ are of the kind just studied, letting x tend to infinity yields

$$\lim_{x\to\pm\infty} G\underline{f}(x) \leqslant \underline{\lim}_{x\to\pm\infty} Gf(x) \leqslant \overline{\lim}_{x\to\pm\infty} Gf(x) \leqslant \lim_{x\to\pm\infty} G\bar{f}(x).$$

Thus

$$c_{\pm}\underline{\sigma} \leqslant \underline{\lim}_{x\to\pm\infty} Gf(x) \leqslant \overline{\lim}_{x\to\pm\infty} Gf(x) \leqslant c_{\pm}\bar{\sigma},$$

which completes the proof.

We now proceed to compute c_+ and c_-. The features are different according to whether there exists or does not exist a first moment for μ, that is whether $\int |x| \, d\mu(x) < \infty$ or not; if it exists we shall call the *mean* of X the number

$$\lambda = \int x \, d\mu(x).$$

We recall that λ cannot be zero, since the random walk is transient (ch. 3 §4).

Proposition 3.4. *If the first moment exists, the random walk is of type II. If $\lambda > 0$, then $c_- = \lambda^{-1}$ and $c_+ = 0$. For $\lambda < 0$, the symmetric conclusion holds.*

Proof. Set $g(x) = \mu([-x, \infty[)$ for $x \leqslant 0$, and $g(x) = -\mu(]-\infty, -x[)$ for $x > 0$. Let k be the function defined on $\mathbf{R}^2$ by $k(x, y) = 1$ for $x < 0$, $y \geqslant -x$, $k(x, y) = -1$ for $x \geqslant 0$, $y < -x$ and $k(x, y) = 0$ elsewhere.

Since

$$\int_{\mathbf{R}^2} |k(x, y)| \, dm(x) \, d\mu(y) = \int_{-\infty}^{+\infty} |y| \, d\mu(y) < +\infty,$$

we may apply the Fubini theorem to $\int k(x, y) \, dm(x) \, d\mu(y)$, and get $\int_{-\infty}^{+\infty} g(x) \, dm(x) = \lambda$. The function g is thus R-integrable, hence the potential Gg is bounded. Set $h = 1_{]-\infty, 0[}$; we have $Ph(x) = \mu(]-\infty, -x[)$ and thus $(I - P)\,h = g$. The function $l = h - Gg$ is therefore a bounded harmonic function which is m-a.e. equal to a constant a. We may find two sequences x_n and y_n of points in $\mathbf{R}$ converging respectively to $-\infty$ and $+\infty$ and such that for every

n, $l(x_n) = l(y_n) = a$. Passing to the limit in the equalities

$$h(x_n) = Gg(x_n) + l(x_n), \qquad h(y_n) = Gg(y_n) + l(x_n)$$

yields, by Proposition 3.3, that

$$1 = c_-\lambda + a, \qquad 0 = c_+\lambda + a.$$

We cannot have $c_+ > 0$, since it would imply $c_- = 0$, $a = 1$ and therefore $\lambda < 0$. We thus have $c_+ = 0$; hence $a = 0$ and $c_- = \lambda^{-1}$, which completes the proof.

Remark. The reader will find, in ch. 3, Exercise 3.13, examples of random walks for which the above result is obvious. The reader will also find in Exercise 3.10 why it is natural to consider the function g in the above proof.

The same argument works when μ is carried by $[0, \infty]$. The expression $\lambda = \int_{\mathbf{R}} x \, d\mu(x)$ is then always meaningful, but λ may be $+\infty$ and we will then put $\lambda^{-1} = 0$.

Corollary 3.5. *If μ is carried by $[0, \infty]$, then $c_- = \lambda^{-1}$* (of course we have $c_+ = 0$).

Proof. If $\lambda < \infty$, this is Proposition 3.4. To deal with $\lambda = +\infty$, we use the notation of Proposition 3.4. The function g vanishes on $]0, \infty[$, the potential Gg is the smallest positive solution of the equation $(I - P)f = g$, and consequently $Gg \leqslant h$ (actually $Gg = h$ m-a.e.).

The function $g_n = g\, 1_{[-n,0]}$ is R-integrable and $Gg_n \leqslant Gg = h$. Letting x tend to $-\infty$ in this inequality yields

$$c_- \int_{-\infty}^{+\infty} g_n(x) \, dx \leqslant 1;$$

this cannot be true for every n unless $c_- = 0$.

The foregoing results are now applied to prepare the final proof of this section. However, the following lemmas are of intrinsic interest. We call T_+ the hitting time of $]0, \infty[$.

Lemma 3.6. If $\lambda^+ = \int_0^{+\infty} x \, d\mu(x) = +\infty$, *the measures* $P_{T_+}(x, \cdot)$ *converge vaguely to zero as* x *tends to* $-\infty$.

Proof. Let f be a positive continuous function with compact support contained in $]0, +\infty[$. By the strong Markov property,

$$Gf(x) = \int_0^\infty P_{T_+}(x, dy)\, Gf(y).$$

If μ is carried by $]0, \infty[$, by Corollary 3.5 the function $Gf(x)$ tends to zero as x tends to $-\infty$, and as $Gf \geqslant f$ this implies the desired conclusion.

If μ is not carried by $]0, \infty[$, we define inductively a sequence $\{S_k\}$ of stopping times by

$$S_0 = 0, \dots, S_k = \inf\{n: X_n > X_{S_{k-1}}\}, \dots,$$

where the infimum of the empty set is understood to be $+\infty$. By ch. 1, Exercise 4.9, the random variables $X_{S_k} - X_{S_{k-1}}$ are independent and equidistributed with law $\mu' = P_{S_1}(0, \cdot) = P_{T_+}(0, \cdot)$; indeed S_1 and T_+ are P_0-a.s.equal. The sequence $\{X_{S_k}\}$ is thus a random walk whose law μ' is carried by $]0, \infty[$, and since μ' is clearly larger than the restriction of μ to $]0, \infty[$ we have $\int_0^\infty x \, d\mu'(x) = +\infty$. Now starting from $x < 0$, the hitting distributions of $]0, \infty[$ are the same for X and X' because the first time that X_{S_k} is > 0 is then almost surely equal to the first time X_n is > 0; we then conclude by applying the first part of the proof to μ'.

Lemma 3.7. *If* $\lambda^- = \int_{-\infty}^0 (-x) \, d\mu(x) < \infty$ *and if* $\lambda > 0$, *then* $P_x[T_+ < \infty] = 1$ *for all* $x \in \mathbf{R}$.

Proof. Let us first assume that $\lambda^+ < \infty$; the strong law of large numbers then implies that X_n/n converges P_0-a.s. towards λ. Thus X_n converges a.s. to $+\infty$ and $P_x[T_+ < \infty] = 1$ for every $x \in \mathbf{R}$.

Now if $\lambda^+ = \infty$, we choose a number c such that $\int_0^c x \, d\mu(x) > \lambda^-$, and set $Z_n^c = Z_n$ if $Z_n \leqslant c$ and $Z_n^c = c$ if $Z_n > c$. We may apply the above reasoning to $X_n^c = \sum_1^n Z_k^c$, and as $X_n \geqslant X_n^c$ a.s., this completes the proof.

We are now ready to prove the renewal theorem for the case in which the first moment does not exist.

Theorem 3.8. *If the first moment does not exist, the random walk is of type I.*

Proof. Assume $c_- > 0$, so that also $c_+ = 0$, and let us suppose that $\lambda^+ = +\infty$. By the strong Markov property, if $f \in C_K^+$ vanishes on $]-\infty, 0[$,

$$Gf(x) = \int_0^\infty P_{T_+}(x, dy)\, Gf(y).$$

Since $c_+ = 0$ for every $\varepsilon > 0$, there is a number $a > 0$ such that $y > a$ implies $Gf(y) < \varepsilon$, and hence

$$Gf(x) \leqslant \int_0^a P_{T_+}(x, dy)\, Gf(y) + \varepsilon;$$

if we let x tend to $-\infty$, $P_{T_+}(x, \cdot)$ converges vaguely to zero, and therefore for x sufficiently small we get $Gf(x) \leqslant 2\varepsilon$, which is impossible since c_- is strictly positive. We thus have $\lambda^+ < \infty$ and therefore $\lambda^- = +\infty$.

If we consider $\hat{\mu}$ in place of μ, this is equivalent to supposing that $c_- = 0$, $c_+ > 0$, $\lambda^- < \infty$, $\lambda^+ = \infty$. We then have $P_x[T_+ < \infty] = 1$ for every $x \in \mathbf{R}$; on the other hand $P_{T_+}(x, \cdot)$ converges vaguely to zero as x tends to $-\infty$. We thus get, with f as above,

$$0 = c_- m(f) = \lim_{x \to -\infty} Gf(x)$$

$$= \lim_{x \to -\infty} \int_{\mathbf{R}} P_{T_+}(x, dy)\, Gf(y) = \lim_{y \to \infty} Gf(y) = c_+ m(f);$$

we have a contradiction and the proof is completed.

These results extend to the groups which are isomorphic to $\mathbf{R} \oplus \boldsymbol{K}$ and $\mathbf{Z} \oplus \boldsymbol{K}$, where $\boldsymbol{K}$ is a compact group. As before, we write the proof only in the former case.

Let $\boldsymbol{G} = \mathbf{R} \oplus \boldsymbol{K}$; every point $x \in \boldsymbol{G}$ can be written uniquely as a pair (y, k), with $y \in \mathbf{R}$ and $k \in \boldsymbol{K}$. We call ψ the canonical mapping from $\boldsymbol{G}$ to $\mathbf{R}$: $(y, k) \to y$ and we shall say that x tends to $+\infty$ (resp. $-\infty$) in $\boldsymbol{G}$ if $\psi(x)$ tends to $+\infty$ (resp. $-\infty$) in $\mathbf{R}$. Finally, the Haar measure m_G on $\boldsymbol{G}$ will be taken equal to the product of the Lebesgue measure m on $\mathbf{R}$ and of the normalized Haar measure on $\boldsymbol{K}$.

Theorem 3.9. *If $\boldsymbol{G}$ is isomorphic to $\mathbf{R} \oplus \boldsymbol{K}$ or $\mathbf{Z} \oplus \boldsymbol{K}$, and if the random walk of law μ is of type II, then*

$$\int_G |\psi(x)|\, d\mu(x) < \infty.$$

In that case, if $\lambda = \int_G \psi(x)\, d\mu(x) > 0$, *then in the vague topology,*

$$\lim_{x \to +\infty} G(x, \cdot\,) = 0 \quad \textit{and} \lim_{x \to -\infty} G(x, \cdot\,) = \lambda^{-1} m.$$

For $\lambda < 0$ *the symmetric conclusion holds.*

Proof. By § 2, from every sequence converging to $+\infty$ we can extract a sequence $\{x_n\} = \{(y_n, k_n)\}$ such that $\{G(x_n, \cdot\,)\}$ converges vaguely to a multiple αm_G of the Haar measure of G. Let $f \in C_K^+$ be constant on the co-sets of K. By Lemma 2.3, we have

$$\lim_{x_n \to +\infty} Gf(x_n) = \lim_{y_n \to +\infty} Gf(y_n, 0).$$

Put $\tilde{f}(y) = f(x)$ and $\tilde{\mu} = \psi(\mu)$, and call $\tilde{G}$ the potential of the random walk of law $\tilde{\mu}$ on $\mathbf{R}$. Then $Gf(y_n, 0) = \tilde{G}\tilde{f}(y_n)$, and moreover, applying Propositions 3.4–3.8 to this random walk, we get

$$\lim_{y_n \to +\infty} \tilde{G}\tilde{f}(y_n) = c_+ m(\tilde{f}),$$

where c_+ is non-zero if and only if

$$\int_{-\infty}^{+\infty} |y|\, d\tilde{\mu}(y) = \int_G |\psi(x)|\, d\mu(x) < +\infty$$

and

$$\int_{-\infty}^{+\infty} |y|\, d\tilde{\mu}(y) = \int_G |\psi(x)|\, d\mu(x) < 0.$$

Since furthermore $m(\tilde{f}) = m_G(f)$, the proof is complete.

In the following section we shall show that, up to an isomorphism, these groups are the only abelian ones for which there exist type II random walks.

Exercise 3.10. (1) Under the hypothesis of Corollary 3.5, show that $]-\infty, 0[$ is a transient set and that g is the probability of never returning to $]-\infty, 0[$.

(2) Similarly, show that if $a > 0$, $1_{]-\infty, a[} = Gg(x - a)$. Show that $P_{T_+}(x, [0, a[)$ is the potential of a function vanishing outside $[0, a[$. Compute this function and derive that if A is a relatively compact subset of $]0, +\infty[$ whose boundary is of Haar measure zero, then

$$\lim_{x \to -\infty} P_{T_+}(x, A) = \lambda^{-1} \int_A g(-x)\, dx.$$

[*Hint*: Use the duality between the kernels H_B and P_B described in ch. 2 § 4. This result will be generalized in § 5.]

Exercise 3.11. (1) Set $\mathbf{R}_+ = [0, \infty[$, and show that for a transient random walk the set of measures $\{H_{\mathbf{R}_+}(x, \cdot), x \in \mathbf{R}\}$ is vaguely relatively compact.

(2) Under the hypothesis of Corollary 3.5, prove that $\lim_{x\to-\infty} H_{\mathbf{R}_+}(x, \cdot)$ exists for the vague topology, and compute this limit.

[*Hint*: Consider sequences for which the limit exists and then apply the duality between P_A and H_A.]

Exercise 3.12. Assume the hypothesis of Corollary 3.5, with $\lambda < \infty$, and set

$$N_t(\omega) = \sum_{n=1}^{\infty} 1_{\{X_n(\omega) \leqslant t\}}.$$

(1) Show that N_t is an a.s. finite random variable.

(2) Find the starting measure ν such that the function $t \to E_\nu[N_t]$ is linear.

[*Hint*: Show that this amounts to finding ν such that $\nu * \bar{G} = \alpha m^+$, where $\alpha \in \mathbf{R}_+$ and m^+ is the restriction of the Lebesgue measure to $[0, \infty[$. It turns out that ν has density $\lambda^{-1} \mu([t, \infty[)$.] If $\lambda = +\infty$ there is no such measure.

(3) Deduce from (2) the renewal theorem for this particular set of hypothesis.

Exercise 3.13. Under the hypothesis of Corollary 3.5, with $\lambda < \infty$ and $\sigma^2 = \int x^2 \, d\mu(x) < \infty$, show that

$$\lim_{x\to\infty} G([0, x]) - \lambda^{-1}x = \sigma^2/2\lambda^2.$$

[*Hint*: Use the Poisson equation with second member

$$\lambda^{-1} 1_{[0,\infty[}(x) \int_x^{\infty} \mu([t, \infty[) \, dt.]$$

Exercise 3.14. This exercise is designed to show that in the discrete case the renewal theorem is equivalent to a limit theorem for recurrent chains which will be studied in ch. 6 §2 in greater generality.

(1) Let X be a recurrent irreducible discrete chain and using the notation of ch. 3 §1.11, set $Y_0 = S_x$ and $Y_n = S_{n+1} - S_n$ for $n > 0$. Prove that for every probability measure P_ν the random variables $\{Y_n, n \geqslant 1\}$ are independent and equidistributed of law $\mu = P_x[S_x = \cdot\,]$. Prove that μ is

adapted if and only if X is aperiodic (ch. 3 §1.17).

(2) Suppose henceforth that X is aperiodic and prove that for every pair (x, y) of points in E,

$$\lim_{n\to\infty} P_n(x, y) = (E_y[S_y])^{-1}.$$

[*Hint*: Use the renewal theorem for the random walk S_n on $\mathbf{Z}$.]

(3) Show that the measure m such that $m(y)$ is equal to the limit in (2) is invariant for X and derive that $E_x[S_x]$ is either finite for every x in E or is infinite for every x in E. This is another proof of Theorem **1.9** and Corollary 1.10 in ch. 3.

(4) We now go in the converse direction. Let $\{Z_n\}$ be an adapted random walk on $\mathbf{Z}$ with law μ carried by the integers $n \geqslant 1$. Define

$$X_n(\omega) = n - Z_k(\omega) \quad \text{if} \quad Z_k(\omega) \leqslant n < Z_{k+1}(\omega).$$

Prove that for P_0, the sequence $\{X_n, n \geqslant 0\}$ is a Markov chain with initial distribution ε_0 of the type studied in Exercise 1.16 of ch. **3**. Deduce then the renewal theorem from the convergence result in (2).

4. The renewal theorem

We now return to the case of a general abelian LCCB group $\mathbf{G}$. As before, f is an arbitrary non-zero function in C_K^+.

Proposition 4.1. *If $\mathbf{G}$ has a compactly generated, open, non-compact sub-group $\mathbf{G}_1$ such that $\mathbf{G}/\mathbf{G}_1$ is infinite, all transient random walks on $\mathbf{G}$ are type I.*

Proof. Suppose that a random walk X on $\mathbf{G}$ is of type II and let $\{x_n\}$ be a sequence such that the measures $G(x_n, \cdot)$ converge vaguely to cm $(c > 0)$. We shall work towards a contradiction.

We may assume that the cosets $x_n + \mathbf{G}_1$ are pairwise disjoint. Indeed if no sub-sequence of $\{x_n\}$ has this property, then there is a coset of $\mathbf{G}_1$ which contains infinitely many x_n. This amounts to assuming that $\{x_n\}$ is contained in one and the same co-set. Pick now a sequence $\{z_k\}$, such that the cosets $z_k + \mathbf{G}_1$ are pairwise disjoint. By Lemma 2.3, for every k we have

$$\lim_n Gf(x_n + z_k) = cm(f).$$

For every k, pick an integer n_k such that

$$|Gf(x_{n_k} + z_k) - cm(f)| < 2^{-k},$$

and set $y_k = x_{n_k} + z_k$; then the sequence $\{y_k\}$ has the desired property.

Now, since G_1 is compactly generated but non-compact, there exists a point $x \in G_1$ such that $nx \to \Delta$ as $n \to \infty$. Using Proposition 2.4 and Lemma 2.3 in the same way as in the proof of Theorem 3.1, it is easily seen that $\lim_n Gf(nx) = 0$ or $\lim_n Gf(-nx) = 0$. Without loss of generality we assume the former contingency. Then, for every n, there is a smallest integer h_n such that for $n' \geqslant h_n$

$$Gf(x_n + n'x) \leqslant \tfrac{1}{2}c\, m(f).$$

For otherwise there would be an integer n_0 and a subsequence $\{n'\}$ such that

$$Gf(x_{n_0} + n'x) > \tfrac{1}{2}c\, m(f),$$

and this is impossible since, by Lemma 2.3, $Gf(x_{n_0} + n'x)$ and $Gf(n'x)$ have the same limit, which is zero.

For fixed l, $Gf(x_n + lx)$ tends to $c\, m(f)$; hence $h_n > 1$ for n sufficiently large. Thus, for n sufficiently large,

$$Gf(x_n + h_n x - x) > \tfrac{1}{2}c\, m(f),$$

$$Gf(x_n + h_n x) \leqslant \tfrac{1}{2}c\, m(f).$$

The sequence $\{x_n + h_n x - x\}$ converges to Δ because G_1 being open, G/G_1 is discrete and every compact subset of G is contained in the union of a finite number of cosets of G_1. As a result, $Gf(x_n + h_n x - x)$ converges to $cm(f)$ as n tends to infinity and $Gf(x_n + h_n x)$ to zero which by Lemma 2.3 is a contradiction. The proof is complete.

Corollary 4.2. *If G is isomorphic to $\mathbf{R}^{d_1} \oplus \mathbf{Z}^{d_2} \oplus K$ (K a compact group) and if $d_1 + d_2 > 1$, every transient random walk on G is of type I.*

Proof. By the argument of Theorem 3.9, it suffices to consider the case $G = \mathbf{R}^{d_1} \oplus \mathbf{Z}^{d_2}$. If $d_2 \geqslant 1$, we are done thanks to Proposition 4.1. If $G = \mathbf{R}^{d_1}$, with $d_1 > 1$, then G is the union of an increasing sequence of compact sets whose complements are connected. If the continuous function Gf has a non-zero limit a, every point in $[0, a]$ is also a limit point, which by Theorem 2.6 is a contradiction.

Before we proceed let us recall that a compact sub-semigroup of a group is a compact subgroup.

Proposition 4.3. *If every element of the group* G *is compact, then every transient random walk on* G *is of type I.*

Proof. Let X be a type II random walk, and let $\{z_n\}$ be a sequence such that $\{G(z_n, \cdot)\}$ converges to cm $(c > 0)$. For every n, the closure of the set $\{kz_n, k > 0\}$ is a compact sub-semigroup of G, hence a subgroup of G, and there is a sequence k_p of positive integers such that $\{k_p z_n\}$ converges to $-z_n$ as p tends to infinity.

Thus for every $\varepsilon > 0$, one can find an integer k_n such that

$$Gf(k_n z_n) \leqslant Gf(-z_n) + \varepsilon n^{-1}$$

and since, by Proposition 2.5, $Gf(-z_n)$ tends to zero as n tends to infinity, the sequence $Gf(k_n z_n)$ tends to zero.

Choose $M > \max(||Gf||, \frac{1}{4}c\, m(f))$. Since $c^2\, m(f)^2/4M < cm(f)$, for all sufficiently large values of $n \geqslant 1$, we may pick the largest integer $h_n < k_n$ such that

$$Gf(h_n z_n) \geqslant \frac{c^2\, m(f)^2}{4M}, \qquad Gf((h_n + 1)\, z_n) < \frac{c^2\, m(f)^2}{4M},$$

By the proof of Proposition 2.4, for n sufficiently large,

$$Gf(h_n z_n)\, Gf(z_n) \leqslant M[\varepsilon + Gf((h_n + 1)\, z_n)].$$

Moreover, for n sufficiently large, $Gf(z_n) \geqslant \frac{1}{2}c\, m(f)$, which implies

$$Gf(h_n z_n) \leqslant \frac{2M}{cm(f)} [Gf((h_n + 1)\, z_n) + \varepsilon]$$

$$\leqslant \tfrac{1}{2}c\, m(f) + \frac{2M\varepsilon}{cm(f)} \leqslant \tfrac{1}{2}c\, m(f) + \tfrac{1}{2}\varepsilon,$$

and

$$\frac{c^3\, m(f)^3}{8M^2} - \varepsilon \leqslant Gf((h_n + 1)\, z_n).$$

Choosing ε in an appropriate way, we see that there exist four constants $\alpha, \beta, \gamma, \delta$ such that for n sufficiently large,

$$0 < \alpha \leqslant Gf(h_n z_n) \leqslant \beta < cm(f),$$

$$0 < \gamma \leqslant Gf((h_n + 1)\, z_n) \leqslant \delta < cm(f).$$

Since $\{z_n\}$ tends to Δ when n tends to infinity, at least one of the sequences $\{h_n z_n\}$ and $\{(h_n + 1) z_n\}$ contains a subsequence $\{y_n\}$ converging to Δ. The sequence $Gf(y_n)$ has at most 0 and $cm(f)$ as limit points; we thus have a contradiction.

Theorem 4.4 (renewal theorem). *If there exists on* $\boldsymbol{G}$ *a random walk of type II, the group* $\boldsymbol{G}$ *is isomorphic to* $\mathbf{R} \oplus \boldsymbol{K}$ *or* $\mathbf{Z} \oplus \boldsymbol{K}$*, where* $\boldsymbol{K}$ *is a compact group. The asymptotic behaviour of the measures* $G(x, \cdot\,)$ *is then described by Theorem* 3.9.

Proof. By Proposition 4.3, there exists a non-compact element y. As every group, $\boldsymbol{G}$ has a compactly generated open sub-group $\boldsymbol{G}'$; the subgroup $\boldsymbol{G}_1$ generated by $\boldsymbol{G}'$ and y is non-compact, compactly generated and open. Thus by Proposition 4.1, the group $\boldsymbol{G}/\boldsymbol{G}_1$ is finite; hence $\boldsymbol{G}$ also is compactly generated. The theorems on the structure of abelian groups then assert that $\boldsymbol{G}$ is isomorphic to $\mathbf{R}^{d_1} \oplus \mathbf{Z}^{d_2} \oplus \boldsymbol{K}$, where $\boldsymbol{K}$ is a compact group. By Corollary 4.2, the proof is complete.

5. Refinements and applications

Throughout this section, μ will be taken spread out, and we shall give applications of the renewal theorem to hitting probabilities and to recurrent sets.

The first result is true if $\boldsymbol{G}$ is merely unimodular.

Theorem 5.1. *The measure* $\bar{G}$ *can be written* $\bar{G} = \bar{G}' + \bar{G}''$*, where* $\bar{G}'$ *is an absolutely continuous measure having a bounded continuous density* p *and* $\bar{G}''$ *a finite measure.*

Proof. By hypothesis, there is an integer k and a bounded integrable function ψ such that $\mu^k = \psi m + \alpha$, where α is a finite measure of norm less than 1. Setting $n_0 = 2k$ it is easily seen that one can write $\mu^{n_0} = \varphi m + \nu$, where $\varphi \in C_K^+$ and $\|\nu\| < 1$. The measure $\bar{G}$ can be written

$$\bar{G} = \left(\sum_{n=0}^{n_0-1} \mu^n\right) * \left(\sum_{n=0}^{\infty} \mu^{nn_0}\right).$$

Setting $\tilde{G} = \sum_{n=0}^{\infty} \mu^{nn_0}$, we have

$$\begin{aligned}\tilde{G} &= \varepsilon_e + (\varphi m + \nu) * \tilde{G} \\ &= \varepsilon_e + \varphi m * \tilde{G} + \nu * (\varepsilon_e + (\varphi m + \nu) * \tilde{G})\end{aligned}$$

and inductively, for every $n \geqslant 1$,

$$\tilde{G} = \varepsilon_e + \nu + \cdots + \nu^n + (\varepsilon_e + \cdots + \nu^n) * \varphi m * \tilde{G} + \nu^{n+1} * \tilde{G}. \tag{5.1}$$

If $f \in C_K$, the function $x \to \int f(xx')\, \tilde{G}(\mathrm{d}x')$ is bounded and continuous, since $\tilde{G}$ is the potential measure corresponding to μ^{n_0}. From this remark, we derive easily that $\lim_{n\to\infty} \nu^{n+1} * \tilde{G} = 0$ in the vague topology. On the other hand, $\nu' = \sum_0^\infty \nu^n$ is a finite measure, and letting n tend to infinity in eq. (5.1) yields

$$\tilde{G} = \nu' + \nu' * \varphi m * \tilde{G},$$

which shows that Theorem 5.1 is true for $\tilde{G}$. It is then verified for G with $\bar{G}'' = (\sum_0^{n_0-1} \mu^n) * \nu'$ and $p = \bar{G}'' * \varphi * \tilde{G}$.

As is apparent from the proof, this decomposition is not unique, but the renewal theorem settles the asymptotic behaviour of p.

Theorem 5.2. *If the random walk is of type I, then*

$$\lim_{x\to\Delta} p(x) = 0.$$

If the random walk is of type II with $\lambda > 0$, then

$$\lim_{x\to+\infty} p(x) = \lambda^{-1}, \qquad \lim_{x\to-\infty} p(x) = 0.$$

For $\lambda < 0$, the symmetric conclusion holds.

Proof. If X is of type I, the random walk of law μ^{n_0} is also type I and therefore $\lim(\tilde{G} * \varphi)(x) = 0$, that is $\tilde{G} * \varphi \in C_0$; since $\bar{G}''$ is finite, $p = \bar{G}'' * \varphi * \tilde{G}$ is also in C_0, and the proof is completed in that case.

If X is of type II with $\lambda > 0$, then using the notation of Theorem 3.9,

$$\int_G \psi(x)\, \mathrm{d}\mu^{n_0}(x) = n_0 \lambda,$$

and therefore, by the renewal theorem for $\tilde{G}$,

$$\lim_{x\to-\infty} (\tilde{G} * \varphi)(x) = 0, \qquad \lim_{x\to+\infty} (\tilde{G} * \varphi)(x) = m(\varphi)/n_0\lambda.$$

Since $\bar{G}''(G) = n_0/m(\varphi)$, it follows that

$$\lim_{x\to-\infty} p(x) = 0 \quad \text{and} \lim_{x\to+\infty} p(x) = \lambda^{-1}.$$

Corollary 5.3. *For every bounded Borel function f with compact support,*

$$\lim_{x \to \Delta} Gf(x) = 0$$

if the random walk is of type I, and

$$\lim_{x \to -\infty} Gf(x) = \lambda^{-1} m(f), \qquad \lim_{x \to +\infty} Gf(x) = 0$$

if the random walk is of type II with $\lambda > 0$.

Proof. We have

$$Gf(x) = \langle \varepsilon_x * \bar{G}'', f \rangle + \int_G f(y)\, p(y-x)\, m(dy).$$

The first term tends to zero as x tends to Δ since $\bar{G}''$ is finite, and the result follows by a simple application of the Lebesgue theorem.

We now turn to the study of the asymptotic behaviour of hitting probabilities. As usual, the symbol $\hat{}$ refers to the dual random walk.

Theorem 5.4. *Let B be any relatively compact Borel set, and f any bounded Borel function with compact support. Then if X is type I*

$$\lim_{x \to \Delta} U_B f(x) = 0$$

and if X is type II with $\lambda > 0$,

$$\lim_{x \to +\infty} U_B f(x) = 0, \qquad \lim_{x \to -\infty} U_B f(x) = \lambda^{-1} \int f(y)\, \hat{\gamma}_B(y)\, dy,$$

where $\hat{\gamma}_B = \hat{P}.[S_B = \infty]$.

Proof. Consider first the type II case. Clearly the kernels $U_0(x, \cdot)$ and $G(x, \cdot)$ have the same limits on functions with compact supports as x tends to infinity. Passing to the limit in the identity

$$U_0 f(x) = U_B f(x) + U_0 1_B U_B f(x)$$

yields

$$\lim_{x \to -\infty} U_B f(x) = \lambda^{-1} m(f) - \lambda^{-1} m(1_B U_B f).$$

Using the duality between the kernels U_B and $\hat{U}_B$ (ch. 2, Proposition 4.9) then yields

$$\lim_{x\to-\infty} U_B f(x) = \lambda^{-1}\langle 1 - \hat{U}_B 1_B, f\rangle_m = \lambda^{-1}\int f(y)\,\hat{P}_y[S_B = \infty]\,dy.$$

The proof is then easily completed.

Corollary 5.5. *With the same hypothesis, the hitting distributions $P_B(x, \cdot)$ converge vaguely to zero or to $\lambda^{-1}\hat{g}_B m$.*

Proof. Straightforward since the function $\hat{g}_B$ which was defined in ch. 2 §3 is equal to $1_B\hat{\gamma}_B$.

By ch. 2 §3 and the Choquet–Deny theorem, when μ is spread out every set is either recurrent or transient. This raises the problem of characterizing the recurrent sets. The following result gives a partial but useful answer.

If the group G is isomorphic to $\mathbf{R} \oplus K$ or $\mathbf{Z} \oplus K$, with K a compact group, we denote by G_+ the set of points (x, k) such that $x \geq 0$.

Proposition 5.6. *If the random walk is of type II with $\lambda > 0$, a set A is recurrent if and only if $m(A \cap G_+) = +\infty$.*

Proof. We take $G = \mathbf{R}$, the other groups requiring only obvious changes. The set $G_- = G_+^C$ is transient as is easily seen by applying the law of large numbers (ch. 3, Exercise 3.12). We therefore take $A \subset G_+$. If $m(A) < +\infty$, Theorem 5.2 implies that $G(x, A) < \infty$ for every $x \in G$, and thus A is transient.

Suppose conversely that $m(A) = +\infty$. We can write $A = \sum_1^\infty A_k$, where the sets A_k can be chosen disjoint, of finite measure and such that the sequence $\{m(A_k)\}$ increases to $+\infty$. Each A_k being transient by the preceding argument, to hit A infinitely many times, the random walk must hit infinitely many sets A_k and therefore

$$P_e\left[\overline{\lim_n}\{X_n \in A\}\right] = P_e\left[\overline{\lim_k}\{T_{A_k} < \infty\}\right] \geq \overline{\lim_k}\, P_e[T_{A_k} < \infty].$$

By the strong Markov property, we get easily (see ch. 2, Theorem 1.11)

$$P_e[T_{A_k} < \infty] \geq \bar{G}(A_k)/\sup_{y\in A_k} G(y, A_k).$$

Next, by Theorem 5.1,

$$\sup_{y\in A_k} G(y, A_k) \leq \|\bar{G}''\| + \|p\|\, m(A_k),$$

which implies

$$P_e[T_{A_k} < \infty] \geqslant \bar{G}(A_k)/(\|\bar{G}''\| + \|p\|\, m(A_k)).$$

As k tends to infinity, $\bar{G}(A_k)$ is equivalent to $\lambda^{-1}\, m(A_k)$, by Theorem 5.1, hence

$$\overline{\lim_{k\to\infty}}\, P_e[T_{A_k} < \infty] \geqslant (\lambda\|p\|)^{-1} > 0.$$

The set A is recurrent and the proof is complete.

Corollary 5.7. *A set $A \subset G_+$ is recurrent if and only if $\bar{G}(A) = +\infty$.*

Proof. Obvious.

Remark. For the case in which the first moment does not exist, but $\int_{-\infty}^{0} |x|\, d\mu(x) < \infty$, one could hope for similar results, but in fact they are not true (cf. Exercise 5.12).

Exercise 5.8. Investigate the asymptotic behaviour as x tends to Δ of $P_x[T_A < T_B]$ and of $G^A(x, B)$, where A and B are relatively compact Borel sets.

Exercise 5.9. Try to carry over Theorem 5.4 to non-spread-out laws.

[*Hint*: Use relatively compact Borel sets whose boundaries have zero Haar measure.]

Exercise 5.10. Prove Theorem 5.4 without relying on Theorem 5.1.

Exercise 5.11. Let A be a transient set of infinite Haar measure. Prove that its capacity relative to m (ch. 2, Exercise 4.13) is equal to $+\infty$ if the random walk is of type I and to $|\lambda|$ if the random walk is of type II with mean λ.

Exercise 5.12. If $G = \mathbf{R}$ and $\int_{-\infty}^{0} |x|\, d\mu(x) < \infty$, prove that the following two statements are equivalent:

(i) for every set A such that $m(G_+ \cap A) = \infty$, we have $P_0[S_A < \infty] = 1$;

(ii) the random walk is spread-out, type II and λ is > 0.

[*Hint*: Observe that a set with property (i) is recurrent; then use the fact that for a non-spread-out random walk the σ-algebra $\mathscr{I}$ is not a.s.trivial and that there exists therefore a set which is neither recurrent nor transient.]

Exercise 5.13. Let X be a transient random walk on a discrete group. Prove that $P_e[T_{\{x\}} < \infty] < 1$ for all $x \neq e$ with one and only one exception: the group is isomorphic to $\mathbf{Z}$ and either X is right continuous (ch. **3**, Exercise **3.13**) and the mean λ exists and is > 0, or X is left continuous with the symmetric conditions.

Exercise 5.14. Let $\boldsymbol{H}$ be a Borel sub-group of $\boldsymbol{G}$. Show that $\boldsymbol{H}$ is a recurrent set if and only if $G(e, \boldsymbol{H}) = +\infty$.

CHAPTER 6

ERGODIC THEORY OF HARRIS CHAINS

In this chapter, we aim at proving limit theorems for Harris chains. For this purpose we shall need a certain amount of potential theory, which will be introduced in §§ 4 and 5, but potential theory for Harris chains will really be dealt with in ch. 8. Throughout the chapter, the state space $(E, \mathscr{E})$ is separable.

1. The zero-two laws

In this section we do not deal with Harris chains but with general chains for which we introduce several notions which will prove useful for Harris chains as well as in other situations.

Definition 1.1. The σ-algebra

$$\mathscr{A} = \bigcap_{n \geq 0} \theta_n^{-1}(\mathscr{F}) = \bigcap_{n \geq 0} \sigma(X_m, m \geq n)$$

is called the *asymptotic* or *tail* σ-algebra of the canonical Markov chain X. A random variable measurable with respect to $\mathscr{A}$ is said to be an *asymptotic* or *tail* random variable and in particular, a set belonging to $\mathscr{A}$ is an *asymptotic* or *tail* event.

Clearly, an event A is asymptotic if and only if for every n there exists an event $A_n \in \mathscr{F}$ such that $A = \theta_n^{-1}(A_n)$. The σ-algebra $\mathscr{I}$ of invariant events is thus contained in $\mathscr{A}$. Moreover, the sets A_n may be chosen in a canonical fashion. The shift operator θ is not one-to-one, so that θ^{-1} is not a mapping and $\theta^{-1}(\omega)$ is the set of trajectories ω' such that $\theta(\omega') = \omega$. However, if $Z \in \sigma(X_m, m \geq n)$, this random variable does not depend on the first n coordinates and one can define unambiguously $Z \circ \theta^{-1}$ and $Z \circ \theta^{-p}$ for $p \leq n$. For $\omega = (x_0, x_1, \ldots)$,

$$Z \circ \theta^{-1}(\omega) = Z(y, x_0, x_1, \ldots),$$

where y is arbitrary in E. If Z is asymptotic, one can therefore define unambiguously and for every $n \in \mathbf{N}$ another asymptotic random variable Z_n, by setting $Z_n = Z \circ \theta^{-n}$. We then have $Z = Z_n \circ \theta_n$ and $Z_m = Z_{n+m} \circ \theta_n$ for every pair of integers n and m. We shall write θ_{-n} rather than θ^{-n}, in order to make θ_n ($n \in \mathbf{Z}$) appear as a transformation group of $\mathscr{A}$.

Definition 1.2. Two asymptotic random variables Z and Z' are said to be *equivalent* if, for every $n \in \mathbf{Z}$, the asymptotic random variables $Z \circ \theta_n$ and $Z' \circ \theta_n$ are almost surely equal.

Notice that if suffices to have the equality for $n \leqslant 0$, the equality for $n > 0$ following from the Markov property. This notion of equivalence is stronger than the usual almost sure equality; two invariant random variables equivalent in this sense are equivalent in the sense of ch. 2 §3. Finally, the operators θ_n operate on the equivalence classes of asymptotic random variables, and the classes of invariant random variables are those which are invariant under θ_n.

We shall state a result similar to ch. 2, Proposition 3.2, for which we need the following notational device: we define a T.P. denoted by Q on $(E \times \mathbf{N}, \mathscr{E} \otimes \mathscr{P}(\mathbf{N}))$ by setting

$$Q((x, n), A \times \{m\}) = \begin{cases} P(x, A) & \text{if } m = n + 1, \\ 0 & \text{if } m \neq n + 1. \end{cases}$$

For every probability measure P_x on $(\Omega, \mathscr{F})$ and every integer m, the sequence $\{Y_n\} = \{(X_n, n + m)\}$ is a homogeneous Markov chain with T.P. Q, called a *space–time* chain (ch. 1, Exercise 2.17(2)).

Proposition 1.3. *If P is markovian, the formula*

$$h(x, n) = E_x[Z_n] = E_x[Z \circ \theta_{-n}]$$

sets up a one-to-one and onto correspondence between the equivalence classes of asymptotic and bounded random variables and the bounded Q-harmonic functions. Moreover,

$$\lim_{n \to \infty} h(X_n, n) = Z \quad a.s.$$

Proof. If $Z \in b\mathscr{A}$, we have

$$Qh(x, n) = E_x[E_{X_1}[Z_{n+1}]] = E_x[Z_{n+1} \circ \theta_1] = E_x[Z_n] = h(x, n),$$

and h is therefore Q-harmonic. Conversely if h is Q-harmonic and bounded, $h(X_n, n)$ is a bounded martingale which converges P_x-a.s. and in $L^1(P_x)$ for every $x \in E$. The limit Z is almost surely asymptotic, since for every $m \in \mathbf{N}$ the sequence $h(X_n, n+m)$ converges also to a limit Z_m, and obviously $Z = Z_m \circ \theta_m$ a.s. The convergence in the L^1 sense implies clearly that $h(x, n) = E_x[Z_n]$.

Remark. The space $\mathscr{K}$ of bounded Q-harmonic functions is a Banach subspace of $\mathrm{b}(\mathscr{E} \otimes \mathscr{P}(\mathbf{N}))$ and it is easily seen that for $h \in \mathscr{K}$,

$$||h|| = \lim_n ||h(\cdot\,, n)||.$$

Moreover, if we make $Z = \overline{\lim}_n h(X_n, n)$, the random variable Z is in $\mathrm{b}\mathscr{A}$ and it follows from the above formulas that $||h|| = ||Z||$ where the norms are the uniform norms on the respective spaces. The correspondence described above thus becomes a Banach isomorphism.

If h is a bounded Q-harmonic function, by setting $h_n = h(\cdot\,, n)$, $n \in \mathbf{N}$, we get a sequence of bounded functions on E such that $P_n h_{n+k} = h_k$ for $n, k \geqslant 0$. Conversely, if h_0 is a bounded function such that for every n there is a bounded function h_n such that $Ph_1 = h_0$ and $Ph_{n+1} = h_n$ for every $n > 0$, the function defined on $E \times \mathbf{N}$ by $h(\cdot\,, n) = h_n$ is Q-harmonic. We will denote by $\mathscr{D}$ the set of such functions h_0; we clearly have a one-to-one and onto correspondence between $\mathscr{D}$ and the equivalence classes of bounded asymptotic random variables expressed, when P is markovian, by the formula

$$h_0(x) = E_x[Z].$$

We henceforth assume that P is markovian and we are going to express asymptotic properties of $\{P_n\}$ in terms of $\mathscr{A}$. In the following if γ is a bounded measure on $(\Omega, \mathscr{F})$ we shall write $||\gamma||_{\mathscr{A}}$ for the norm of γ when it is restricted to $\mathscr{A}$.

Theorem 1.4. *For any bounded measure ν on $(E, \mathscr{E})$ we have*

$$\lim_n ||\nu P_n|| = ||P_\nu||_{\mathscr{A}}.$$

Proof. The sequence $\{||\nu P_n||\}$ is clearly decreasing hence convergent. We are going to prove that its limit is equal to $\sup\{\int h \, d\nu;\ h \in \mathscr{D},\ ||h|| \leqslant 1\}$ which is equal to $||P_\nu||_{\mathscr{A}}$ thanks to the remark following Proposition 1.3

together with the fact that if $h \in \mathscr{K}$, then $h(\cdot, n)$ is in $\mathscr{D}$ for every n.

By ch. 4 §1 we may choose a probability measure μ such that $\mu P \ll \mu$ and $\nu \ll \mu$ and we set $f = d\nu/d\mu$. Using the operator T on $L^1(E, \mathscr{E}, \mu)$ associated with P we have

$$||\nu P_n|| = ||T^n f||_1 = \sup\left\{\int f \cdot T^{*n} g \, d\mu;\ g \in B\right\}$$

where B is the unit ball in $L^\infty(\mu)$. It follows that

$$\lim_n ||\nu P_n|| \geqslant \sup\left\{\int fg \, d\mu;\ g \in \bigcap_0^\infty T^{*n}B\right\}.$$

On the other hand, for any $\varepsilon > 0$, we can find a sequence $\{g_n\}$ of points in B such that for every n,

$$\int f \cdot T^{*n} g_n \, d\mu \geqslant ||\nu P_n|| - \varepsilon.$$

Since $(E, \mathscr{E})$ is separable, B is $\sigma(L^\infty, L^1)$-sequentially compact so that we can find a subsequence $\{n_k\}$ such that $\{T^{*n_k} g_{n_k}\}$ converges to a limit g_0 in B. The operator T^* being weakly continuous, for each n the function g_0 is in the compact set $T^{*n}B$ and therefore g_0 is in $\bigcap_0^\infty T^{*n}B$. By passing to the limit in the above inequality we get

$$\int fg_0 \, d\mu \geqslant \lim_n ||\nu P_n|| - \varepsilon,$$

hence

$$\lim_n ||\nu P_n|| = \sup\left\{\int fg \, d\mu;\ g \in \bigcap_0^\infty T^{*n}B\right\}.$$

Now T maps the set $A = \bigcap_0^\infty T^{*n}B$ onto itself. Indeed we clearly have $T^*A \subset A$ and if $k \in A$, then for every n there is a function k_n such that $k = T^{*n}k_n$. Each of the sets $T^{*n}B$ being $\sigma(L^\infty, L^1)$-compact the sequence $\{T^{*(n-1)}k_n\}$ has a limit point l which lies in A and as T is $\sigma(L^\infty, L^1)$-continuous we have $T^*l = k$, and T^* is onto. As a result, for any $g \in A$ we can inductively find a sequence $\{\bar{g}_n\}$ of points in B such that $\bar{g}_n = T^*\bar{g}_{n+1}$ and $\bar{g}_0 = g$.

By applying now Proposition 1.6 of ch. 4 to the transition probability Q and to the measure $\mu \otimes m$ where m is the counting measure on the integers, we see that there is a one-to-one correspondence between classes of equivalent functions in A and functions in $\mathscr{D}$. The proof is then complete.

We now set, for $k \in \mathbf{Z}$,

$$\alpha_k(x) = \lim_n \|P_{n+k}(x, \cdot) - P_n(x, \cdot)\|.$$

The limit exists since the sequence involved is clearly decreasing.

Lemma 1.5. *For every k, the function α_k is measurable and $P\alpha_k \geqslant \alpha_k$.*

Proof. To prove that α_k is measurable we shall prove the stronger property: the positive part $N^+(x, \cdot)$ of the bounded kernel N is itself a kernel. We shall use the set-up of Lemma 5.3 in ch. 1. Let x be fixed and choose for probability measure λ the suitable multiple of $|N(x, \cdot)|$; the functions f_n defined by

$$f_n(y) = N(x, E_y^n)/\lambda(E_y^n) \quad \text{if} \quad \lambda(E_y^n) > 0, \qquad f_n(y) = 0 \quad \text{otherwise},$$

are bounded uniformly and converge almost surely to the density of $N(x, \cdot)$ with respect to λ. Consequently $\{f_n^+\}$ converges a.s. to the density of $N^+(x, \cdot)$ and because of the uniform boundedness, for every $A \in \mathscr{E}$,

$$N^+(x, A) = \lim_n \int_A f_n^+ \, d\lambda.$$

But if $A \in \mathscr{P}^k$ and $n > k$, then $\int_A f_n^+ \, d\lambda = N_n^+(x, A)$ where N_n^+ is the positive part of the restriction of N to $\mathscr{P}^n$ and the map $x \to N_n^+(x, A)$ is easily seen to be measurable. Consequently $N^+(\cdot, A)$ is measurable for every A in $\bigcup_k \mathscr{P}^k$ and by the monotone class theorem N^+ is a kernel.

To prove the second assertion put $\gamma_n(x, \cdot) = P_{n+k}(x, \cdot) - P_n(x, \cdot)$ and $\delta_n(x) = \|\gamma_n(x, \cdot)\|$. For $A \in \mathscr{E}$, we have

$$\delta_n(x) \geqslant \gamma_n(x, A) - \gamma_n(x, A^c),$$

hence

$$P\, \delta_n(x) \geqslant \gamma_{n+1}(x, A) - \gamma_{n+1}(x, A^c).$$

By taking for (A, A^c) a Jordan decomposition for $\gamma_{n+1}(x, \cdot)$ we see that $P\, \delta_n \geqslant \delta_{n+1}$, whence $P\alpha_k \geqslant \alpha_k$ follows readily.

Clearly, $0 \leqslant \alpha_k \leqslant 2$, but one can actually say much more as will be seen in the next result. Let us first observe that to say that $\mathscr{A} = \mathscr{I}$ a.s. is to say that for any $Z \in b\mathscr{A}$, there is a $Z' \in b\mathscr{I}$ such that $Z = Z'$ a.s. or equivalently that $Z = Z \circ \theta$ a.s. By Proposition 1.3 and Proposition 3.2 in ch. 2, this is also equivalent to saying that the bounded Q-harmonic functions h do not depend on the variable n, or in other words that $h(x, n) = h(x)$ with h a harmonic function for P. An example of an asymptotic event which is not a.s.equal to an invariant event is provided by $\overline{\lim}\{X_n = n\}$ where X is the chain of translation on integers. Another example is to be found in Exercise 1.9.

Theorem 1.7 (first zero-two law). *For every k, the number $\sup_{x \in E} \alpha_k(x)$ is either zero or two. For $k = 1$, the former case occurs if and only if $\mathscr{A} = \mathscr{I}$ a.s. and then for every probability measure ν on $\mathscr{E}$,*

$$\lim_n \|\nu P_n - \nu P_{n+1}\| = 0.$$

Proof. It is enough to prove the result for $k = 1$ because it can then be applied to the T.P. $P_{|k|}$ to get the result for all k's.

If $\mathscr{A} = \mathscr{I}$ a.s. then by the Markov property P_ν and $P_{\nu P}$ agree on $\mathscr{A}$ and the "if" part follows from Theorem 1.4.

Suppose conversely that $\mathscr{A}$ is not equal to $\mathscr{I}$ a.s. There is then a probability measure ν and a set $F \in \mathscr{A}$ such that $P_\nu[F \,\Delta\, \theta^{-1}(F)] > 0$, and by replacing if necessary ν by $\sum_0^\infty 2^{-n-1}\nu P_n$ we may assume that $\nu P \ll \nu$. If $P_\nu[F \backslash \theta^{-1}(F)] > 0$ we set $G = F \backslash \theta^{-1}(F)$; the set G is in $\mathscr{A}$, $P_\nu[G]$ is > 0 and $G \cap \theta^{-1}(G) = (F \backslash \theta^{-1}(F)) \cap (\theta^{-1}(F) \backslash \theta^{-2}(F)) = \emptyset$. If $P_\nu[F \backslash \theta^{-1}(F)] = 0$, then $P_\nu[\theta^{-1}(F) \backslash F] > 0$; but this is equal to $P_{\nu P}[F \backslash F_1]$ and therefore we have $P_\nu[F \backslash F_1] > 0$. Then again the set $G = F \backslash F_1$ is in $\mathscr{A}$, has positive P_ν-measure and $G \cap \theta^{-1}(G) = (F \backslash F_1) \cap (\theta^{-1}(F) \backslash F) = \emptyset$.

In each case we now set $Z = 1_G - 1_G \circ \theta$ and $h_n(x) = E_x[Z_n]$. By Proposition 1.3 we have

$$\lim_n h_n(X_n) = Z, \qquad \lim_n h_{n+1}(X_n) = Z_1 \quad P_\nu\text{-a.s.}$$

But $|Z - Z_1| = 2$ on G and therefore, for any $\varepsilon > 0$, there is an $x \in E$ and an integer l such that $|h_{l+1}(x) - h_l(x)| > 2 - \varepsilon$. Since $\|h_n\| \leqslant 1$ and $P_k h_{n+k} = h_n$ it follows that for every n,

$$\begin{aligned} \|P_n(x, \cdot) - P_{n+1}(x, \cdot)\| &\geqslant |P_n h_{l+n+1}(x) - P_{n+1} h_{l+n+1}(x)| \\ &= |h_{l+1}(x) - h_l(x)| > 2 - \varepsilon. \end{aligned}$$

We consequently get $\alpha_1(x) > 2 - \varepsilon$, which completes the proof.

Before we proceed, let us recall that a sub-σ-algebra $\mathscr{G}$ of $\mathscr{F}$ is said to be a.s. trivial if for every $A \in \mathscr{G}$ the function $P.[A]$ is identically zero or one; this is a more stringent condition than to be P_x-trivial for every $x \in E$.

Theorem 1.8 (second zero-two law). *For any chain, the number*

$$\sup_{x,y\in E} \lim_n \|P_n(x,\cdot) - P_n(y,\cdot)\|$$

is either zero or two. The former case holds if and only if the following equivalent conditions hold:

(i) *$\mathscr{A}$ is a.s.trivial;*

(ii) *the bounded Q-harmonic functions are constant;*

(iii) *for every pair (ν_1, ν_2) of probability measures on $(E, \mathscr{E})$*

$$\lim_n \|(\nu_1 - \nu_2) P_n\| = 0;$$

(iv) *$\alpha_1 = 0$ and the bounded harmonic functions are constant.*

Proof. The equivalence of (i) and (ii) follows readily from Proposition 1.3. If $\mathscr{A}$ is a.s.trivial, then $\|P_{\nu_1-\nu_2}\|_{\mathscr{A}} = 0$ and (iii) follows from (i) by Theorem 1.4. If (iii) holds, by taking $\nu_1 = \varepsilon_x$, $\nu_2 = P(x, \cdot)$ we get $\alpha_1 = 0$. Furthermore if h is bounded and harmonic,

$$|h(x) - h(y)| = |P_n h(x) - P_n h(y)| \leqslant \|h\| \, \|P_n(x,\cdot) - P_n(y,\cdot)\|$$

for every n, so that h is constant. If (iv) holds, then since $\alpha_1 = 0$, we see from the last theorem that $\mathscr{A} = \mathscr{I}$ a.s. and since the bounded harmonic functions are constant, $\mathscr{I}$ and hence $\mathscr{A}$, is a.s.trivial.

It remains to prove that if $\mathscr{A}$ is not a.s.trivial we are in the second case of the first sentence in the statement. But we can then find a set $F \in \mathscr{A}$ and a probability measure ν such that $0 < P_\nu[F] < 1$; set $Z = 1_F - 1_{F^c}$ and $g_n(x) = E_x[Z_n]$. Since $\lim_n g_n(X_n) = Z$ P_ν-a.s., for any $\varepsilon > 0$ there exist an integer n and points x and y in E such that $g_n(x) > 1 - \varepsilon$ and $g_n(y) < -1 + \varepsilon$. As in the proof of Theorem 1.7 we then get

$$\lim_k \|P_k(x,\cdot) - P_k(y,\cdot)\| \geqslant |g_n(x) - g_n(y)| > 2 - \varepsilon$$

and the proof is now easily completed.

Exercise 1.9. Prove that for the chain of Exercise 1.24 of ch. 2 the σ-algebra

$\mathscr{A}$ is strictly larger than $\mathscr{I}$ when $\prod_0^\infty p_x > 0$. Give explicitly an event which is in $\mathscr{A}$ and is not a.s.equal to an invariant event. The renewal chain of ch. **3** §1.16 also provides examples of the same kind.

Exercise 1.10. For any T.P. R and $0 < \alpha < 1$ show that for the chain with T.P. $P = \alpha I + (1 - \alpha) R$ the algebras $\mathscr{A}$ and $\mathscr{I}$ are a.s.equal.

Exercise 1.11. (1) With the notation of Theorem 1.4 prove that

$$\lim_n n^{-1} \left\| \sum_1^n \nu P_k \right\| = \|P_\nu\|_{\mathscr{I}} = \sup_{h \in \mathscr{H}} \left\{ \int h \, d\nu \right\}$$

where $\mathscr{H}$ is the set of bounded harmonic functions h such that $\|h\| \leqslant 1$.

(2) The number $\sup_{x,y}(\lim_n n^{-1}\|(\varepsilon_x - \varepsilon_y) \sum_1^n P_k\|)$ is equal to either zero or two. The first case occurs if and only if $\mathscr{I}$ is a.s.trivial and then

$$\lim_n n^{-1} \left\| (\nu_1 - \nu_2) \sum_1^n P_k \right\| = 0$$

for any pair (ν_1, ν_2) of probability measures on $(E, \mathscr{E})$.

Exercise 1.12. Suppose that P induces a conservative contraction on $L^1(\mu)$. Prove that α_1 takes on only the values 0 and 2 μ-almost-surely.

[*Hint*: The set $A = \{a < \alpha_1 < b\}$, where $0 < a < b < 2$ is in $\mathscr{C}$ hence almost-surely equal to an absorbing set $B \subset A$; apply Theorem 1.7 to the trace chain.]

Exercise 1.13. Prove that the equivalent conditions in Theorem 1.8 are also equivalent to the following statement: for any $B \in \mathscr{F}$ and any initial measure ν,

$$\lim_{n \to \infty} \sup_{A \in \mathscr{A}_n} |P_\nu(A \cap B) - P_\nu(A) P_\nu(B)| = 0,$$

where $\mathscr{A}_n = \theta_n^{-1}(\mathscr{F})$.

Exercise 1.14. Let μ be a probability measure on a group $\boldsymbol{G}$, and $\boldsymbol{H}$ the smallest closed normal subgroup containing the support of $\mu * \hat{\mu}$ (cf. ch. 3, Exercise 3.15).

(1) Prove that the set of bounded measures on $\boldsymbol{G}$ such that $\lim_n \|\mu^n * \nu\| = 0$ is a right ideal (in the convolution algebra of $\boldsymbol{G}$) contained in the kernel of the canonical mapping from $\mathrm{b}\mathscr{M}(\boldsymbol{G})$ into $\mathrm{b}\mathscr{M}(\boldsymbol{G}/\boldsymbol{H})$.

(2) Show that one can extend uniquely Q-harmonic functions on $E \times \mathbf{N}$ to Q-harmonic functions on $E \times \mathbf{Z}$, and restate Proposition 1.3 in that case.

(3) Compute the sub-group $\boldsymbol{H}'$ of periods of continuous Q-harmonic functions in terms of $\boldsymbol{H}$ and prove that $\boldsymbol{G} \times \mathbf{Z}/\boldsymbol{H}'$ is isomorphic to $\boldsymbol{G}/\boldsymbol{H}$.

(4) Assume that $\boldsymbol{G}$ is abelian and that μ is spread-out and prove that the inclusion in (1) is then an equality. If in addition $\boldsymbol{G}$ is connected, for every bounded measure ν such that $\nu(\boldsymbol{G}) = 0$,

$$\lim_n \|\mu^n * \nu\| = 0.$$

2. Cyclic classes and limit theorems for Harris chains

In this section we take up the study of Harris chains for which we want to prove that the limit behaviour of Theorem 1.8 occurs. Actually simple examples show that this cannot hold when the chain has a periodic behaviour and this leads to the notion of cyclic classes (see ch. **3**, Exercise 1.17).

Our study will rest on the following

Proposition 2.1. *For each integer k either $\alpha_k(x) = 0$ for every x in E, or $\alpha_k(x) = 2$ for m-almost every x in E.*

Proof. By Lemma 1.5, the functions $2 - \alpha_k$ are superharmonic hence constant m-a.e. and larger than this constant everywhere. The result follows therefore immediately from Theorem 1.7.

We shall need the following notation. According to Lemma 5.3 in ch. 1, there exists for every k an $\mathscr{E} \otimes \mathscr{E}$-measurable function p_k such that the Lebesgue decomposition of P_k with respect to m may be written

$$P_k(x, \mathrm{d}y) = p_k(x, y)\, m(\mathrm{d}y) + P_k^1(x, \mathrm{d}y).$$

We shall write $P_k^0(x, \cdot\,)$ for the first part of this decomposition. Because of the irreducibility, for every x there is an n such that $P_n^0(x, E) > 0$ and actually, it is easily proved (ch. **3**, Exercise 2.14) that $P_n^1(\,\cdot\,, E)$ decreases to zero as n increases to infinity.

Theorem 2.2. *For a Harris chain there are only two possibilities:*

(i) *either for any pair (ν_1, ν_2) of probability measures on $\mathscr{E}$*

$$\lim_{n\to\infty} \|\nu_1 P_n - \nu_2 P_n\| = 0,$$

(ii) *or there exist an integer* d *and a measurable partition* $C_1, C_2, \ldots, C_d, F$ *of* E *such that* $m(F) = 0$ *and* $P(\cdot, C_i) = 1$ *on* C_{i-1}, $P(\cdot, C_1) = 1$ *on* C_d.

Proof. It is easily shown that $\alpha_{k+l} \leqslant \alpha_k + \alpha_l$ and it follows that the set $G = \{k: \alpha_k \equiv 0\}$ is a subgroup of the group $\mathbf{Z}$. We consequently have $G = \{0\}$, $G = \mathbf{Z}$ or $G = d\mathbf{Z}$ for an integer $d > 1$. We are going to prove that the first case is impossible and that the latter two correspond to the two situations described above.

Let us first suppose that $G = \{0\}$; by discarding if necessary a set of measure zero we may assume that $||P_{n+k}(x, \cdot) - P_n(x, \cdot)|| = 2$ for every x in E and every integers $n, k > 0$. Pick a point x and an integer n such that $P_n^0(x, E) > 0$ and set $A = \{y: p_n(x, y) > 0\}$. Let B be a set of measure zero such that $\sum_0^\infty P_n^1(x, B^c) = 0$. Because the chain is Harris, we can find $k > n$ such that $P_k(x, A\backslash B) > 0$, hence such that $p_k(x, \cdot)$ is strictly positive on a subset of $A\backslash B$ of positive m-measure. It follows that

$$||P_n(x, \cdot) - P_k(x, \cdot)|| < 2,$$

which is a contradiction. The case $G = \{0\}$ is impossible.

If $G = \mathbf{Z}$ we have in particular $\alpha_1 = 0$ and since the bounded harmonic functions are constant it follows at once from Theorem 1.8 that (i) holds.

We now turn to the case $G = d\mathbf{Z}$ with $d > 1$. By discarding an m-null set which will eventually be lumped with the set F of the statement we may assume that for every x in E,

$$\lim_n ||P_{n+d}(x, \cdot) - P_n(x, \cdot)|| = 0, \qquad ||P_{n+j}(x, \cdot) - P_n(x, \cdot)|| = 2$$

for every $n > 0$ and every j which is not a multiple of d.

Pick an arbitrary point x_0 in E, and for $j = 1, 2, \ldots, d$, set

$$B_j = \bigcup_{n=0}^{\infty} \{y: p_{nd+j}(x_0, y) > 0\}.$$

For every pair (i, j), we have $m(B_i \cap B_j) = 0$; otherwise one could find integers n_1 and n_2 such that

$$m\{y: p_{n_1 d+i}(x_0, y) > 0 \text{ and } p_{n_2 d+j}(x_0, y) > 0\} > 0$$

and it would follow that $||P_{n_1 d+i}(x_0, \cdot) - P_{n_2 d+j}(x_0, \cdot)|| < 2$ which is impossible. On the other hand, because of the recurrence property and the fact that $\sum_0^\infty P_n^1(x_0, \cdot)$ is singular with respect to m, the union of the sets

B_j is equal to E up to a set of measure zero. We can therefore assume that the sets B_j are pairwise disjoint and that their union has full measure.

For $j \neq i+1$ (modulo d), set $D = \{y \in B_i \colon P(y, B_j) > 0\}$ and suppose $m(D) > 0$. There exists an integer n such that $P^0_{nd+i}(x_0, D) > 0$, hence such that

$$P^0_{nd+i+1}(x_0, B_j) \geqslant \int_D P^0_{nd+i}(x_0, \mathrm{d}y)\, P(y, B_j) > 0.$$

But this is impossible since $B_j \cap B_{i+1}$ is empty. As a result $m(D) = 0$ and $P(\cdot\,, B_{i+1}) = 1$ m-a.e. on B_i. By defining

$$C_i = B_i \backslash \{y \in B_i \colon P(y, B_{i+1}) < 1\}$$

and $F = (\bigcup_i C_i)^c$ we get the situation (ii).

Definition 2.3. In the first case of Theorem 2.2 the chain is said to be *aperiodic*. In the second case it is said to be *periodic* and the integer d is called the *period* of X. The sets C_i are called the *cyclic classes* and the partition of E into $C_1, C_2, \ldots, C_d, F$ is a *cyclic decomposition* of E. This cyclic decomposition is unique up to equivalence.

It is worth recording that the limit property in part (i) of Theorem 2.2 characterizes aperiodic Harris chains among conservative and ergodic chains (see Exercise 2.12). In that case the asymptotic σ-algebra is a.s.trivial as we have just seen. In the periodic case these properties have a weaker version.

Proposition 2.4. *If X is periodic with period d and if ν_1 and ν_2 are two probabilities which agree on F and satisfy $\nu_1(C_i) = \nu_2(C_i)$ for every $i \leqslant d$, then*

$$\lim_n \|(\nu_1 - \nu_2)P_n\| = 0.$$

Furthermore the σ-algebra $\mathscr{A}$ is atomic in the sense that every bounded asymptotic random variable can be written, up to equivalence,

$$Z = \sum_{i=1}^{d} \alpha_i 1_{\underline{\lim}_{n\to\infty} \{X_{nd} \in C_i\}}$$

where the α_i's are arbitrary real numbers.

Proof. The first sentence is an easy consequence from the fact that each C_i

is absorbing for P_d and that P_d restricted to C_i is Harris and aperiodic.

To prove the second sentence we first restrict ourselves to P_x for x in F^c. The events $\Gamma_i = \underline{\lim}\,\{X_{nd} \in C_i\}$ are in $\mathscr{A}$ and $P_x[\Gamma_i] = 0$ or 1 according as x belongs to C_i or not. In other words P_x is carried by Γ_i if $x \in C_i$.

Let now Γ be an arbitrary event of $\mathscr{A}$ and h the associated bounded Q-harmonic function; from the inequality

$$|h(x, n) - h(x', n)| \leqslant ||h||\,||P_k(x, \cdot) - P_k(x', \cdot)||$$

and the convergence property of the first sentence it follows that, for every n, $h(\cdot, n)$ is constant on each C_i. If $x \in C_i$, $h(X_n, n)$ is thus constant P_x-a.s. for every n and consequently, its almost-sure limit 1_Γ is constant P_x-a.s. This constant can only be zero or one, hence either $\Gamma = \emptyset$ P_x-a.s. or $\Gamma = \Gamma_i$ P_x-a.s.

For every bounded asymptotic random variable Z there are therefore real numbers α_i such that $Z = \sum_{i=1}^d \alpha_i 1_{\Gamma_i}$ P_x-a.s. for every x in F^c. Call h the space-time harmonic function associated with Z. For x in F^c, we have

$$h(x, n) = E_x\left[\left(\sum_{i=1}^d \alpha_i 1_{\Gamma_i}\right) \circ \theta_{-n}\right];$$

if x is in F, since F is transient, we eventually have P_x-a.s.

$$\begin{aligned} h(X_n, n) &= E_{X_n}\left[\left(\sum_{i=1}^d \alpha_i 1_{\Gamma_i}\right) \circ \theta_{-n}\right] \\ &= E_x\left[\sum_{i=1}^d \alpha_i 1_{\Gamma_i} \mid \mathscr{F}_n\right] \end{aligned}$$

which converges to $\sum_{i=1}^d \alpha_i 1_{\Gamma_i}$ P_x-a.s. As $h(X_n, n)$ converges also to Z P_x-a.s. the proof is complete.

Remark. The reader will find in Exercise 2.20 another description of $\mathscr{A}$ in the periodic case.

As in ch. 3 §1 we lay down the following

Definition 2.4. The chain X is said to be *positive* or *ergodic* if $m(E) < \infty$ (and we then always make $m(E) = 1$) and *null* if $m(E) = \infty$.

Our next result says that aperiodic positive chains evolve towards a stationary state.

Proposition 2.5. If X is positive and aperiodic,

$$\lim_n ||\nu P_n - m|| = 0$$

for any probability measure ν on $\mathscr{E}$.

Proof. Make $\nu_1 = \nu$ and $\nu_2 = m$ in Theorem 2.2(i).

Null chains and positive chains have quite different asymptotic behaviours. We have just seen that, for a positive and aperiodic chain, $\lim_n P_n f(x) = m(f)$ for every $x \in E$ and $f \in b\mathscr{E}$. The following result, which does not require X to be aperiodic, shows that a null chain is "less recurrent" than a positive chain. Other evidence to support this statement is given in ch. 3, Corollary 1.10 and in Exercises 2.10–2.11.

Theorem 2.6. *If X is null, for every function $f \in L^1(m) \cap b\mathscr{E}$ and every $x \in E$*

$$\lim_{n\to\infty} P_n f(x) = 0.$$

Proof. We shall show the more stringent property: for every $x \in E$ and $\varepsilon > 0$,

$$\lim_n \{P_n f(x)/(m(f) + \varepsilon)\} = 0,$$

the convergence being uniform in $f \in \mathscr{U}_+ \cap L^1(m)$.

We first assume that X is aperiodic. If our statement were false, there would exist a sequence $\{n_k\}$ tending to infinity, a sequence of functions f_k in $\mathscr{U}_+ \cap L^1(m)$, a point x_0, and two numbers ε, $\delta > 0$ such that

$$P_{n_k} f_k(x_0)/(m(f_k) + \varepsilon) \geqslant \delta$$

for every k. Since $m(E) = \infty$, Egoroff's theorem and Theorem 2.2 allow us to pick a set B in $\mathscr{E}$ such that $m(B) > 2\delta^{-1}$ and such that

$$||P_{n_k}(x_0, \cdot) - P_{n_k}(y, \cdot)||$$

tends to zero uniformly in y on B. Hence

$$m(f_k) \geqslant \int_B m(\mathrm{d}x)\, P_{n_k} f_k(x) \geqslant \int_B m(\mathrm{d}x)\, (P_{n_k} f_k(x_0) - \tfrac{1}{2}\varepsilon\delta)$$

if k is sufficiently large to have

$$||P_{n_k}(x_0, \cdot) - P_{n_k}(y, \cdot)|| < \tfrac{1}{2}\varepsilon\delta$$

on B. These two relations together imply

$$1 > m(f_k)/(m(f_k)+\varepsilon) \geqslant \int_B m(\mathrm{d}x)\,(\delta - (\tfrac{1}{2}\varepsilon\delta)/(m(f_k)+\varepsilon)) \geqslant \tfrac{1}{2}\delta m(B),$$

which is a contradiction.

Suppose that X is periodic and let $C_1, \ldots, C_d, F$ be a cyclic decomposition of E. If $x \in C_i$, then $P(f 1_F)(x) = 0$. Considering the chain on C_i of T.P. P_d, we conclude from above that $P_n f(x)/(m(f)+\varepsilon)$ tends to zero uniformly on $\mathscr{U}_+ \cap L^1(m)$. If $x \in F$, then for $r < n$

$$\frac{P_n f(x)}{m(f)+\varepsilon} = \sum_{j=1}^{r} \int_F \cdots \int_F \int_{E\setminus F} P(x, \mathrm{d}x_1) \cdots P(x_{j-1}, \mathrm{d}x_j) \frac{P_{n-j} f(x_j)}{m(f)+\varepsilon}$$
$$+ \int_F \cdots \int_F P(x, \mathrm{d}x_1) \cdots P(x_{r-1}, \mathrm{d}x_r) \frac{P_{n-r} f(x_r)}{m(f)+\varepsilon}.$$

Since F is transient, $P_n(x, F)$ tends to zero as n tends to infinity, and one can choose r large enough to make the second term on the right as small as we please uniformly in n and f. After r has been chosen, since $P_{n-j} f(x_j)/(m(f)+\varepsilon)$ tends to zero on $E\setminus F$, and is dominated by the integrable function $1/\varepsilon$, we can make the first term as small as we choose by taking n sufficiently large.

Exercise 2.7. A point transformation θ on $(E, \mathscr{E})$ (ch. 1, Examples 1.4) possessing an invariant probability measure ν, is said to be *strongly mixing* if, for every pair (A, B) of sets in $\mathscr{E}$,

$$\lim_n \nu(\theta^{-n}(A) \cap B) = \nu(A)\nu(B).$$

(1) Prove that if the limit property holds for all A's and B's in a semi-algebra generating $\mathscr{E}$, then θ is strongly mixing. Observe also that a strongly mixing θ is ergodic (ch. 4, Exercise 3.13).

(2) Show that the shift operator θ of an aperiodic positive Harris chain is strongly mixing on $(\Omega, \mathscr{F}, P_m)$. If X is a Markov chain with invariant probability measure m, θ is ergodic if and only if $P_n f$ converges to $m(f)$ in the $\sigma(L^1, L^\infty)$ sense for every $f \in L^1(m)$.

Exercise 2.8. Prove by an example that an aperiodic Harris chain may induce on a subset a periodic trace chain.

Exercise 2.9. If X is periodic and in duality with a chain $\hat{X}$, then $\hat{X}$ is periodic with the same period and cyclic classes (up to equivalence) but these are visited in reverse order.

Exercise 2.10. If X is positive prove that $m(A) > 0$ implies that $E_x[S_A]$ is finite for m-almost every x in E.

[*Hint*: Start with the equality $mH_A = m$.]

Exercise 2.11. Suppose that E is an LCCB space and that m is Radon. Prove that there exists a sequence $\{n_k\}$ of integers such that $\{X_{n_k}\}$ converges almost-surely to Δ if and only if X is null.

Exercise 2.12. Let ν be a σ-finite measure such that $\nu P \ll \nu$ and suppose that

(i) the induced operator on $L^1(\nu)$ is conservative;

(ii) for every pair (x, y) of points in E,

$$\lim_n \|P_n(x, \cdot) - P_n(y, \cdot)\| = 0;$$

prove then that P is an aperiodic Harris chain.

[*Hint*: Use the results of ch. 4 §4.]

Exercise 2.13. Show that, if X is Harris and positive, then for every $f \in L^p(m)$, the sequence $P_n f$ converges to $m(f)$ in the L^p sense $(1 \leqslant p < \infty)$.

Exercise 2.14. Prove that if X is Harris, for every pair (x, y) of points in E the sequence of functions $p_n(x, \cdot) - p_n(y, \cdot)$ converges to zero in $L^1(m)$. If X is positive $p_n(x, \cdot)$ converges to 1 in $L^1(m)$.

Exercise 2.15. If X is null then $P_n f$ converges to zero for all functions f which belong to the closure of $b\mathscr{E} \cap L^1(m)$ in the topology of uniform convergence. If E is LCCB and P maps C_K into C_0, then $P_n f$ converges to zero for all functions f in C_0.

Exercise 2.16. With the notation of Exercise 1.14 prove that if the left random walk of law μ is Harris then $\boldsymbol{H}$ is open, $\boldsymbol{G}/\boldsymbol{H}$ is finite, the cosets of $\boldsymbol{H}$ are the cyclic classes and the inclusion of Exercise 1.14 is an equality. What can be said along the same line if X is not Harris but merely recurrent?

Exercise 2.17. Let m be an invariant probability measure for P. Show that the following two conditions are equivalent:

(i) for $m \otimes m$-almost every (x, y) there exists an integer n such that $P_n(x, \cdot)$ and $P_n(y, \cdot)$ are not singular with respect to one another;

(ii) for m-almost every x

$$\lim_n \|P_n(x, \cdot) - m\| = 0.$$

Exercise 2.18. (1) Prove that there exist Harris random walks on the group of rigid motions of the Euclidean plane.

(2) Let μ be the law of such a random walk. For every pair (ν_1, ν_2) of probability measures on the plane,

$$\lim_n \|\mu^n * \nu_1 - \mu^n * \nu_2\| = 0.$$

(Of course the plane is to be considered as a homogeneous space of the group of rigid motions.)

Exercise 2.19. For resolvents recurrent in the sense of Harris (ch. 3, Exercise 2.19), state and prove results similar to Theorems 2.2 and 2.6.

[*Hint*: One should look for the asymptotic behaviour of αV_α as α tends to zero. For $\alpha < 1$ one can write

$$\alpha V_\alpha = \sum_{n=0}^{\infty} (1 - \alpha)^n (V_1)^{n+1}.]$$

Exercise 2.20. If X is periodic and the set F is empty, $\mathscr{A}$ is a.s.equal to the σ-algebra generated by the events $\{X_0 \in C_i\}$.

Exercise 2.21. Extend the results of this section to Harris chains with a non-separable state space.

[*Hint*: Use admissible σ-algebras.]

3. Quasi-compact transition probabilities and the strong ergodic theorem

Following the usual terminology for convergence of operators on a Banach space, one defines, for the contractions P_n of $b\mathscr{E}$, several types of convergence.

Definition 3.1. The operators P_n are said to converge to a bounded operator T

(i) *uniformly* if

$$\lim_{n\to\infty} \|P_n - T\| = \lim_{n\to\infty} \sup_{\|f\|\leq 1} \sup_{x\in E} |P_n f(x) - T f(x)| = 0;$$

(ii) *strongly* if, for every $f \in b\mathscr{E}$,

$$\lim_{n\to\infty} \|P_n f - Tf\| = \lim_{n\to\infty} \sup_{x\in E} |P_n f(x) - T f(x)| = 0;$$

(iii) *weakly* if, for every $f \in b\mathscr{E}$ and every $\nu \in b\mathscr{M}(\mathscr{E})$,

$$\lim_{n\to\infty} \langle \nu, P_n f - Tf \rangle = 0.$$

The weak convergence is equivalent to pointwise convergence on E of functions $P_n f$. Let us also observe that if P is a T.P. and $\{P_n\}$ converges uniformly to T, then T is itself a T.P.

Theorem 2.4 asserts that if X is positive recurrent, the operators P_{nd} converge weakly to the operator with finite range $\sum_{i=1}^{d} 1_{C_i} \otimes m_i$. The present section is devoted to the study of the necessary and sufficient conditions for this convergence to be uniform. For this purpose we need the following

Definition 3.2. The transition probability P is said to be *quasi-compact* if there exists a sequence $\{U_n\}$ of compact operators on $b\mathscr{E}$ such that

$$\lim_n \|P_n - U_n\| = 0.$$

In the following section, we shall show that with each Harris chain we can associate plenty of quasi-compact operators. We begin with another definition of quasi-compactness.

Proposition 3.3. *The T.P. P is quasi-compact if and only if there exist a compact operator U and an integer n_0 such that*

$$\|P_{n_0} - U\| < 1.$$

Proof. The necessity is obvious. To prove the sufficiency, let us first recall that the set of compact operators is a closed two-sided ideal in the algebra of bounded operators.

Assume that $P_{n_0} = U + Q$, where U is compact and $\|Q\| < 1$. An integer $n > n_0$ may be written $n = mn_0 + l$ in only one way with $m \geqslant 1$ and $0 \leqslant l < n_0$; so

$$P_n = (U + Q)^m P_l = U_n + Q^m P_l,$$

where U_n is compact. Since

$$\|Q^m P_l\| \leqslant \left(\sup_{l<n_0} \|P_l\|\right) \|Q^m\| \underset{n\to\infty}{\longrightarrow} 0,$$

the T.P. P is quasi-compact.

Remark. In Definition 3.2 and Proposition 3.3, one can put "operator with finite range" instead of "compact operator" without altering their validity.

Theorem 3.4. *The following two statements are equivalent:*

(i) $\{P_n\}$ *converges uniformly to an operator of rank one;*

(ii) P *is Harris, aperiodic and quasi-compact.*

If these equivalent conditions hold, then the chain is positive and the rank one operator is $1 \otimes m$ *where* m *is the invariant probability measure.*

Proof. If (i) holds, then P is obviously quasi-compact. Moreover the rank one operator is a T.P., hence is of the form $g \otimes m$. But $g = \lim_n P_n 1 = 1$ and m is clearly invariant. The chain is now obviously m-irreducible, and by 2.4 and the results in ch. 4 § 4, it is easily seen that P is Harris and aperiodic. We leave to the reader as an exercise the task of filling in the details.

Suppose conversely that (ii) holds. Since P is quasi-compact, for any $f \in b\mathscr{E}$, the sequence $\{P_n f\}$ is relatively compact in $b\mathscr{E}$. Indeed, if we fix $\varepsilon > 0$, we may pick a k such that $||P_k - U|| < \varepsilon/2$, where U is a compact operator. A finite number of balls of radius $\varepsilon/2$ cover the set $U(\mathscr{U})$; the balls with the same centers and radius ε cover the set $P_k(\mathscr{U})$, which, since ε is arbitrary, proves our claim.

Now using a diagonal process, from any sequence of integers we may extract a subsequence $\{n_j\}$ such that for every $r \geqslant 0$, $\{P_{n_j-r} f\}$ converges in $b\mathscr{E}$ to a function g_r. Obviously $P_r g_{s+r} = g_s$ for any pair (r, s), but since X is aperiodic all these functions are equal to the same constant α. For $n > n_j$ we consequently have

$$||P_n f - \alpha|| = ||P_{n-n_j}(P_{n_j} f - \alpha)|| \leqslant ||P_{n_j} f - \alpha||.$$

It follows that $\{P_n f\}$ converges in $b\mathscr{E}$ to a constant Sf.

The mapping S is a contraction and $PS = SP = S$. The sequence $\{P_n\}$ converges strongly to S and we now show that this convergence is actually uniform. Let ε and k be as above. For $f \in \mathscr{U}$, we have

$$||P_{n+k} f - Sf|| = ||(P_n - S)\, P_k f|| \leqslant \varepsilon + ||(P_n - S)\, Uf||;$$

since $U(\mathscr{U})$ is relatively compact, the convergence

$$\lim_n ||(P_n - S)\, Uf|| = 0$$

is uniform on $\mathscr{U}$. It follows upon letting n tend to infinity and then ε to zero that the convergence of $\{P_n\}$ to S is uniform. As a result S is a Transition Probability and the proof is now easily completed.

We will now remove the assumption of aperiodicity in the above result.

Theorem 3.5. *The following three conditions are equivalent:*

(i) *P is Harris and quasi-compact;*

(ii) *there is a bounded invariant probability measure m, the bounded harmonic functions are constant and*

$$(I - P)\, b\mathscr{E} = b^\circ\mathscr{E}$$

where $b^\circ\mathscr{E} = \{f \in b\mathscr{E} : m(f) = 0\}$;

(iii) *the sequence $n^{-1} \sum_0^{n-1} P_k$ converges uniformly to an operator of rank one.*

If these equivalent conditions hold the chain is Harris and positive and the rank one operator is equal to $1 \otimes m$.

Proof. (i) ⇒ (ii). By the preceding result, if X is aperiodic and P quasi-compact the chain is positive and for n sufficiently large the norm of the restriction of P_n to $b^\circ\mathscr{E}$ is strictly less than one. As a result $I - P$ is invertible on $b^\circ\mathscr{E}$, hence $(I - P)\, b^\circ\mathscr{E} = b^\circ\mathscr{E}$ and the proof is complete in that case.

If X is periodic the T.P., cU_c is easily seen to be Harris (ch. **3**, Proposition 2.9), aperiodic and quasi-compact and m is still the invariant measure which is thus bounded. Consequently $(I - cU_c)\, b\mathscr{E} = b^\circ\mathscr{E}$. But on the other hand the resolvent equation shows that

$$I - cU_c = (I - P)\,(I + (1 - c)\, U_c),$$

and therefore $(I - P)\, b\mathscr{E} \supset (I - cU_c)\, b\mathscr{E}$ which completes the proof.

(ii) ⇒ (iii). The hypothesis entails that $(I - P)$ is one-to-one and onto from $b^\circ\mathscr{E}$ into $b^\circ\mathscr{E}$. By Banach's theorem it has therefore a continuous inverse Q. For $h \in \mathscr{U}$, we have therefore

$$\left\| n^{-1} \sum_0^{n-1} P_k h - (1 \otimes m)\, h \right\| = \left\| n^{-1} \sum_0^{n-1} P_k (h - m(h)) \right\|$$

$$= \left\| n^{-1} \sum_0^{n-1} P_k (I - P)\, Q(h - m(h)) \right\| \leqslant 4n^{-1} \|Q\|$$

which completes the proof.

(iii) ⇒ (i). The proof follows exactly the same pattern as for the proof of (i) ⇒ (ii) in the preceding result.

The remainder of this section may be skipped without hampering the understanding of the sequel. It will be devoted to a more thorough discussion of quasi-compactness so as to see how it relates to Harris recurrence in general but the results thus obtained will not be used in the following sections. We first recall a few definitions and facts pertaining to the general theory of Riesz spaces.

Definition 3.6. A continuous linear functional on $b\mathscr{E}$ is called a *mean*.

Every mean has a minimal decomposition as a sum of a finite measure and of a *pure mean*; a pure mean is a mean that is singular with respect to all measures. A mean φ is said to be *invariant* if for every $f \in b\mathscr{E}$, we have $\langle \varphi, f \rangle = \langle \varphi, Pf \rangle$.

Theorem 3.7. *If X is Harris the conditions of Theorem 3.5 are equivalent to each of the following conditions:*

(iv) *there exists no invariant pure mean;*

(iv)$'$ *there exists no positive invariant pure mean;*

(v) *the measure m is bounded and $\overline{(I - P)\, b\mathscr{E}} = b^\circ\mathscr{E}$;*

(vi) *the measure m is bounded and is the only invariant mean.*

Proof. We first observe that (iv)$'$ is equivalent to (iv). Indeed, if φ is an invariant pure mean, call φ^+ and φ^- the positive and negative part of its minimal decomposition. From the equality $\varphi^+ - \varphi^- = \varphi^+P - \varphi^-P$, it follows that $\varphi^+ \leqslant \varphi^+P$ and $\varphi^- \leqslant \varphi^-P$; but since $P1 = 1$ these inequalities are in fact equalities. Thus φ^+ and φ^- are positive invariant pure means, and the result follows.

(iii) $\Rightarrow$ (iv). If φ is a pure mean, there exists a set $A \in \mathscr{E}$ such that $m(A) > 0$ and $\varphi(A) = 0$. If φ were invariant, for every n we should have

$$\left\langle \varphi, n^{-1} \sum_{1}^{n} P_k(\cdot, A) \right\rangle = 0,$$

and this is impossible since the sequence on the right converges in $b\mathscr{E}$ to the constant $m(A)$.

(iv) $\Rightarrow$ (i). We begin with the following remark. If P has no invariant pure mean, then P_k, for any integer k, has the same property. Indeed if φ is a pure mean and $\varphi P_k = \varphi$, then the mean $\sum_0^{k-1} \varphi P_n$ is invariant by P, hence is not pure. Consequently, there is an integer $0 < n < k$ such that φP_n is not pure, and then $\varphi P_k = \varphi P_n P_{k-n}$ is not pure either, which yields a contradiction.

By § 1 we know that, X being Harris, there exists an integer n such that P_n is the sum of an integral kernel K and of another kernel. By ch. 1, Theorem 5.7 there exist sets $A_k \in \mathscr{E}$ of arbitrarily large measure such that $I_{A_k}K$ is a compact operator on b$\mathscr{E}$. We may therefore write $P_n = Q + R$, where Q is compact and $m\{R1 < 1\} > 0$. Moreover, we may pick n outside the set of multiples of d in order that m be the only measure invariant by P_n. We will be finished if we show that there exists a k such that $\|R^k\| < 1$.

Let φ be an invariant mean for R and $\varphi = \varphi^+ - \varphi^-$ its minimal decomposition into positive and negative parts. From $\varphi R = \varphi$, we derive $\varphi^+ \leqslant \varphi^+ R \leqslant \varphi^+ P_n$, and since $P_n 1 = 1$, $\varphi^+ = \varphi^+ P_n$. Let $\varphi^+ = \varphi_1 + \varphi_2$, where φ_1 is a measure and φ_2 a pure mean. The equality $\varphi_1 + \varphi_2 = \varphi_1 P_n + \varphi_2 P_n$ implies that $\varphi_1 \geqslant \varphi_1 P_n$, and since X is Harris, $\varphi_1 = \varphi_1 P_n$, and thus $\varphi_2 = \varphi_2 P_n$. Since P_n has no invariant pure mean, $\varphi_2 = 0$ and φ^+ is zero or a multiple of m. In the latter case, since $m\{R1 < 1\} > 0$, we have

$$\langle \varphi^+, 1 \rangle > \langle \varphi^+, R1 \rangle,$$

which is a contradiction. In the same way we prove that $\varphi^- = 0$, so that the kernel R has no invariant mean.

Consequently, by the Hahn–Banach theorem, we have $\overline{(I - R)\, \mathrm{b}\mathscr{E}} = \mathrm{b}\mathscr{E}$, since a mean which is zero on $(I - R)\, \mathrm{b}\mathscr{E}$ is R-invariant. For every $\varepsilon > 0$, there exists therefore a function $g \in \mathrm{b}\mathscr{E}$ such that $\|g - Rg - 1\| < \varepsilon$. Hence

$$\left\| n^{-1} \sum_{k=1}^{n} R^k 1 \right\| \leqslant \left\| n^{-1} \sum_{k=1}^{n} R^k(1 - g + Rg) \right\| + \left\| n^{-1} \sum_{k=1}^{n} R^k(g - Rg) \right\|$$

$$\leqslant \varepsilon + 2\|g\|\, n^{-1};$$

it follows that $n^{-1} \sum_{k=1}^{n} R^k 1$ converges uniformly to zero, and since $R^k 1$ is a decreasing function of k, this proves that $\|R^k 1\| < 1$ for k sufficiently large.

(ii) $\Rightarrow$ (v). Obvious.

(v) $\Rightarrow$ (vi). The invariant means take on the value zero on $\overline{(I - P)\, \mathrm{b}\mathscr{E}}$. If this subspace is equal to $\mathrm{b}^\circ\mathscr{E}$, the measure m is the only invariant mean.

(vi) $\Rightarrow$ (iv). Obvious.

We now turn to a direct study of quasi-compactness. Plainly, if some subsequence of $\{P_n\}$ converges uniformly to an operator with finite range, the T.P. P is quasi-compact. We are going to prove that this condition is also sufficient; this will be done independently of the results we have already obtained. The space $\mathscr{K}$ used below was introduced in the remark following Proposition 1.3.

Proposition 3.8. *The following two statements are equivalent:*

(i) *the σ-algebra of equivalence classes of asymptotic events is a.s. finite;*

(ii) *the dimension of $\mathscr{K}$ is finite.*

If these equivalent conditions are satisfied, there exists a finite number of integers d_ρ $(\rho = 1, 2, \ldots, r)$ and functions $U_{\rho,\delta}$ in $\mathscr{E}_+$, where δ is an integer taken modulo d_ρ, such that every element f in $\mathscr{K}$ has a unique representation

$$f(\cdot, n) = \sum_{\rho=1}^{r} \sum_{\delta=1}^{d_\rho} C_{\rho,\delta}\, U_{\rho,\delta+n}$$

where $C_{\rho,\delta} \in \mathbf{R}$. If we set $U_\rho = \sum_{\delta=1}^{d_\rho} U_{\rho,\delta}$, every bounded harmonic function f has a unique representation

$$f = \sum_{\rho=1}^{r} b_\rho U_\rho.$$

Moreover one may choose the functions $U_{\rho,\delta}$ in one and only one way in order that

$$\sum_{\rho=1}^{r} \sum_{d=1}^{d_\rho} U_{\rho,\delta} = 1.$$

Proof. The equivalence of (i) and (ii) is an obvious consequence of Proposition 1.3. Assume now that these conditions are satisfied, and consider a finite partition $\mathscr{P}$ of Ω consisting of atoms of $\mathscr{A}$. For every $A \in \mathscr{P}$ the event $\theta_n^{-1}(A)$ is also an atom of $\mathscr{A}$; otherwise we would have $1_A \circ \theta_n = 1_{B_1} + 1_{B_2}$ with $B_1 \cap B_2 = \emptyset$, hence

$$1_A = 1_{B_1} \circ \theta_n^{-1} + 1_{B_2} \circ \theta_n^{-1},$$

and since these two sets are disjoint we would have a contradiction.

Consequently the group θ_n $(n \in \mathbf{Z})$ induces a one-to-one group of transformations of $\mathscr{P}$. One can therefore partition $\mathscr{P}$ into r classes invariant under θ and within each, label the atoms as $A_{\rho,\delta}$ in such a way that $\theta_n^{-1}(A_{\rho,\delta}) = A_{\rho,\delta-n}$.

It remains to set $U_{\rho,\delta} = P_\cdot[A_{\rho,\delta}]$ to get the desired result.

Proposition 3.9. *If P is quasi-compact, then $\mathscr{K}$ is finite dimensional.*

Proof. A Banach space is finite dimensional if and only if it is locally compact. It suffices therefore to prove that the unit ball $\mathscr{K}_1$ of $\mathscr{K}$ can be covered, for any $\varepsilon > 0$, by a finite number of balls of radius less than ε. This amounts to showing that there exist at most a finite number of points in $\mathscr{K}_1$ such that their pairwise distance is greater than ε.

Fix $\varepsilon > 0$ and choose n such that $\|P_n - U\| < \frac{1}{4}\varepsilon$ for a compact operator U. There exist a finite number N of balls of radius $\frac{1}{4}\varepsilon$, with centres f_k ($k = 1, 2, \ldots, N$) which cover $U(\mathscr{K}_1)$. Therefore the N balls with centre f_k and radius $\frac{1}{2}\varepsilon$ cover $P_n(\mathscr{K}_1)$ and one cannot find more than N points in $P_n(\mathscr{K}_1)$ whose pairwise distance is greater than ε. Then the same property is true for $\mathscr{K}_1$, since otherwise there would exist $N + 1$ points g_k in $\mathscr{K}_1$ such that, for $k \neq k'$ and m sufficiently large,

$$\|g_k(\cdot, m) - g_{k'}(\cdot, m)\| > \varepsilon,$$

but as

$$g_k(\cdot, m) = P_n g_k(\cdot, n + m) \in P_n(\mathscr{K}_1),$$

this is impossible.

If P is quasi-compact, we are therefore in the situation described in Proposition 3.8, the notation of which we use below.

Theorem 3.10. *If P is quasi-compact, there exist $d_\rho \geq 1$ ($\rho = 1, \ldots, r$) bounded measurable functions $U_{\rho,\delta}$ ($\delta = 1, 2, \ldots, d$) such that $\sum_\rho \sum_\delta U_{\rho,\delta} = 1$, and probability measures $m_{\rho,\delta}$ carried by the pairwise disjoint sets $E_{\rho,\delta} = \{U_{\rho,\delta} = 1\}$ such that, if d denotes the least common multiple of the d_ρ, then for every k,*

$$\lim_{n\to\infty} \left\| P_{nd+k} - \sum_{\rho=1}^{r} \sum_{\gamma=1}^{d_\rho} U_{\rho,\delta-k} \otimes m_{\rho,\delta} \right\| = 0.$$

The sets $E_\rho = \bigcup_\delta E_{\rho,\delta}$ are absorbing sets, and the chain induced on E_ρ is a Harris chain with invariant measure $m_\rho = \sum_\delta m_{\rho,\delta}$. The sets $E_{\rho,\delta}$ and the number d_ρ are the corresponding cyclic classes and period.

Proof. We have already shown in the proof of Theorem 3.4 that for $f \in b\mathscr{E}$ the sequence $\{P_n f\}$ is relatively compact in $b\mathscr{E}$. Therefore using the diagonal procedure there exists a sequence n_j such that $g(\cdot, m) = \lim_j P_{n_j d - m} f$ exists for every m in the sense of convergence in the norm of $b\mathscr{E}$. Clearly the function g is in $\mathscr{K}_1$ and by virtue of Proposition 3.8, $g(\cdot, nd + m) = g(\cdot, m)$ for every n and therefore $P_{nd-m} g(\cdot, 0) = g(\cdot, m)$ as long as $nd \geq m \geq 0$.

Consequently, if $nd \geqslant n_j d + m$,

$$\|P_{nd-m}f - g(\cdot, m)\| = \|P_{(n-n_j)d-m}(P_{n_j}f - g(\cdot, 0))\| \leqslant \|P_{n_j d}f - g(\cdot, 0)\|.$$

Letting n, and then n_j, tend to infinity, we get that $P_{nd-m}f$ converges to $g(\cdot, m)$ in $b\mathscr{E}$, and thus the operators P_{nd-m} converge strongly. This allows us to define a contraction S of $b\mathscr{E}$ by setting $Sf = \lim_n P_{nd}f$. For every integer m, $SP_{md} = P_{md}S = S$. Exactly as in the proof of Theorem 3.4 we can then prove that the convergence $\lim_n P_{nd} = S$ is uniform. This implies that P_{nd+m} also converges uniformly to $P_m S = SP_m$.

By virtue of Proposition 3.8, for every $f \in b\mathscr{E}$ there exist uniquely determined numbers $C_{\rho,\delta}(f)$ such that

$$Sf = \sum_{\rho=1}^{r} \sum_{\delta=1}^{d_\rho} C_{\rho,\delta}(f)\, U_{\rho,\delta}.$$

Since S is the uniform limit of transition probabilities, it is easily seen that the mappings $f \to C_{\rho,\delta}(f)$ are defined by probability measures $m_{\rho,\delta}$, and thus S can be written

$$S = \sum_{\rho=1}^{r} \sum_{\delta=1}^{d_\rho} U_{\rho,\delta} \otimes m_{\rho,\delta}.$$

It follows from $P_d U_{\rho,\delta} = U_{\rho,\delta}$, that $SU_{\rho,\delta} = U_{\rho,\delta}$; hence $\langle m_{\rho',\delta'}, U_{\rho,\delta}\rangle = 1$ or 0 according as $\rho = \rho'$ and $\delta = \delta'$ (mod d_ρ) or not. Finally, the equalities

$$\sum_\rho \sum_\delta (U_{\rho,\delta} \otimes m_{\rho,\delta})\, P_n = SP_n = P_n S = \sum_\rho \sum_\delta P_n U_{\rho,\delta} \otimes m_{\rho,\delta}$$

$$= \sum_\rho \sum_\delta U_{\rho,\delta-n} \otimes m_{\rho,\delta} = \sum_\rho \sum_\delta U_{\rho,\delta} \otimes m_{\rho,\delta+n}$$

imply that $m_{\rho,\delta} P_n = m_{\rho,\delta+n}$.

Now denote $E_{\rho,\delta} = \{U_{\rho,\delta} = 1\}$. Clearly $m_{\rho,\delta}(E_{\rho,\delta}) = 1$, and by Proposition 3.8 the sets $E_{\rho,\delta}$ are pairwise disjoint. From the equality

$$1 - U_{\rho,\delta}(x) = \int P(x, dy)\, (1 - U_{\rho,\delta+1}(y))$$

it follows that $E_{\rho,\delta} = \{x : P(x, E_{\rho,\delta+1}) = 1\}$. The sets E_ρ are therefore absorbing sets, the measure m_ρ is clearly invariant and by the uniform convergence the restriction of X to E_ρ is m_ρ-irreducible and therefore recurrent in the sense of Harris. The last statement is obvious.

We finally record that quasi-compactness has been used in another guise.

If we denote by ν any convex combination of the probability measures $m_{\rho,\delta}$, it is easily seen that P satisfies the following condition, known as *Doeblin's condition*: *there exist an integer n, two real numbers $\theta < 1$ and $\eta > 0$, and a probability measure ν on $(E, \mathscr{E})$ such that, for $A \in \mathscr{E}$,*

$$\nu(A) > \theta \qquad \textit{implies} \qquad P_n(\cdot, A) \geqslant \eta \quad \textit{on } E.$$

This condition is in fact also sufficient for P to be quasi-compact (see Exercise 3.17 and the Notes and comments).

Exercise 3.11. If E is finite, all T.P.'s on E are quasi-compact. Show by an example that the union of the sets $E_{\rho,\delta}$ may be a proper subset of E.

Exercise 3.12. Set $U_\rho = \sum_{\delta=1}^{d_\rho} U_{\rho,\delta}$, and prove that

$$\lim_{n\to\infty} \left\| n^{-1} \sum_{1}^{n} P_k - \sum_{\rho=1}^{r} U_\rho \otimes m_\rho \right\| = 0.$$

State and prove a similar result for positive Harris chains.

Exercise 3.13. With the notations of Proposition 3.8 and Theorem 3.10 show that

$$A_{\rho,\delta} = \lim_n \{X_n \in E_{\rho,\delta+n}\} \quad \text{a.s.}$$

Prove a similar result for invariant events. Prove that

$$U_\rho(x) = P_x[T_{E_\rho} < \infty].$$

Can one give a similar result for $U_{\rho,\delta}$?

Exercise 3.14. In the situation of Proposition 3.8, we set

$$\lambda_{\rho,k} = \exp(2i\pi k/d_\rho), \qquad f_{\rho,k} = \sum_\delta (\lambda_{\rho,k})^\delta \, U_{\rho,\delta},$$

where k is an integer modulo d_ρ. Prove that $Pf_{\rho,k} = \lambda_{\rho,k} f_{\rho,k}$. Show that for every complex number λ of modulus 1, the space of bounded measurable functions f such that $Pf = \lambda f$ is finite dimensional and admits as a basis the $f_{\rho,k}$ corresponding to those of the $\lambda_{\rho,k}$ equal to λ.

[*Hint*: Notice that $g(\cdot, n) = \lambda^{-n} f$ is Q-harmonic.]

Exercise 3.15. In Theorem 3.5(iii), one can use strong convergence instead of uniform convergence.

Exercise 3.16. If a T.P. satisfies Theorem 3.7(iv), then there exists a bounded invariant measure.

[*Hint*: Make P operate on the convex compact set (for $\sigma(\mathrm{b}\mathscr{E}^*, \mathrm{b}\mathscr{E})$) of positive invariant means, and apply a fixed point theorem.]

Exercise 3.17. For a Harris chain prove that quasi-compactness is equivalent to Doeblin's condition.

Exercise 3.18. Let X be Harris and positive and aperiodic and B the set of Lemma 2.4 in ch. 3. Prove that P is quasi-compact if and only if there is a bounded operator G on B such that for every $\nu \in B$

$$\lim_n \left\| \sum_0^n \nu P_k - \nu G \right\| = 0$$

the norm being that of B. Then $(I - P)\, G = I - 1 \otimes m$; G is thus a potential operator for measures (see §5).

Exercise 3.19. Let $\boldsymbol{M}$ be a compact homogeneous space of a group $\boldsymbol{G}$. We assume that there exists a continuous cross-section $s\colon \boldsymbol{M} \to \boldsymbol{G}$. Prove that, if μ is a spread-out probability measure on $\boldsymbol{G}$, the T.P. $\mu * \varepsilon_x$ $(x \in \boldsymbol{M})$ is quasi-compact, and derive a convergence theorem. (See example in ch. 5 § 1.10.) In particular, if $\boldsymbol{G}$ is a compact connected group with normalized Haar measure m, and μ a spread-out probability measure on $\boldsymbol{G}$, then

$$\lim_{n\to\infty} \sup_{x\in \boldsymbol{G}} \|\mu^n * \varepsilon_x - m\| = \lim_{n\to\infty} \sup_{x\in \boldsymbol{G}} \|\varepsilon_x * \mu^n - m\| = 0.$$

Exercise 3.20. Let (V_α), $\alpha > 0$, be a resolvent such that V_1 is quasi-compact. Prove that, as α tends to zero, αV_α converges uniformly to an operator of finite rank. (See also Exercises 2.19 and 4.14.)

4. Special functions

We proceed with the study of Harris chains. Our goal is twofold: we want to build a potential theory suitable for the recurrent case – in particular we want to solve the Poisson equations – this will be initiated in the next section and achieved in chs. 8 and 9; on the other hand we want to show a general quotient limit theorem generalizing ch. 4, Exercise 4.10, and this will be done in § 6. Throughout this study the class of functions to be defined below will play a prominent role.

Definition 4.1. A function $f \in \mathscr{E}_+$ is said to be *special* if for every non-negligible function $g \in \mathscr{U}_+$, the function $U_g(f)$ is bounded. A set $A \in \mathscr{E}$ is called *special* if its characteristic function 1_A is special.

Proposition 4.2. *The set $\mathscr{S}$ of special functions is a convex hereditary sub-cone of $L^1_+(m)$.*

Proof. Let $g \in \mathscr{U}_+$ and $0 < m(g) < \infty$; by ch. 3, Proposition 2.9,

$$m(f) = m(gU_g(f)) \leqslant m(g)\,||U_g(f)|| < \infty.$$

Moreover, every function dominated by a special function is obviously special.

To give a first characterization of special functions as well as for other important purposes, we need the following:

Proposition 4.3. *There exists a strictly positive function $h \in \mathscr{U}_+$ such that $m(h) < 1$ and $U_h > 1 \otimes m$.*

Proof. By ch. 3, Proposition 2.7, there is a strictly positive function $h_0 \in \mathscr{U}_+$ and a measure m_0 equivalent to m such that $U_{h_0} \geqslant 1 \otimes m_0$. Let f be a strictly positive function such that $m_0 = fm$. The function $h = h_0 f/(1+f)$ is a strictly positive function of $\mathscr{U}_+$, and

$$\begin{aligned} U_h &> U_{h_0} I_{h_0 - h} U_h \\ &\geqslant 1 \otimes ((h_0 - h)\, m_0)\, U_h = 1 \otimes (hm)\, U_h = 1 \otimes m \end{aligned}$$

by virtue of ch. 3, Proposition 2.9. Since $1 = U_h(h) > m(h)$, the proof is complete.

Proposition 4.4. *Let h_0 be a strictly positive function in $\mathscr{U}_+$ such that $U_{h_0} \geqslant 1 \otimes m_0$ for a positive measure m_0. Then a function $f \in \mathscr{E}_+$ is special if and only if $U_{h_0}(f)$ is bounded on E.*

Proof. The necessity is obvious. Conversely suppose $U_{h_0}(f) \leqslant a$; for $0 < h \leqslant h_0$, we have

$$U_{h_0}(h_0 - h) = 1 - U_{h_0}(h) \leqslant 1 - m_0(h) < 1,$$

and therefore

$$U_h(f) = \sum_{n \geqslant 0} (U_{h_0} I_{h_0 - h})^n\, U_{h_0} f \leqslant a/m_0(h).$$

If h is the function of Proposition 4.3, the function $h' = h \wedge h_0$ has the same properties as h, since $U_{h'} \geqslant U_h$ and $0 < h' \leqslant h_0$, so that $U_{h'}(f) \leqslant a' = a/m_0(h')$. The same argument as above then yields that, for $g \leqslant h'$ and $m(g) > 0$,

$$U_g(f) \leqslant a'/m(g) < \infty.$$

Finally, if g is any non-negligible function in $\mathscr{U}_+$, the function $g \wedge h'$ satisfies the preceding conditions, and since $U_g(f) \leqslant U_{g\wedge h'}(f)$, we get the desired result.

Remark. Special functions are integrable by the measure m_0.

Corollary 4.5. *The cone $\mathscr{C}$ of special functions is invariant under the operators $I_h U_h$, where $h \in \mathscr{U}_+$ and $m(h) > 0$. In particular, it is invariant under P.*

Proof. Let $f \in \mathscr{C}$; by the resolvent equation, we have

$$\begin{aligned} U_{h_0}(hU_h(f)) &\leqslant U_{h_0\wedge h}(hU_h(f)) \\ &= U_{h_0\wedge h}(f) - U_h(f) + U_{h_0\wedge h}((h \wedge h_0)\, U_h(f)) \\ &\leqslant \|U_{h_0\wedge h}(f)\| + \|U_h(f)\|, \end{aligned}$$

and this is finite since f is special.

The following corollary shows that there exist special functions, and that E is the union of an increasing sequence of special sets.

Corollary 4.6. *Every strictly positive function $h_0 \in \mathscr{U}_+$ such that $U_{h_0} \geqslant 1 \otimes m_0$ for a non-zero measure m_0 is special.*

Proof. Since $U_{h_0}(h_0) = 1$ this follows immediately from Proposition 4.4.

It follows from ch. 3, Lemma 2.4 and ch. 6 §3 that the T.P. $U_{h_0}I_{h_0}$ is quasi-compact. We shall now show a converse to this result, which will lead to another characterization of special functions.

Lemma 4.7. *Let $h \in \mathscr{U}_+$, c be a real number in $[0, 1[$ and $A \in \mathscr{E}$. For every T.P. on $(E, \mathscr{E})$ the following inequality holds on $\{U_A(1_A) = 1\}$:*

$$U_{ch}(1_A) \geqslant (1 - c)^{U_A(h)}.$$

Proof. By ch. 2, Definition 2.4,

$$U_{ch}1_A(x) = E_x\left[\sum_{n>0} (1 - ch(X_1)) \cdots (1 - ch(X_{n-1}))\, 1_A(X_n)\right]$$

$$\geqslant E_x[(1 - ch(X_1)) \cdots (1 - ch(X_{S_A-1}))].$$

If $U_A 1_A(x) = 1$, the stopping time S_A is P_x-a.s. finite. The inequality $1 - ch(x) \geqslant (1 - c)^{h(x)}$, which holds since c and $h(x)$ are in $[0, 1]$, and Jenssen's inequality then imply

$$U_{ch}1_A(x) \geqslant E_x[(1 - c)^{(h(X_1)+\cdots+h(X_{S_A-1}))}]$$

$$\geqslant (1 - c)^{E_x[h(X_1)+\cdots+h(X_{S_A-1})]}.$$

Since

$$U_A h(x) = E_x[h(X_1) + \cdots + h(X_{S_A})],$$

the proof is complete.

Proposition 4.8. *For every non-negligible bounded special function f and every positive real number c strictly less than $\|f\|^{-1}$, there is a measure μ_c equivalent to m such that $U_{cf} \geqslant 1 \otimes \mu_c$.*

Proof. We may suppose $\|f\| < 1$. By the proof of Proposition 2.2 in ch. 3, for $0 < c < 1$, there are two strictly positive functions a and b such that $U_c \geqslant a \otimes bm$, and consequently

$$U_{cf} \geqslant U_{cf}(c(1 - f))\, U_c \geqslant U_{cf}(c(1 - f)\, a) \otimes bm.$$

The function $c(1 - f)\, a$ being strictly positive on E, there is a real $\varepsilon > 0$ and a set A of positive m-measure such that $\varepsilon 1_A \leqslant c(1 - f)\, a$. By Lemma 4.7, we have therefore

$$U_{cf}(c(1 - f)\, a) \geqslant \varepsilon U_{cf}(1_A) \geqslant \varepsilon(1 - c)^{U_A f}.$$

Since f is special, $U_A(f)$ is bounded; hence

$$U_{cf} \geqslant 1 \otimes (\varepsilon(1 - c)^{\|U_A f\|})\, bm,$$

and it remains to call this measure μ_c to get the desired result.

If we argue as in ch. 3, Theorem 2.5, the preceding proposition implies that $cU_{cf}I_f$ is quasi-compact. This will be used in the following important results.

Proposition 4.9. *If X is Harris, the following two statements are equivalent:*
(i) *P is quasi-compact;*
(ii) *every positive bounded function is special.*

Proof. If f is special for every integer n, the function $f_n = n^{-1} \sum_{k=1}^{n} P_k f$ is special by virtue of Corollary 4.5. If P is quasi-compact the sequence $\{f_n\}$ converges uniformly to $m(f)$. If $m(f) > 0$, there is therefore an integer n such that f_n majorizes a positive constant, and consequently all bounded functions are special.

Conversely if the function 1 is special, Proposition 4.8 implies that for $c \in]0, 1[$ the T.P. cU_c is quasi-compact; hence, by Theorem 3.5, and its proof P is quasi-compact.

Remark. The last proof could be written using Theorem 3.7 instead of Theorem 3.5. In that way one would avoid the use of the results in ch. 4 §4.

Theorem 4.10. *A non-negligible function $f \in \mathscr{U}_+$ is special if and only if the transition function $U_f I_f$ is quasi-compact.*

Proof. It is known (ch. 3, Proposition 2.9) that the T.P. $Q = U_f I_f$ is Harris, and that for $h \in \mathscr{U}_+$, the kernel U_h^Q is equal to $U_{hf} I_f$. Let $g \in b\mathscr{E}_+$; fg is special for P and therefore $U_h^Q(g) = U_{hf}(fg)$ is bounded on E. The bounded functions are therefore special for Q; hence Q is quasi-compact.

Conversely, if Q is quasi-compact and if $h_0 \in \mathscr{U}_+$ is such that $U_{h_0} \geq 1 \otimes m_0$, then

$$U_{h_0}(f) \leqslant U_{h_0 f}(f) = U_{h_0}^Q(1),$$

which is a bounded function. Thus f is special and the proof is complete.

Exercise 4.11. (1) Prove that, if E is countable, all functions with finite support are special. More generally if E is LCCB and if P is strong Feller, the bounded functions with compact support are special. The result can even be shown for Feller transition probabilities.

[This is solved in ch. 8 §4.]

(2) For a random walk recurrent in the sense of Harris, all bounded functions with compact support are special.

Exercise 4.12. If $f \in \mathscr{U}_+$ let $Q = U_f I_f$, as in Theorem 4.10. Show that if g is special for Q then fg is special for P. Is the converse true?

Exercise 4.13. Prove that, for every bounded special function f, there is constant c such that $U_{cf} I_{cf}$ is 1-quasi-compact, i.e. there is a compact operator K such that $\|U_{cf} I_{cf} - K\| < 1$.

Exercise 4.14. Let the resolvent $\{V_\alpha\}$ be Harris.

(1) Prove that the set of special functions is the same for all the chains X^α and is equal to the set of functions f such that $V_h(f)$ is bounded whenever $V_h > 1 \otimes m$.

We shall call these functions *the special functions of the resolvent* $\{V_\alpha\}$.

(2) If αV_α is quasi-compact for an $\alpha > 0$, it is quasi-compact for every $\alpha > 0$. (See also Exercise 3.20.)

(3) Prove that f is special if and only if $V_f I_f$ is quasi-compact.

Exercise 4.15. Let P be a T.P. enjoying the two following properties:

(i) There is a measure ν such that $\nu P \ll \nu$ and P induces a conservative and ergodic contraction of $L^1(\nu)$.

(ii) There is a function $f \in \mathscr{U}_+$ such that $\nu(f) > 0$ and $U_f I_f$ is quasi-compact.

Then prove that there exists an absorbing set F such that $\nu(F^c) = 0$ and that the restriction of P to F is Harris.

Exercise 4.16. For $A \in \mathscr{E}$, $m(A) > 0$, a function f is said to be A-special if for every $h \in \mathscr{U}_+$ with $m(h) > 0$ we have

$$\sup_A U_h(f) < \infty.$$

(1) Prove that if f is A-special, bounded and non-negligible there exists a measure m' equivalent to m and a number $c < \|f\|^{-1}$ such that $U_{cf} \geqslant 1_A \otimes m'$.

(2) Prove that f is A-special if and only if $U_A f$ is bounded on A and is special for the T.P. $Q = U_A I_A$. If A is special, f is A-special if and only if $U_A f$ is bounded on A.

(3) If f is A-special, bounded, and vanishes outside A, then f is special.

5. Potential kernels

In ch. 2 it was shown that for transient chains, the kernel G gives rise to a satisfactory potential theory, and in particular permits us to solve the Poisson equation. In the present situation of Harris chains, the kernel G is no longer proper, and we shall look for another kernel which could be used for similar purposes.

Definition 5.1. A *charge* is a function $f \in b\mathscr{E}$ such that $|f|$ is special and $m(f) = 0$. The set of charges will be denoted by $\mathscr{N}$. A proper kernel Γ is said to be a *potential kernel* if it maps special functions into finite functions and if, for every $f \in \mathscr{N}$,

$$(I - P)\,\Gamma f = f.$$

Theorem 5.2. *With each function h in $\mathscr{U}_+$ such that $U_h > 1 \otimes m$, one can associate a positive proper kernel W such that, for every non-negligible function $g \in \mathscr{U}_+$,*

$$U_g + U_g I_g W = W + \frac{1}{m(h)}\, U_g(h) \otimes m, \tag{5.1}$$

$$U_g + W I_g U_g = W + \frac{1}{m(h)} \otimes (hm)\, U_g. \tag{5.2}$$

The positive proper kernel $\Gamma = I + W$ is a potential kernel.

Proof. For clarity we divide the proof into four parts.

(1) We may write $U_h = V + 1 \otimes m$ with V a strictly positive kernel. Since $U_h(h) = 1$ and $(hm)\,U_h = m$, we have

$$Vh = 1 - m(h), \qquad (hm)\,V = (1 - m(h))\,m.$$

Recalling the formula

$$U_0 = \sum_{n \geq 0} (U_h I_h)^n\, U_h,$$

we define

$$W = \sum_{n \geq 0} (V I_h)^n\, V = \sum_{n \geq 0} V(I_h V)^n.$$

The kernel W is clearly a positive kernel, and setting $q = (1 - m(h))/m(h)$, we have

$$Wh = \sum_{n \geq 0} (1 - m(h))^{n+1} = q,$$

and $(hm)\,W = qm$.

(2) It is easily seen that

$$V = P + P I_{1-h} V - Ph \otimes m.$$

Thus the obvious identity $W = V + VI_hW$ may be written

$$W = P + PI_{1-h}V - Ph \otimes m + PI_hW + PI_{1-h}VI_hW - Ph \otimes (hm)\, W.$$

But since $(hm)\, W = qm$, it follows that

$$W = P + PI_hW + PI_{1-h}(V + VI_hW) - Ph \otimes (1 + q)\, m,$$

and consequently

$$W = P + PW - (Ph/m(h)) \otimes m, \tag{5.3}$$

which is eq. (5.1) in the special case $g = 1$. From this equation, it is easily derived by induction that, for every integer n,

$$\sum_1^n P_k + P_nW = W + \left(\sum_1^n P_k(h) \Big/ m(h)\right) \otimes m.$$

Let c be a real number with $0 < c < 1$; from the last displayed equality we get

$$\sum_{n\geq 1} (1-c)^{n-1}\left(\sum_1^n P_k\right) + U_cW = c^{-1}W + \sum_{n\geq 1} (1-c)^{n-1}\left(\left(\sum_1^n P_kh \Big/ m(h)\right) \otimes m\right).$$

Now it is easily calculated that

$$\sum_{n\geq 1} (1-c)^{n-1}\left(\sum_1^n P_k\right) = c^{-1}U_c,$$

and consequently

$$U_c + cU_cW = W + (U_ch/m(h)) \otimes m, \tag{5.4}$$

which is eq. (5.1) in the special case $g = c$.

In analogous fashion, we get the following special cases of eq. (5.2):

$$P + WP = W + m(h)^{-1} \otimes (hm)\, P, \tag{5.5}$$

$$U_c + cWU_c = W + m(h)^{-1} \otimes (hm)\, U_c. \tag{5.6}$$

(3) Now let g be a non-negligible function in $\mathscr{U}_+$, with $\|g\| < 1$. We may write eq. (5.3) in the form

$$P(I + I_gW) + PI_{1-g}W = W + (Ph/m(h)) \otimes m.$$

We repeat p times the following computation: multiply through the above equality to the left by PI_{1-g} and add $P(I + I_gW)$ to both sides. We get

$$\sum_{n\leq p} (PI_{1-g})^n\, P(I + I_gW) + (PI_{1-g})^{p+1}\, W$$

$$= W + \left(\sum_{n \leqslant p} (PI_{1-g})^n\, Ph/m(h)\right) \otimes m.$$

Letting p tend to infinity in the resulting inequality

$$\sum_{n \leqslant p} (PI_{1-g})^n\, P(I + I_g W) \leqslant W + \left(\sum_{n \leqslant p} (PI_{1-g})^n\, Ph/m(h)\right) \otimes m$$

yields

$$U_g(I + I_g W) \leqslant W + (U_g h/m(h)) \otimes m. \tag{5.7}$$

Pick a number c such that $\|g\| < c < 1$; the function $h' = cU_c h$ is strictly positive, and if we apply h' to the right of both sides of eq. (5.7) we get, on account of eq. (5.6),

$$cU_g U_c h + U_g I_g (Wh - U_c h + m(h)^{-1}\,(hm)\, U_c h)$$

$$\leqslant Wh - U_c h + m(h)^{-1}\,(hm)\, U_c h + U_g h.$$

Using the fact that Wh is constant, this inequality reduces, after cancellation, to

$$cU_g U_c h - U_g I_g U_c h \leqslant U_g h - U_c h;$$

but the left-hand side is equal to $U_g I_{c-g} U_c h$, and by the resolvent equation of ch. 2 § 2, this inequality is in fact an equality. As a result, the inequality (5.7) is an equality, and eq. (5.1) holds in that case.

Finally, if $\|g\| = 1$, we deduce from the preceding discussion and the resolvent equation

$$U_g = U_{cg} + (1 - c)\, U_g I_g U_{cg},$$

where $0 < c < 1$, that eq. (5.1) is true in full generality.

We now turn to the proof of eq. (5.2). Following the same pattern as above, with right multiplications instead of left multiplications, we get, for any non-negligible function g in $\mathscr{U}_+$,

$$U_g + WI_g U_g \leqslant W + m(h)^{-1} \otimes (hm)\, U_g.$$

We claim that this inequality is in fact an equality. From eq. (5.6), it follows that $Wh' = cWU_c h \leqslant Wh + \langle h, cU_c h\rangle/m(h)$ is a bounded function. Consequently, if $0 \leqslant f \leqslant h'$, the function

$$k = U_g f + WI_g U_g f - Wf - \langle h, U_g f\rangle/m(h)$$

is bounded. Moreover, k is harmonic; this follows from writing the equalities

$$\begin{aligned} P + P(U_g + WI_gU_g) &= P + (PI_{1-g}U_g + PI_gU_g) + PWI_gU_g \\ &= (P + PI_{1-g}U_g) + (P + PW)\, I_gU_g \\ &= U_g + WI_gU_g + (Ph/m(h)) \otimes m \end{aligned}$$

and

$$P + P(W + m(h)^{-1} \otimes (hm)\, U_g) = W + m(h)^{-1} \otimes (hm)\, U_g + (Ph/m(h)) \otimes m$$

and taking their difference. By ch. 3 § 2, the function k is thus constant. On the other hand, combining the two equalities

$$(hm)\,(U_g + WI_gU_g) = (hm)\, U_g + qm$$

$$(hm)\,(W + m(h)^{-1} \otimes (hm)\, U_g) = qm + (hm)\, U_g$$

proves, since the measure hm is bounded, that $\langle hm, k\rangle = 0$ and therefore that k is identically zero. The proof of eq. (5.2) is thus complete.

(4) We have already noticed that Wh' is bounded; since h' is strictly positive this implies that W is a proper kernel. Furthermore, if f is special, eq. (5.2) written for $g = h$ yields

$$\begin{aligned} Wf &\leqslant U_h(f) + W(hU_hf) \\ &\leqslant \|U_h(f)\|\,(1 + W(h)) = \|U_hf\|/m(h). \end{aligned}$$

The kernel W thus maps special functions into bounded functions and $\Gamma = I + W$ maps special functions into finite functions.

Moreover, if $f \in \mathcal{N}$, then by eq. (5.1),

$$f + Wf = f + Pf + PWf,$$

which reads

$$(I - P)\,\Gamma f = f.$$

We observe that Γf is then bounded.

Corollary 5.3. *Let f be a charge; then the Poisson equation*

$$(I - P)\, g = f$$

has a bounded solution, unique up to addition of a constant function.

Proof. By the preceding result, Γf is a bounded solution of the Poisson equation. The difference between two such solutions is a bounded harmonic function, hence a constant function.

The potential kernel Γ is not unique and there are always many potential kernels. Indeed, if φ is a finite function and ν a measure taking finite values on special functions, then the kernel $\Gamma + \varphi \otimes m + 1 \otimes \nu$ is also a potential kernel. In the following result, we prove that in this way we obtain all possible potential kernels. We notice that some of these kernels are not positive.

Proposition 5.4. *Let Γ and Γ' be two potential kernels; then there is a finite function φ and a measure ν taking finite values on special functions such that*

$$\Gamma' = \Gamma + \varphi \otimes m + 1 \otimes \nu.$$

Proof. If $f \in \mathcal{N}$, then Γf and $\Gamma' f$ are two bounded solutions of the Poisson equation and thus differ by a constant. Pick x_0 in E and set

$$\nu = \Gamma'(x_0, \cdot\,) - \Gamma(x_0, \cdot\,);$$

the measure ν takes finite values on special functions and

$$\Gamma' f - \Gamma f = \nu(f).$$

Now pick a special function f_0 such that $m(f_0) = 1$. For every special function f, the function $f - m(f)\, f_0$ is a charge, and consequently

$$\Gamma'(f - m(f)\, f_0) - \Gamma(f + m(f_0)) = \nu(f) - m(f)\, \nu(f_0),$$

hence

$$\Gamma' f = \Gamma f + m(f)\, (\Gamma' f_0 - \Gamma f_0 - \nu(f_0)) + \nu(f).$$

It remains to put $\varphi = \Gamma' f_0 - \Gamma f_0 - \nu(f_0)$ to get the desired result, since it is easily seen that two measures equal on special functions are equal.

The case in which Γ and Γ' are associated with positive special functions by Theorem 5.2 is of special interest, since in that case one can compute φ and ν. When necessary to avoid possible misinterpretation, we shall write W_h and Γ_h for the kernels associated in Theorem 5.2 with the function h.

Corollary 5.5. *If h_1 and h_2 are two positive special functions such that $U_{h_i} > 1 \otimes m$ $(i = 1, 2)$, then the kernels W_{h_i} $(i = 1, 2)$ are related by the formula*

$$W_{h_1} + \frac{1}{m(h_1)} W_{h_2}(h_1) \otimes m = W_{h_2} + \frac{1}{m(h_2)} \otimes (h_2 m)\, W_{h_1}.$$

Proof. Within the course of this proof let us abbreviate W_{h_i} to W_i. By Proposition 5.4, there is a function φ and a measure ν such that

$$W_1 + \varphi \otimes m = W_2 + 1 \otimes \nu.$$

Since $W_1(h_1) = q_1$ and $(h_2 m) W_2 = q_2 m$, by multiplying this equality to the left by $(h_2 m)$ and to the right by h_1 we get

$$\varphi = m(h_1)^{-1} (W_2(h_1) + a), \qquad a \in \mathbf{R},$$

$$\nu = m(h_2)^{-1} ((h_2 m) W_1 + bm), \qquad b \in \mathbf{R},$$

and we may write

$$W_1 + m(h_1)^{-1} W_2(h_1) \otimes m = (W_2 + m(h_2)^{-1} \otimes (h_2 m) W_1) + k1 \otimes m,$$

where k is a real number. As

$$(h_2 m) (W_1 + m(h_1)^{-1} W_2(h_1) \otimes m) (h_1) = m(h_2)\, q_1 + q_2\, m(h_1)$$

and

$$(h_2 m) (W_2 + m(h_2)^{-1} \otimes (h_2 m) (h_1) = q_2\, m(h_1) + m(h_2)\, q_1,$$

it follows that $k = 0$, and we are done.

The existence of potential kernels gives yet another characterization of special functions.

Proposition 5.6. *A positive function f is special if and only if Γf is bounded for any potential kernel*

$$\Gamma = I + W + \varphi \otimes m + 1 \otimes \nu$$

such that φ is bounded.

Proof. The necessity follows from Theorem 5.2 and Proposition 5.4. To prove the sufficiency, we start by noticing that if Γf is bounded, then Wf is bounded, and since (hm) is a bounded measure we have

$$m(f) = q^{-1}(hm)\, Wf < \infty.$$

The result then follows from the inequality

$$U_g(f) \leqslant Wf + m(h)^{-1}\, U_g(h)\, m(f).$$

The next result, another version of which is given in Exercise 6.16, may be seen as a refinement of the ratio-limit theorem of ch. 4, Theorem 4.2 and will be used to give a sharper form of this theorem in the following section.

Proposition 5.7. *If $f \in \mathcal{N}$, then*

$$\sup_n \left\| \sum_{m=0}^{n} P_m f \right\| < +\infty.$$

Proof. From $(I - P)\,\Gamma f = f$, it is easily derived that

$$\sum_{m=0}^{n} P_m f = \Gamma f - P_{n+1}\Gamma f;$$

therefore

$$\left\| \sum_{m=0}^{n} P_m f \right\| \leqslant 2\|\Gamma f\|.$$

We now turn to the problem of solving the Poisson equation for measures. For this purpose, we lay down the following

Definition 5.8. A positive measure μ on $(E, \mathscr{E})$ is said to be *special* if there is a bounded measure σ and a non-negligible function $g \in \mathscr{U}_+$, such that $\mu \leqslant \sigma U_g$.

Proposition 5.9. *Let h be a strictly positive function in $\mathscr{U}_+$ such that $U_h > 1 \otimes m$. Then a measure μ is special if and only if there is a bounded measure ρ such that $\mu \leqslant \rho U_h$.*

Proof. The sufficiency is obvious. Conversely assume that $\mu \leqslant \sigma U_g$; since $U_g \leqslant U_{g\wedge h}$, we may assume $g \leqslant h$.

The measure

$$\rho = \sum_{n\geqslant 0} \sigma(U_h I_{h-g})^n$$

is bounded because

$$\rho(E) \leqslant \sigma(E) \sum_{n\geqslant 0} (1 - U_h(g))^n \leqslant \sigma(E) \sum_{n\geqslant 0} (1 - m(g))^n = \sigma(E)/m(g) < \infty.$$

By the resolvent equation, we get $\rho U_h = \sigma U_g$, which proves the necessity.

Remark. The measure m is special since $m = (hm)\, U_h$.

Corollary 5.10. *Special measures take finite values on special functions. They are thus σ-finite.*

Proof. Let μ and f be special; then

$$\mu(f) \leqslant \langle \sigma U_h, f \rangle \leqslant \sigma(E) \| U_h(f) \| < \infty.$$

The convex cone of special measures is obviously hereditary to the left and we have the following

Proposition 5.11. *The cone of special measures is invariant by the operators $I_h U_h$.*

Proof. Left to the reader as an exercise.

Proposition 5.12. *For every bounded measure ν such that $\nu(E) = 0$, there is a measure μ such that $|\mu - \nu|$ is special and*

$$\nu = \mu(I - P).$$

The measure μ is unique up to addition of a multiple of m.

Proof. If ρ is a bounded measure, then ρW is special since

$$\rho W \leqslant \left(\sum_{n \geqslant 0} \rho (V I_h)^n \right) U_h,$$

and

$$\sum_{n \geqslant 0} \rho (V I_h)^n (E) = \rho(E)/m(h) < \infty.$$

The measure

$$\mu = \nu \Gamma = \nu + \nu^+ W - \nu^- W$$

is such that $|\mu - \nu|$ is special, and by Theorem 5.2 is the desired measure.

We turn to show the latter half of the statement. Let μ' be another measure with the same properties. Then $\lambda = \mu - \mu'$ is such that $|\lambda|$ is special and $\lambda = \lambda P$, which implies $\lambda^{\pm} \leqslant \lambda^{\pm} P$. It remains to show that the multiples of m are the only special measures λ such that $\lambda \leqslant \lambda P$.

If $\lambda \leqslant \lambda P$, we may obtain inductively the inequality

$$\lambda \leqslant \lambda I_h P \left(\sum_{l < n} (I_{1-h} P)^l \right) + \lambda (I_{1-h} P)^n.$$

Then if $\lambda \leqslant \rho U_h$, we have

$$\lambda(I_{1-h}P)^n \leqslant \rho U_h(I_{1-h}P)^n = \sum_{l \geqslant n} \rho P(I_{1-h}P)^l;$$

as $\rho U_h(h) = \rho(E) < +\infty$, it follows that $\lim_n (I_{1-h}P)^n h = 0$, and since h is strictly positive, passing to the limit in the next to the last displayed inequality, we get $(h\lambda) \leqslant (h\lambda) I_h U_h$. These two measures are bounded since

$$\lambda(h) \leqslant \rho U_h(h) = \rho(E) < \infty$$

and have the same total variation, hence they are equal. The measure $h\lambda$ is thus $U_h I_h$-invariant, and therefore by ch. 3 § 2.9, there is a constant c such that $h\lambda = chm$. Since h is strictly positive, $\lambda = cm$.

Remark. The condition that $|\mu - \nu|$ is special is essential to have the uniqueness property as will be seen from examples in ch. 9 (see also ch. 3, Exercise 2.20). The reader will also find in Exercise 6.10 the condition under which the solution ν may be chosen positive.

Exercise 5.13. Prove that the measure $(hm)\, U_g$, which appears in eq. (5.2), is majorized by a multiple of m.

Exercise 5.14. With the notation used at the beginning of §2 prove that the singular part W^1 of W with respect to m is equal to $\sum_1^\infty P_n^1$.

Exercise 5.15. Let ϑ be the set of bounded special functions such that $m(f) = 1$.

(1) Prove that for any $f \in \vartheta$ there is at least one proper kernel Γ_f such that

(i) $\Gamma_f g$ is bounded for any special and bounded g;

(ii) $(I - P)\, \Gamma_f = I - f \otimes m$.

(2) If λ and ν are two probability measures show that $\nu \prec \lambda$ if and only if $\Lambda_f = ((\lambda - \nu)\, \Gamma_f)^-$ is absolutely continuous with respect to m for one (hence for all) $f \in \vartheta$.

Exercise 5.16. For every special set K there is a constant C_K such that, for every bounded charge f vanishing outside K,

$$\sup_n \left\| \sum_{m=0}^{n} P_m f \right\| \leqslant C_K \|f\|.$$

Exercise 5.17. Assume X aperiodic, and let f be a charge. Prove that for any pair (ν_1, ν_2) of probability measures on $(E, \mathscr{E})$,

$$\lim_{n\to\infty} \left\langle \nu_1 - \nu_2, \sum_{k=0}^{n} P_k f \right\rangle = (\nu_1 - \nu_2)\, \Gamma f,$$

where Γ is any potential kernel. Derive that the sequence $WP_n f$ converges pointwise to zero. What can be said for positive periodic X?

Exercise 5.18. The kernel $I + WI_h$ satisfies the reinforced complete maximum principle. The kernel W satisfies the positive maximum principle, that is,

$$\sup_{x\in E} W f(x) = \sup_{x\in\{f>0\}} W f(x).$$

Exercise 5.19. In the situation of Theorem 4.10, prove that $I + WI_f$ is a potential kernel for Q. If f is special, it is then a bounded kernel. More precisely, $I + W$ is bounded if and only if P is quasi-compact.

Exercise 5.20 (Potential kernels for Harris resolvents). As for chains, a charge is a function f such that $|f|$ is special and $m(f) = 0$. A potential kernel is a proper kernel W which maps special functions into finite ones, and which is such that, for every charge f and every $\alpha > 0$,

$$(I + \alpha W)\, V_\alpha f = Wf = V_\alpha (I + \alpha W)\, f.$$

(1) Prove that there is a whole family of potential kernels and describe their pairwise relationship.

(2) Define special measures and prove a result similar to Theorem 5.12.

6. The ratio-limit theorem

We now want to sharpen ch. 4, Theorem 4.2, which was obtained by a straightforward application of the Chacon–Ornstein theorem along the line of ch. 4, Exercise 4.10. The most general result that one could hope for is the following: for every pair (f, g) of integrable functions and every pair (ν_1, ν_2) of probability measures on $(E, \mathscr{E})$,

$$\lim_{n\to\infty} \left(\sum_{m=0}^{n} \nu_1 P_m f \Big/ \sum_{m=0}^{n} \nu_2 P_m g \right) = m(f)/m(g).$$

Unfortunately, it turns out that this result is not true, but below we shall give several theorems which come very close to it.

Proposition 6.1. *If f and g are special functions and $m(g) > 0$, for every probability measure ν on $\mathscr{E}$,*

$$\lim_{n\to\infty}\left(\sum_{m=0}^{n} \nu P_m f \Big/ \sum_{m=0}^{n} \nu P_m g\right) = m(f)/m(g).$$

Proof. The function $m(g)\, f - m(f)\, g$ is a charge; thus, by Proposition 5.7, there is a number M such that, for every integer n,

$$\left| m(g) \sum_{m=0}^{n} \nu P_m f - m(f) \sum_{m=0}^{n} \nu P_m g \right| \leqslant M < \infty.$$

Since $\sum_{m=0}^{n} \nu P_m g > 0$ for n sufficiently large,

$$\left| \sum_{m=0}^{n} \nu P_m f \Big/ \sum_{m=0}^{n} \nu P_m g - m(f)/m(g) \right| \leqslant M \left(m(g) \sum_{m=0}^{n} \nu P_m g \right)^{-1},$$

and this converges to zero as n tends to infinity.

In the sequel, we shall use the kernel W associated, as in Theorem 5.2, with a strictly positive function h. We shall denote by λ the bounded measure hm, and put

$$a_n = \|\lambda\|^{-1} \left(\sum_{m=1}^{n} \lambda P_m h \right).$$

Lemma 6.2. *For every $\varepsilon > 0$, there is a constant c_ε such that, for every integer n,*

$$\sum_{m=1}^{n} P_m h \leqslant c_\varepsilon + (1 + \varepsilon)\, a_n.$$

Proof. Set

$$A_n = \left\{ \sum_{1}^{n} P_m h < (1 + \varepsilon)\, a_n \right\}.$$

Integrating with respect to λ the two terms of the inequality

$$\sum_{1}^{n} P_m h \geqslant (1 + \varepsilon)\, a_n 1_{A_n^c}$$

yields

$$\|\lambda\|\, a_n > (1 + \varepsilon)\, a_n \lambda(A_n^c),$$

and therefore

$$\lambda(A_n) \geqslant \|\lambda\| \,\varepsilon/(1+\varepsilon).$$

For $f \in \mathscr{U}_+$, it is easy to derive inductively the formula

$$\sum_1^n P_m + U_f P_n = U_f I_f \left(\sum_1^n P_m\right) + U_f. \tag{6.1}$$

By applying eq. (6.1) to the function $f_n = h1_{A_n}$, we get

$$\sum_1^n P_m h \leqslant U_{f_n}\left(f_n\left(\sum_1^n P_m h\right)\right) + U_{f_n}(h).$$

But

$$f_n\left(\sum_1^n P_m h\right) < (1+\varepsilon)\, a_n f_n,$$

and, since $U_{f_n}(f_n) \leqslant 1$,

$$\sum_1^n P_m h \leqslant (1+\varepsilon)\, a_n + U_{f_n}(h).$$

Now, because of the resolvent equation,

$$\begin{aligned} U_{f_n}(h) &= \sum_{k\geqslant 0} (U_h I_{h-f_n})^k\, U_h(h) \\ &\leqslant \sum_{k\geqslant 0} (1 - m(f_n))^k = m(f_n)^{-1} = \lambda(A_n)^{-1} \leqslant (1+\varepsilon)\,(\varepsilon\|\lambda\|)^{-1}. \end{aligned}$$

The proof is now complete.

Lemma 6.3. *For every $\delta \in \,]0, 1[$ and every probability measure ν, there are two numbers $\alpha, \beta > 0$ such that*

(i) $U_{\alpha h} I_{\alpha h} \geqslant (1-\delta)\, \|\lambda\|^{-1} \otimes \lambda$;

(ii) $\nu U_{\alpha h+\beta}(\alpha h) \geqslant 1 - \delta$.

Proof. If $\alpha < 1$, we have

$$U_{\alpha h} I_{\alpha h} = \sum_{n\geqslant 0} \alpha(1-\alpha)^n\, (U_h I_h)^{n+1};$$

that is, with the notation of § 5,

$$U_{\alpha h} I_{\alpha h} = \sum_{n\geqslant 0} \alpha(1-\alpha)^n\, (1 \otimes \lambda + V I_h)^{n+1}.$$

Since

$$(VI_h)\,(1\otimes\lambda) = (1\otimes\lambda)\, VI_h = (1 - \|\lambda\|)\,(1\otimes\lambda),$$

easy computations prove that

$$U_{\alpha h} I_{\alpha h} = (1 - (1 - \alpha)(1 - \|\lambda\|))^{-1} \otimes \lambda + \sum_{n \geqslant 0} \alpha (1 - \alpha)^n (V I_h)^{n+1}.$$

It is then easily seen that (i) is satisfied for sufficiently small α.

The number α being chosen, when β decreases to zero the operator $U_{\alpha h+\beta}$ increases to $U_{\alpha h}$ and therefore

$$\lim_{\beta \to 0} \nu U_{\alpha h+\beta}(\alpha h) = \nu U_{\alpha h}(\alpha h) = 1;$$

The condition (ii) is thus satisfied for sufficiently small β.

Lemma 6.4. *For every probability measure* ν *on* $(E, \mathscr{E})$,

$$\lim_n \left(\sum_{m=1}^n \nu P_m h \Big/ \sum_{m=1}^n \lambda P_m h \right) = \|\lambda\|^{-1}.$$

Proof. Denote by R_n the ratio which occurs in the statement. Because of Lemma 6.2, we have $\overline{\lim}_n R_n \leqslant \|\lambda\|^{-1}$, and we are going to prove the reverse inequality for the inferior limit.

Let $\delta \in]0, 1[$ and set

$$B_n = \left\{ \sum_1^n P_m h \geqslant (1 - \delta) a_n \right\}.$$

The inequality

$$\sum_1^n P_m h \geqslant (1 - \delta) a_n 1_{B_n},$$

together with eq. (6.1), implies

$$\sum_1^n P_m h \geqslant U_f I_f \left(\sum_1^n P_m h \right) - U_f P_n h$$

$$\geqslant (1 - \delta) a_n U_f(f 1_{B_n}) - U_f P_n h. \tag{6.2}$$

With the notation of Lemma 6.3, put $f = \alpha h + \beta$; it follows that

$$\nu U_f(f 1_{B_n}) \geqslant \nu U_{\alpha h+\beta}(\alpha h 1_{B_n}) \geqslant \nu U_{\alpha h}(\alpha h 1_{B_n}) - \delta,$$

since by Lemma 6.3(ii),

$$\nu(U_{\alpha h} I_{\alpha h} - U_{\alpha h+\beta} I_{\alpha h})(E) \leqslant \delta.$$

By Lemma 6.3(i), it then follows that

$$\nu U_f(f1_{B_n}) \geqslant (1-\delta)\,\|\lambda\|^{-1}\,\lambda(B_n) - \delta.$$

Let us study the asymptotic behaviour of $\lambda(B_n)$. By virtue of Lemma 6.2, we have

$$\sum_1^n P_m h \leqslant (1-\delta)\,a_n \qquad \text{on } B_n^c,$$

$$\sum_1^n P_m h \leqslant c_\varepsilon + (1+\varepsilon)\,a_n \quad \text{on } B_n;$$

taking the integral with respect to λ yields

$$a_n\|\lambda\| \leqslant (1-\delta)\,a_n(\|\lambda\| - \lambda(B_n)) + (c_\varepsilon + (1+\varepsilon)\,a_n)\,\lambda(B_n),$$

hence

$$\delta a_n\|\lambda\| \leqslant (c_\varepsilon + (\delta+\varepsilon)\,a_n)\,\lambda(B_n).$$

Since $\{a_n\}$ converges to infinity, we find

$$\varliminf_n \lambda(B_n) \geqslant \delta\|\lambda\|/(\delta+\varepsilon),$$

and since ε is arbitrary, $\varliminf_n \lambda(B_n) \geqslant \|\lambda\|$; hence $\lim_n \lambda(B_n) = \|\lambda\|$. It then follows that

$$\varliminf_n \nu U_f(f1_{B_n}) \geqslant 1 - 2\delta.$$

Finally as $f \geqslant \beta$ and $P_n h \leqslant 1$, for every n we have

$$U_f P_n h \leqslant U_\beta 1 = \beta^{-1}.$$

Returning to eq. (6.2), we see that

$$\varliminf_n R_n \geqslant \|\lambda\|^{-1}\,(1-\delta)\,(1-2\delta),$$

and since δ is arbitrary, the proof is complete.

Theorem 6.5. *For every pair (f, g) of special functions with $m(g) > 0$ and every pair (ν_1, ν_2) of probability measures on $(E, \mathscr{E})$*

$$\lim_{n\to\infty} \left(\sum_{m=0}^n \nu_1 P_m f \Big/ \sum_{m=0}^n \nu_2 P_m g \right) = m(f)/m(g).$$

Proof. Beginning the summations at 0 instead of 1 causes no trouble, since the denominator diverges. By applying Lemma 6.4 with ν_1, and then with

ν_2, and taking the ratio, we get

$$\lim_{n\to\infty}\left(\sum_{m=0}^{n}\nu_1 P_m h \Big/ \sum_{m=0}^{n}\nu_2 P_m h\right) = 1.$$

We then apply Proposition 6.1 twice, first to ν_1, f, h, then to ν_2, g, h, and writing

$$\frac{\sum_{m=0}^{n}\nu_1 P_m f}{\sum_{m=0}^{n}\nu_2 P_m g} = \frac{\sum_{m=0}^{n}\nu_1 P_m f}{\sum_{m=0}^{n}\nu_1 P_m h} \times \frac{\sum_{m=0}^{n}\nu_1 P_m h}{\sum_{m=0}^{n}\nu_2 P_m h} \times \frac{\sum_{m=0}^{n}\nu_2 P_m h}{\sum_{m=0}^{n}\nu_2 P_m g},$$

we get the desired result.

In the same way we may obtain

Theorem 6.6. *Let f, g be two functions in $L^1(m)$ with $m(g) > 0$. There exist two negligible sets N_f and N_g depending only on f and g and such that for $x \notin N_f$ and $y \notin N_g$*

$$\lim_{n\to\infty}\left(\sum_{m=0}^{n} P_m f(x) \Big/ \sum_{m=0}^{n} P_m g(y)\right) = m(f)/m(g).$$

Proof. It can be performed by following the pattern of Theorem 6.5, except that instead of Proposition 6.1 we use ch. 4, Theorem 4.2. The set N_f is the negligible set which appears in ch. 4, Theorem 4.2, as applied to f and h.

Remark. One could think of enlarging the scope of validity of the preceding theorems to functions f such that $P_n f$ is special. Exercise 6.7 shows that nothing can be gained in that way.

Exercise 6.7. If f is such that $U_g(f)$ is special for a function $g \in \mathscr{U}_+$, then f is special.

Exercise 6.8. Prove that if there is a Harris random walk on the group G, then G is unimodular.

[*Hint*: Apply Theorem 6.5 to the left and right random walks with $\nu_1 = \nu_2 = \varepsilon_e$.]

Exercise 6.9. Carry over the result of this section to the case of Harris resolvents.

Exercise 6.10 (continuation of Exercise 5.15). Retain the notation of ch. 2 §5 and suppose that X is Harris.

(1) Prove that either the measure σ is σ-finite or that $\sigma(A) = \infty$ as soon as $m(A) > 0$. Prove that in the former case $\nu \prec \lambda$ and

$$\sigma = \sigma(f)\, m + (\lambda - \nu)\, \Gamma_f$$

for any $f \in \vartheta$.

(2) Prove that the Poisson equation $\eta = \eta P + \lambda - \nu$ has a σ-finite and positive solution if and only if $\nu \prec \lambda$ and $\phi_f = \mathrm{d}\Lambda_f/\mathrm{d}m$ is bounded m-a.e. in which case σ is the minimal positive solution and $\sigma(f) = ||\phi_f||_\infty$ for any $f \in \vartheta$. In particular if P is quasi-compact and $\nu \prec \lambda$ and if T is the R.S.time of the filling scheme then $E_\lambda[T] = ||\phi_1||_\infty$.

CHAPTER 7

MARTIN BOUNDARY

The purpose of Martin boundary theory is twofold. From the potential theoretic point of view it is to give an integral representation of harmonic functions which turns them into "potentials" of charges carried by the boundary of the space. We thus wish to generalize the classical Poisson integral representation of harmonic functions in the unit disc, namely the one-to-one correspondence between the non-negative harmonic functions h and the Borel measures μ^h on the circle S^1 given by (with classical notation)

$$h(re^{i\theta}) = \int_{S^1} \frac{1-r^2}{1-2r\cos(\theta-t)+r^2}\, d\mu^h(t).$$

The function 1 corresponds to the Lebesgue measure m on S^1, and if f is the density of μ^h with respect to m, then Fatou's theorem states that $h(re^{i\theta})$ converges to $f(t)$ whenever $re^{i\theta}$ converges to t non-tangentially.

From the probabilistic point of view we wish to study the asymptotic behaviour of transient Markov chains. If the potential kernel G is proper, the space E is the union of an increasing sequence of transient sets, and one can thus say, loosely speaking, that the chain eventually leaves the space. For example, a transient random walk on $\mathbf{R}$ with positive first moment converges towards $+\infty$; we can say that it leaves the space on the right. We wish to describe in general how a transient chain leaves its state space.

1. Regular functions

In this section we shall investigate more thoroughly some of the notions introduced in ch. 2 §3. For $f \in \mathscr{E}$ we set

$$R(f) = \overline{\lim_n}\, f(X_n),$$

and put $R(1_A) = R(A)$, to be consistent with the notation of ch. 2 §3. The random variable $R(f)$ is invariant.

Definition 1.1. A function f in $\mathscr{E}$ is said to be *regular* if $\lim_n f(X_n)$ exists a.s. A set $A \in \mathscr{E}$ is said to be *regular* if 1_A is a regular function. Two regular functions f and g are said to be equivalent if $R(f) = R(g)$ a.s.

Finite superharmonic functions and bounded harmonic functions are regular. Absorbing sets are regular; transient sets are regular and equivalent to the empty set.

Proposition 1.2. *The set $\mathscr{R}$ of regular functions is an algebra (for ordinary pointwise multiplication) and a lattice. The collection of regular sets is a Boolean algebra of sets which will also be denoted by $\mathscr{R}$.*

Proof. Obvious.

We call $b\mathscr{R}$ the sub-algebra of bounded regular functions, $\bar{\mathscr{R}}$ and $b\bar{\mathscr{R}}$ the collections of equivalence classes in $\mathscr{R}$ and $b\mathscr{R}$. We observe that the family of transient sets is an ideal in the Boolean algebra $\mathscr{R}$ and $\bar{\mathscr{R}}$ is the quotient set of $\mathscr{R}$ by this ideal.

Proposition 1.3. *The map $A \to R(A)$ is an isomorphism of the Boolean algebra $\bar{\mathscr{R}}$ onto the Boolean algebra of equivalence classes of invariant events.*

Proof. Plainly $R(A^c) = R(A)^c$ and $R(A \cup B) = R(A) \cup R(B)$ if A, B are regular; moreover $R(A) = \emptyset$ a.s. if and only if A is transient. Finally, if $\Gamma \in \mathscr{I}$, the set $A = \{x\colon P_x[\Gamma] > \frac{1}{2}\}$ is regular and $\Gamma = R(A)$, because $\lim_n P_{X_n}[\Gamma] = 1_\Gamma$ according to the reasoning in ch. 2 §3.

Proposition 1.4. *A function f in $b\mathscr{E}$ is regular if and only if it is the uniform limit of simple functions over the Boolean algebra $\mathscr{R}$.*

Proof. The sufficiency is obvious. Conversely, let $f \in b\mathscr{R}$ and pick an invariant, simple random variable $Y = \sum \alpha_i 1_{\Gamma_i}$ such that $\|Y - \mathscr{R}(f)\| < \varepsilon$. Let A_i be regular sets such that $R(A_i) = \Gamma_i$ a.s., and set $g = \sum \alpha_i 1_{A_i}$; the set $A_\varepsilon = \{|f - g| > \varepsilon\}$ is easily seen to be transient. Let h be any bounded simple function such that $|h - f| < \varepsilon$ on A_ε and set $f_\varepsilon = g\, 1_{A_\varepsilon^c} + h\, 1_{A_\varepsilon}$; then f_ε is a simple and regular function, and $\|f - f_\varepsilon\| < \varepsilon$, which yields the desired result.

In ch. 2 §3, we have seen how to establish a one-to-one correspondence between bounded harmonic functions and equivalence classes of a.s. bounded invariant random variables. We may summarize all these results in the following statement.

Theorem 1.5. *If P is markovian, the following three vector spaces are isomorphic:*
(i) *the space of bounded harmonic functions;*
(ii) *the space of equivalence classes of a.s. bounded invariant functions;*
(iii) *the space* $\mathrm{b}\bar{\mathscr{R}}$.

This theorem has the following corollary. We recall that a positive harmonic function h is said to be *extremal* if, for every harmonic function h' such that $0 \leqslant h' \leqslant h$, there is a real number c such that $h' = ch$.

Corollary 1.6. *The extremal bounded harmonic functions are in one-to-one correspondence with the atoms of the Boolean algebra* $\bar{\mathscr{R}}$.

We are now going to exhibit a particular class of regular functions which will be of paramount importance in the following sections.

We assume henceforth that P is markovian and that G is proper. There is therefore an increasing sequence of sets E_k in $\mathscr{E}$ with union E, such that the potentials $G(\cdot\,, E_k)$ are bounded, and if we write L_k for the last hitting time of E_k (see ch. 1, Exercise 3.18) then L_k is a.s.-finite, $L_k \leqslant L_{k+1}$ and $\lim_k L_k = +\infty$ a.s. We define $Y_k = X_{L_k}$ and $\mathscr{H}_k = \sigma(Y_l, l \geqslant k)$.

Proposition 1.7. *The σ-algebra $\mathscr{I}$ of invariant events is a.s. equal to the σ-algebra* $\bigcap_k \mathscr{H}_k$.

Proof. Let $\Gamma \in \mathscr{I}$ and set $A = \{x: P_x[\Gamma] > \frac{1}{2}\}$; since P is markovian we know from ch. 2 §3 that

$$\Gamma = \lim_n \{X_n \in A\} \quad \text{a.s.},$$

and by the afore-mentioned properties of times L_k we also have

$$\Gamma = \lim_k \{Y_k \in A\} \quad \text{a.s.}$$

Conversely, it is easily seen that $L_k \circ \theta = L_k - 1$, hence $Y_k \circ \theta = Y_k$ on the set $\{0 < L_k < \infty\}$; consequently, if $\Gamma \in \mathscr{H}_k$, then

$$\theta^{-1}(\Gamma \cap \{0 < L_k < \infty\}) = \Gamma \cap \{1 < L_k < \infty\}.$$

If $\Gamma \in \bigcap \mathscr{H}_k$, it follows from the properties of times L_k that $\theta^{-1}(\Gamma) = \Gamma$ a.s., which is the desired conclusion.

Let ν be a probability measure on $\mathscr{E}$; the measure νG is σ-finite and for any probability measure γ on $\mathscr{E}$, we write $\gamma G = \gamma K \cdot \nu G + \nu G^1$ where γK is a function in $\mathscr{E}_+$ and νG^1 is singular with respect to νG. We denote by A_γ a set such that $\nu G(A_\gamma^c) = 0$ and $\nu G^1(A_\gamma) = 0$; clearly $P_\nu[\bigcup_n \{X_n \in A_\gamma^c\}] = 0$. We are going to prove that γK is regular and we begin with some lemmas.

Lemma 1.8. *Let $A_0 \subset A_\gamma, A_1, \ldots, A_n, B$ be in $\mathscr{E}$ and for $n \geqslant 1$, set*

$$\Lambda = \{X_{L_B - n} \in A_0, X_{L_B - n+1} \in A_1, \ldots, X_{L_B} \in A_n, n \leqslant L_B < \infty\}.$$

Then

$$P_\gamma[\Lambda] = E_\nu[\gamma K(X_{L_B - n})\, 1_\Lambda].$$

Proof. We have

$$P_\gamma[\Lambda] = \sum_{p=n}^{\infty} P_\gamma[X_{p-n} \in A_0, \ldots, X_p \in A_n \cap B, X_q \notin B, q > p]$$

$$= \sum_{p=n}^{\infty} \int_{A_0} \gamma P_{p-n}(\mathrm{d}x_0)\, P_{x_0}[\Gamma],$$

where

$$\Gamma = \{X_1 \in A_1, \ldots, X_n \in A_n \cap B, X_q \notin B, q > n\}.$$

Since $\gamma G^1(A_0) = 0$, it follows that

$$P_\gamma[\Lambda] = \int_{A_0} \gamma G(\mathrm{d}x_0)\, P_{x_0}[\Gamma] = \int_{A_0} \gamma K(x_0)\, P_{x_0}[\Gamma]\, \nu G(\mathrm{d}x_0)$$

$$= \sum_{p=n}^{\infty} \int_{A_0} \nu P_{p-n}(\mathrm{d}x_0)\, \gamma K(x_0)\, P_{x_0}[\Gamma]$$

$$= E_\nu[\gamma K(X_{L_B - n})\, 1_\Lambda].$$

Let μ and ν be two probability measures. The Radon–Nikodym derivative of the restriction of μ to the sub-σ-algebra $\mathscr{B}$ with respect to the restriction of ν to $\mathscr{B}$ will be denoted $(\mathrm{d}\mu/\mathrm{d}\nu)_{\mathscr{B}}$.

Lemma 1.9. *For every k we have*

$$\gamma K(Y_k) = (\mathrm{d}P_\gamma/\mathrm{d}P_\nu)_{\mathscr{H}_k} \quad P_\nu\text{-a.s.}$$

Proof. Let $A_0, A_1, \ldots, A_n$ be in $\mathscr{E}$, and let us compute

$$P_\gamma[\Lambda] = P_\gamma[Y_k \in A_0, Y_{k+1} \in A_1, \ldots, Y_{k+n} \in A_n]$$

$$= \sum_{p \geqslant 0} P_\gamma[L_k = p, Y_k \in A_0, Y_{k+1} \in A_1, \ldots, Y_{k+n} \in A_n].$$

For $p > 0$, one has $L_{k+n} = p + L_{k+n} \circ \theta_p$, hence $Y_{k+n} = Y_{k+n} \circ \theta_p$, on the event $\{L_k = p\}$. By an application of Markov property, we get

$$P_\gamma[\Lambda] = \sum_{p \geqslant 0} \int_E \gamma P_p(\mathrm{d}y)\, P_y[\Gamma],$$

where

$$\Gamma = \{L_k = 0, X_0 \in A_0, Y_{k+1} \in A_1, \ldots, Y_{k+n} \in A_n\}.$$

As in the above lemma, it follows that

$$P_\gamma[\Lambda] = \int_E \nu G(\mathrm{d}y)\, \gamma K(y)\, P_y[\Gamma] + \int_{A_\gamma^{\mathrm{c}}} \gamma G(\mathrm{d}y)\, P_y[\Gamma]$$

$$= E_\nu[\gamma K(Y_k)\, 1_\Lambda] + E_\gamma\left[\sum_p 1_{\{X_p \in A_\gamma^{\mathrm{c}}\}}\, P_{X_p}[\Gamma]\right]$$

$$= E_\nu[\gamma K(Y_k)\, 1_\Lambda] + E_\gamma\left[\left(\sum_p 1_{\{X_p \in A_\gamma^{\mathrm{c}}\}}\right) 1_\Lambda\right];$$

since the event $\bigcup \{X_p \in A_\gamma^{\mathrm{c}}\}$ has zero P_ν-measure, we get the desired result.

Corollary 1.10. *One has*

$$\lim_n \gamma K(Y_n) = (\mathrm{d}P_\gamma/\mathrm{d}P_\nu)_{\mathscr{I}} \quad P_\nu\text{-}a.s.$$

Proof. This is a well-known result in the theory of martingales.

We turn to the main result of this section.

Theorem 1.11. *For every probability measure γ on $\mathscr{E}$, we have*

$$\lim_n \gamma K(X_n) = (\mathrm{d}P_\gamma/\mathrm{d}P_\nu)_{\mathscr{I}}. \quad P_\nu\text{-}a.s.$$

Proof. Set $L = L_B$, where B is any set in $\mathscr{E}$, and define the random variables

Z_n by

$$Z_n = \gamma K(X_{L-n}) \quad \text{on } \{n \leqslant L < \infty\},$$

$$Z_n = 0 \qquad\qquad \text{on } \{L < n\} \cup \{L = \infty\}.$$

We claim that the sequence $\{Z_n\}$ is a supermartingale with respect to the σ-algebras $\mathscr{B}_n = \sigma(X_{L-k}, k \leqslant n)$. Indeed $\mathscr{B}_{n-1}$ is generated by the sets

$$\Lambda = \{X_{L-n+1} \in A_1, \ldots, X_{L-n+j} \in A_j\},$$

where $A_i \in \mathscr{E}$, and we prove that

$$E_\nu[Z_n 1_\Lambda] \leqslant E_\nu[Z_{n-1} 1_\Lambda].$$

Since $P_\nu[\bigcup \{X_n \in A_\gamma^c\}] = 0$, we may take $A_i \subset A_\gamma$ for every i, and by Lemma 1.8 we then have

$$E_\nu[Z_n 1_\Lambda] = E_\nu[Z_n 1_\Lambda 1_{\{n\leqslant L<\infty\}}] = E_\gamma[1_\Lambda 1_{\{n\leqslant L<\infty\}}] \leqslant E_\gamma[1_\Lambda 1_{\{n-1\leqslant L<\infty\}}]$$
$$= E_\nu[Z_{n-1} 1_\Lambda 1_{\{n-1\leqslant L<\infty\}}] = E_\nu[Z_{n-1} 1_\Lambda].$$

Now let a, b be two real numbers with $a < b$, and let $\xi_a^b(L)$ denote the number of upcrossings of the interval $]a, b[$ by $\{Z_n\}$. It is known from martingale theory that

$$E_\nu[\xi_a^b(L)] \leqslant (b-a)^{-1}\left(a + \sup_k E_\nu[Z_k]\right),$$

which implies that

$$E_\nu[\xi_a^b(L)] \leqslant (b-a)^{-1}(a + E_\nu[Z_0]) = (b-a)^{-1}(a + P_\gamma[L < \infty])$$
$$\leqslant (b-a)^{-1}(a+1).$$

The number ξ_a^b of upcrossings of $]a, b[$ by the sequence $\{\gamma K(X_n)\}$ is the increasing limit of the sequence $\{\xi_a^b(L_k)\}$. Consequently $E_\nu[\xi_a^b] < \infty$ for every pair (a, b). By the classical argument in martingale theory it follows that $\lim_n \gamma K(X_n)$ exists P_ν-a.s., and the limit variable is identified by means of Corollary 1.10.

Remark. Exercise 1.16 hints at the idea which lies beneath the above proof.

We may now state

Corollary 1.12. *If* $P_x \ll P_\nu$ *on* $\mathscr{I}$ *for every* $x \in E$, *then* γK *is a regular function for every probability measure* γ.

Proof. The event

$$\Lambda = \left\{\lim_n \gamma K(X_n) = (\mathrm{d}P_\gamma/\mathrm{d}P_\nu)_{\mathscr{I}}\right\}$$

is plainly an invariant event, and by the previous results $P_\nu[\Lambda^c] = 0$. If $P_x \ll P_\nu$ on $\mathscr{T}$, we have $P_x[\Lambda^c] = 0$ for every x in E, which is the desired result.

The following result gives a useful criterion to decide whether the condition in the above corollary is satisfied or not. We write, according to ch. 1 § 5,

$$P_n(x, \cdot) = P_n^0(x, \cdot) + P_n^1(x, \cdot)$$

for the Lebesgue decomposition of P_n with respect to ν. If $\nu P \ll \nu$, it is easily checked that $P_{n+m}^1 \leqslant P_n^1 P_m^1$, and therefore that the sequence $\{P_n^1(\cdot, E)\}$ decreases to a measurable function g. We then have

Theorem 1.13. *If* $\nu P \ll \nu$, *then for any* x *in* E *the following two conditions are equivalent*: (i) $P_x \ll P_\nu$ *on* $\mathscr{I}$; (ii) $g(x) = 0$.

Proof. We first show that (ii) implies (i). Let Γ be an invariant event such that $P_\nu[\Gamma] = 0$. The bounded harmonic function $P.[\Gamma]$ vanishes ν-a.e., and therefore for every n,

$$P_x[\Gamma] = \int_E P_n(x, \mathrm{d}y)\, P_y[\Gamma] = \int_E P_n^1(x, \mathrm{d}y)\, P_y[\Gamma] \leqslant P_n^1(x, E).$$

It follows that if $g(x) = 0$, then $P_x \ll P_\nu$ on $\mathscr{I}$.

To prove the converse let us denote by A^n a set in $\mathscr{E}$ such that $\nu(A^n) = 0$ and $P_n^1(x, (A^n)^c) = 0$, and set $A = \bigcup_n A^n$. We have

$$g(x) = \lim_n P_x[X_n \in A^n] \leqslant P_x[\overline{\lim}\{X_n \in A\}].$$

The event $\Lambda = \overline{\lim}\,\{X_n \in A\}$ is invariant, and since $\nu(A) = 0$ we have $\nu G(A) = 0$; hence $P_\nu[\Lambda] = 0$. If $P_x \ll P_\nu$, it follows that $P_x[\Lambda] = 0$, hence that $g(x) = 0$ and thus (i) implies (ii).

Remark 1.14. It is now readily seen that the condition of Corollary 1.12 is satisfied if $P(x, \cdot) \ll \nu$ for every $x \in E$, hence for discrete chains; it is also satisfied for spread-out random walks if ν is equivalent to a Haar measure.

We now apply the above results in a way which will be useful in the next section. We assume henceforth that X is in duality with a chain $\hat{X}$ with respect to a measure m.

Proposition 1.15. *For every pair f, g of functions in $\mathscr{L}^1_+(m)$, we have*

$$\lim_n \hat{G}f(X_n)/\hat{G}g(X_n) = (\mathrm{d}P_{fm}/\mathrm{d}P_{gm})_{\mathscr{I}} \quad P_m\text{-a.s.},$$

and hence P_ν-a.s. for every $\nu \ll m$.

If, in addition, $P_x \ll P_m$ on $\mathscr{I}$ for every x in E, the convergence holds almost surely.

Proof. Let g' be a strictly positive function in $\mathscr{L}^1(m)$ with $m(g') = 1$, and let f be any function in $\mathscr{L}^1(m)$ with $m(f) = 1$. In Theorem 1.11, we can take $\nu = g'm$ and $\gamma = fm$, and then $\gamma K = \mathrm{d}(fm)\, G/\mathrm{d}(g'm)\, G$. But $(g'm)\, G \ll m$, and $\hat{G}g'$ is a version of the Radon–Nykodim derivative $\mathrm{d}(g'm)\, G/\mathrm{d}m$. By the well-known properties of these derivatives we thus have $\gamma K = \hat{G}f/\hat{G}g'$ and it follows that

$$\lim_n \{\hat{G}f(X_n)/\hat{G}g'(X_n)\} = (\mathrm{d}P_{fm}/\mathrm{d}P_{g'm})_{\mathscr{I}} \quad P_{g'm}\text{-a.s.},$$

and hence P_m-a.s., since g' is strictly positive. Now if g is any function in $\mathscr{L}^1_+(m)$, by considering the ratios $(\hat{G}f/\hat{G}g')/(\hat{G}g/\hat{G}g')$ we get the general result. The last statement is obtained in the same way as Corollary 1.12.

Exercise 1.16. Assume X to be discrete, and retain the notation of Theorem 1.11. Prove that the sequence of random variables $\{\hat{X}_n\}$ defined by

$$\hat{X}_n = X_{L-n} \quad \text{on } \{n \leqslant L < \infty\}, \qquad \hat{X}_n = \Delta \quad \text{on } \{L < n\} \cup \{L = \infty\}$$

is a homogeneous Markov chain for the probability measure P_ν with starting measure $P_\nu[\hat{X}_0 = x] = \nu G(x)\, g_A(x)$ (cf. ch. 2, Proposition 3.5) and transition probability $\hat{P}$ defined by the identity

$$\nu G(x)\, \hat{P}(x, y) = \nu G(y)\, P(y, x).$$

The reader will observe that P and $\hat{P}$ are in duality with respect to νG.

Exercise 1.17. Let ν be a probability measure such that $P_x \ll P_\nu$ on $\mathscr{T}$ for every x in E.

(1) Prove that the space $\mathscr{L}$ of a.s. bounded invariant random variables is a Banach algebra when endowed with the norm

$$||Z|| = \sup_{x \in E} ||Z||_{L^\infty(\Omega, \mathscr{I}, P_x)}.$$

(2) We endow the space b$\mathscr{H}$ of bounded harmonic functions with the following multiplication

$$(h * h')(x) = E_x[R(h)\, R(h')].$$

Prove that b$\mathscr{H}$ is a Banach algebra for the supremum norm. Prove that the map $h \to R(h)$ is a norm-preserving isomorphism between the Banach algebras b$\mathscr{H}$ and $\mathscr{L}$.

Exercise 1.18. Let X be a right random walk of law μ on a group $\boldsymbol{G}$.

(1) Prove that the results in the preceding exercise are still true even if μ is not spread-out.

(2) Prove that the space $\mathscr{A}$ of left uniformly continuous, harmonic functions is a closed subalgebra of b$\mathscr{H}$, hence a C^*-algebra.

(3) The spectrum of $\mathscr{A}$ is denoted by Π_μ and called the *Poisson space* of μ. It is a compact and metrizable $\boldsymbol{G}$-space. There is an isometry j from $C(\Pi_\mu)$ onto $\mathscr{A}$. If we set $\langle \nu, f \rangle = j(f)(e)$, then ν is a probability measure on Π_μ called the Poisson kernel of μ and

$$j(f)(g) = \int_{\Pi_\mu} f(gx)\, \nu(\mathrm{d}x), \qquad f \in C(\Pi_\mu).$$

The measure ν is invariant by μ ($\mu * \nu = \nu$).

(4) If μ is spread out, then ν is absolutely continuous with respect to some quasi-invariant measure η. Prove that η may be chosen such that the integral formula of (3) defines an isomorphism between the Banach spaces b$\mathscr{H}$ and $L^\infty(\Pi_\mu, \eta)$. If $\boldsymbol{G} = \mathrm{SL}(2, \boldsymbol{R})$ and $T_\mu = \boldsymbol{G}$ then Π_μ is equal to $\boldsymbol{G}/\boldsymbol{K} = \boldsymbol{S}^1$, where $\boldsymbol{K}$ is the compact sub-group of rotations.

Exercise 1.19. In the setting of Theorem 1.13 prove that $g(x)$ is equal to the norm of the singular part of the restriction of P_x to $\mathscr{I}$ with respect to the restriction of P_ν to $\mathscr{I}$.

2. Convergence to the boundary

In this section the following set of hypotheses will be in force:

(i) the space E is an LCCB;

(ii) there exists a transition probability $\hat{P}$ in duality with P with respect to a Radon measure m;

(iii) for any $f \in C_b$, the function $\hat{P}f$ is continuous;

(iv) $\hat{G}f$ is bounded and continuous whenever f is in C_K;

(v) $P1 = 1$, in other words 1 is a harmonic function.

Let us observe that if $\hat{P}1 = 1$, it is enough that $\hat{P}f \in C_b$ for $f \in C_K^+$ to entail that $\hat{P}_n g \in C_b$ for any $g \in C_b$ and every n, and in particular property (iii).

These hypotheses may always be achieved if E is countable, since in that case we can take $m = \nu G$ for any probability measure ν such that $\nu G(x) > 0$ for every x in E, and set

$$\hat{P}(x, y) = \nu G(y)\, P(y, x)/\nu G(x).$$

They are also in force for transient random walks on groups.

Definition 2.1. A *reference function* is a positive m-integrable function r such that $\hat{G}r$ is bounded, continuous and strictly positive.

By condition (iv) above, there are plenty of reference functions, but we shall prove a more precise result. We recall that a Borel function f is said to be *locally integrable* if, for every compact K, we have $\langle m, |f|\, 1_K \rangle < \infty$. If E is countable, every finite function is locally integrable. For a spread-out random walk all m-a.e. finite superharmonic functions are locally integrable. (See Exercise 2.13.)

Lemma 2.2. *For every locally integrable superharmonic function f, there is a reference function r such that $m(fr) < \infty$.*

Proof. Let $\{\varphi_n\}$ be an increasing sequence in C_K^+ with limit 1 at every point in E. Then let C_n be the maximum among the numbers $m(\varphi_n)$, $m(\varphi_n f)$, $\|\hat{G}\varphi_n\|$; it is easily checked that

$$r = \sum_{n=1}^{\infty} C_n^{-1}\, 2^{-n}\, \varphi_n$$

is the required reference function.

Remark. The particular reference function just constructed is continuous and strictly positive. All rm-integrable functions are then locally integrable.

Proposition 2.3. *If r is a reference function, then for every $f \in C_K^+$, the function $\hat{K}f = \hat{G}f/\hat{G}r$ is bounded and continuous. Furthermore if $\{f_n\}_{n \geqslant 1}$ is dense in C_K*

for the topology of uniform convergence on compacta then $\{\hat{K}f_n\}_{n\geqslant 1}$ *is dense in* $\{\hat{K}f, f \in C_K\}$ *for the topology of uniform convergence.*

Proof. That $\hat{K}f$ is continuous is clear. Since we have $\hat{G}r > 0$, there exists a constant $a > 0$ such that $\hat{G}r \geqslant a$ on $\{f > 0\}$. Consequently

$$\hat{G}f \leqslant a^{-1}\|\hat{G}f\|\, \hat{G}r \quad \text{on } \{f > 0\},$$

and hence everywhere, by the complete maximum principle.

Let now f be in C_K and $\{f_k\}$ a subsequence of $\{f_n\}$ converging uniformly to f and such that the sets $\{f_k > 0\}$ are contained in a fixed compact set K. Since

$$\|\hat{G}(f - f_k)\| \leqslant \|f - f_k\|\, \hat{G}(\cdot, K),$$

for any $\varepsilon > 0$ we may find k such that $\|\hat{G}(f - f_k)\| < \varepsilon$. The latter half of the statement is then easily proved by the same argument as the former half.

We shall denote by $\hat{K}(x, \cdot)$ the kernel $\hat{G}(x, \cdot)/\hat{G}r(x)$. The measures $\hat{K}(x, \cdot)$ are co-excessive Radon measures such that $\langle \hat{K}(x, \cdot), r\rangle = 1$, and it follows from the above proposition that the set $\{\hat{K}(x, \cdot), x \in E\}$ is vaguely relatively compact.

We are now ready to construct the Martin boundary of E. We first choose a metric δ on E, compatible with the topology of E, and such that the Cauchy completion of (E, δ) is obtained by adjoining the Alexandrov's point Δ.

We next choose a reference function r such that $m(r) = 1$, and a sequence $\{f_n\}_{n\geqslant 1}$ dense in C_K. For x, y in E, we set

$$d(x, y) = \sum_{n=1}^{\infty} W_n |\hat{K} f_n(x) - \hat{K} f_n(y)|,$$

where the positive real numbers W_n are chosen such that $\sum_1^\infty W_n\|\hat{K}f_n\| < \infty$. Then $\delta + d$ is a metric on E.

Definition 2.4. We define E^* to be the Cauchy completion of the metric space $(E, \delta + d)$ and set $M = E^*\backslash E$. The set E^* is called the *Martin space* for X started with distribution rm, the set M the *Martin exit boundary* for X started with distribution rm.

The pseudo-metric d generates the uniformity relative to which all the functions $\hat{K}f$, $f \in C_K$, are uniformly continuous; consequently the space E^* depends neither on the choice of $\{f_n\}$ nor on the choice of $\{W_n\}$, and each function $\hat{K}f$ may be extended continuously to E^*; the extended function will still be denoted by $\hat{K}f$. Let x be a point in M; a sequence $\{x_n\}$ of points in E converges to x if and only if $x_n \to \Delta$ in the original topology of E and $\hat{K}f(x_n) \to \hat{K}f(x)$ for every $f \in C_K$.

Proposition 2.5. *The space E^* is compact; E is a dense open subset of E^* and its relative topology coincides with the original topology.*

Proof. By the choice of δ, and since the functions $\hat{K}f_n$ are bounded, from every sequence of points in E^* one may, by use of Cantor's diagonal process, extract a subsequence $\{x_k\}$ which is Cauchy for δ and such that $\{\hat{K}f_n(x_k)\}$ is Cauchy in $\boldsymbol{R}$ for every n. The subsequence $\{x_k\}$ is thus Cauchy in $(E^*, \delta + d)$, hence it is convergent, and E^* is a compact metric space.

That E is dense in E^* follows from the definition of E^*; furthermore, since $d + \delta \geqslant \delta$, the original topology is coarser than the relative topology, and E is open in E^*. On the other hand, the functions $\hat{K}f$ being continuous, a sequence of points converging to a point in E in the original topology converges in the relative topology, which proves that the latter is finer than the former. The proof is thus complete.

Remark. The space E^* may depend on the choice of the reference function r; in other words, it may happen that the spaces E^* obtained for two different functions are not homeomorphic.

It is now plain that Borel sets in E are Borel sets in E^*. Furthermore, for every point x in M the map $f \to \hat{K}f(x)$, $f \in C_K$, is clearly a Radon measure on E which will be denoted by $\hat{K}(x, \cdot)$. The extended function $\hat{K}$ is a kernel on E^*, the measures $\hat{K}(x, \cdot)$ vanishing outside E. Finally, if $\{x_n\}$ is a sequence of points in E converging to a point x in M, the measure $\hat{K}(x, \cdot)$ is the limit of the measures $\hat{K}(x_n, \cdot)$ in the vague topology on E. This entails that $\hat{K}r(x) \leqslant 1$ for every x in E^*, and also

Proposition 2.6. *The measures $\hat{K}(x, \cdot)$, $x \in M$, are co-excessive.*

Proof. For $f \in C_K^+$, $\hat{P}f$ is continuous; hence if $\{\mu_n\}$ is a sequence of Radon measures converging vaguely to a measure μ, then

$$\langle \mu, \hat{P}f \rangle \leqslant \varliminf_n \langle \mu_n, \hat{P}f \rangle.$$

The proposition then follows from the previous discussions.

As a result, if σ is a probability measure on the Borel sets of M, the measure $\sigma\hat{K}$ is a co-excessive Radon measure and $\langle\sigma\hat{K}, r\rangle \leqslant 1$. This will be of significance in the forthcoming results, which are the main results of this section.

We shall call Λ the set of trajectories ω such that the sequence $\{X_n(\omega)\}$ has a limit in E^* as $n \to \infty$. Since E^* is compact and metrizable, it is easily seen that Λ is measurable, and moreover, it is an invariant event. Let us pick an arbitrary point x in M and set

$$X_\infty(\omega) = \lim_n X_n(\omega) \quad \text{if } \omega \in \Lambda, \qquad X_\infty(\omega) = x \quad \text{if } \omega \in \Lambda^c.$$

We thus define an invariant random variable X_∞, and in the sequel the statement $\{X_n\}$ converges to X_∞ P_ν-a.s. will mean that $P_\nu[\Lambda] = 1$. We are now ready to state

Theorem 2.7. *The sequence $\{X_n\}$ converges P_m-a.s. to X_∞ and the image σ of P_{rm} by X_∞ is carried by M. Furthermore $m = \sigma\hat{K}$; in other words, for every $f \in C_K$*

$$m(f) = \int_M \hat{K} f(s)\, \mathrm{d}\sigma(s).$$

Finally if $P_x \ll P_m$ on $\mathscr{I}$ for every x in E, the above convergence holds almost surely.

Proof. Since the potentials of compact sets are bounded it follows that $\{X_n\}$ converges almost surely to Δ in the topology of E, hence for the metric δ. The convergence in E^* P_m-a.s. is then a consequence of Proposition 1.15 and of the definition of E^*. The last sentence of the statement is shown in exactly the same way. The equality $m = \sigma\hat{K}$ is shown by the following set of equations, where f ranges through C_K^+:

$$\int_M \hat{K} f\, \mathrm{d}\sigma = \int_\Omega \hat{K} f(X_\infty)\, \mathrm{d}P_{rm} = \int_\Omega (\mathrm{d}P_{fm}/\mathrm{d}P_{rm})_{\mathscr{I}}\, \mathrm{d}P_{rm} = m(f).$$

Finally σ vanishes on E because $X_\infty \in M$ P_{rm}-a.s.

The preceding theorem may be generalized as follows. In the remainder of this chapter, we shall use "harmonic function" to mean "positive harmonic function", but at variance with the convention of ch. 2, we will allow these functions to take infinite values on a negligible set. We denote by $\mathscr{H}_r$ the set of harmonic functions h such that $m(rh) \leqslant 1$. For such a function we set $E^h = \{0 < h < \infty\}$. We define a new T.P. on $(E^h, E^h \cap \mathscr{E})$ by setting for $x \in E^h$ and $A \in E^h \cap \mathscr{E}$,

$$P^h(x, A) = h(x)^{-1} \int_A P(x, \mathrm{d}y)\, h(y).$$

Plainly, $(P^h)^n = I_{h^{-1}} P_n I_h$, and a function f is P^h-(super) harmonic if and only if hf is P-(super) harmonic. In particular 1 is P^h-harmonic. It is also easily seen that P^h is in duality with $\hat{P}$ with respect to the measure hm. The functions $\hat{G}f/\hat{G}r$ are therefore the same for P and P^h, and the preceding theorem will be applied to the chains with transition probability P^h.

Such a chain may be constructed on the space $(\Omega, \mathscr{F})$ of the chain X. For any positive measure ν on E^h, we call P^h_ν the measure on $(\Omega, \mathscr{F})$ for which $\{X_n\}$ is a homogeneous Markov chain with transition probability P^h and starting measure $h\nu$. For any finite collection A_i, $0 \leqslant i \leqslant n$, of sets in $\mathscr{E}$, one has

$$P^h_\nu[X_0 \in A_0, X_1 \in A_1, \ldots, X_n \in A_n]$$
$$= \int_{A_0} \nu(\mathrm{d}x_0)\, E_{x_0}[1_{A_1}(X_1) \cdots 1_{A_n}(X_n)\, h(X_n)].$$

In this formula, E_{x_0} is the usual expectation associated with the probability measure P_{x_0}; in other words, if $h = 1$, then $P^h = P$ and $P^h_\nu = P_\nu$. It is clear that $P^h_\nu[X_n \notin E^h] = 0$ for every integer n. Finally, if h and k are two harmonic functions, then $P^{h+k}_\nu = P^h_\nu + P^k_\nu$.

Theorem 2.8. *Let h be a locally integrable function in $\mathscr{H}_r$. The sequence $\{X_n\}$ converges P^h_m-a.s. to X_∞; the image σ^h of P^h_{rm} by X_∞ is carried by M and $hm = \sigma^h \hat{K}$. Finally if $P^h_z \ll P^h_m$ on $\mathscr{I}$ for every x in E, the convergence holds P^h_x-a.s. for every x in E^h.*

Proof. We cannot immediately apply the preceding theorem, because E^h is no longer LCCB.

By Proposition 1.15 we still have the convergence P^h_{rm}-a.s. of the sequences $\hat{K}f(X_n)$. It remains therefore to prove that X_n converges P^h_{rm}-a.s. to Δ in E. Let K be any compact subset of E; calling G^h the potential kernel of P^h, it is easily seen that $h\, G^h(\cdot, K) = G(h\, 1_K)$, and therefore, since h is locally integrable,

$$\langle hr, G^h 1_K \rangle_m = \langle r, G(h\, 1_K) \rangle_m = \langle \hat{G}r, h\, 1_K \rangle_m < \infty.$$

As a result, $P^h_{rm}[R(K)] = 0$ and $\{X_n\}$ converges P^h_{rm}-a.s. to Δ. The first part of the statement then follows from the definitions of E^* and M; the equality $hm = \sigma^h \hat{K}$ is derived as was the equality $m = \sigma \hat{K}$ of Theorem 2.7.

Remarks. (1) The measure σ of Theorem 2.7 thus appears as a special case of σ^h obtained when $h = 1_E$.

(2) The map $h \to \sigma^h$ is an affine map from the cone of locally integrable functions in $\mathscr{H}_r$ into the space of bounded positive measures on M, and if $h \leqslant k$ rm-a.e., then $\sigma^h \leqslant \sigma^k$.

We conclude this section by giving the probabilistic version of the classical Fatou's theorem in the unit disc and some of its applications.

Theorem 2.9. *If Gg is a finite potential, then $\lim_n Gg(X_n) = 0$ a.s. If h is a locally integrable harmonic function in $\mathscr{H}_r$, and if $\sigma^h = u\sigma + \sigma^h_s$ is the Lebesgue decomposition of σ^h with respect to σ, then*

$$\lim_n h(X_n) = u(X_\infty) \quad P_m\text{-}a.s.$$

This convergence holds a.s. if $P_x \ll P_m$ on $\mathscr{I}$ for every x in E.

Proof. Since $Gg(X_n)$ is a positive supermartingale, the limit exists a.s. by martingale theory. Now by Fatou's lemma, for every x in E,

$$E_x[\lim Gg(X_n)] \leqslant \lim E_x[Gg(X_n)] = \lim P_n Gg(x) = 0;$$

the proof of the first part of the statement is thus complete.

For the second part, let us observe that, by the definition of σ^h, we have for any f in C_K^+,

$$m(fh) \geqslant \int_M \hat{K}f(s)\, u(s)\, \sigma(\mathrm{d}s) = \int_\Omega \hat{K}f(X_\infty)\, u(X_\infty)\, \mathrm{d}P_{rm}$$
$$= \int_\Omega u(X_\infty)\, \mathrm{d}P_{fm}.$$

Set $h' = E_\cdot[u(X_\infty)]$; the function h' is harmonic and since h is locally integrable, $u(X_\infty)$ is P_{fm}-integrable and consequently $h'(X_n)$ converges P_{fm}-a.s. and in $L^1(P_{fm})$ to $u(X_\infty)$. Since this is true for every $f \in C_K^+$, we get

$$\lim h'(X_n) = u(X_\infty) \quad P_m\text{-a.s.}$$

Furthermore

$$m(fh') = m(f \cdot P_n h') = E_{fm}[h'(X_n)],$$

and because of the L^1-convergence above, we may pass to the limit to get

$$m(fh') = E_{fm}[u(X_\infty)] = \int_\Omega (\mathrm{d}P_{fm}/\mathrm{d}P_{rm})_{\mathscr{I}}\, u(X_\infty)\, \mathrm{d}P_{rm}$$

$$= \int_M \hat{K}f \cdot u\, \mathrm{d}\sigma;$$

since the functions $\hat{K}f$, $f \in C_K^+$, separate M we thus obtain that $\sigma^{h'} = u\sigma$. As a result $h' \leqslant h$ m-a.e.

Now the function $h' \wedge h$ is superharmonic; let $Gg + h''$ be its Riesz decomposition. Since $m\{h' > h\} = 0$, it is easily derived that

$$\lim_n (h' \wedge h)(X_n) = u(X_\infty) \quad P_m\text{-a.s.},$$

and by the first part of the theorem $\lim_n Gg(X_n) = 0$; consequently

$$\lim_n h''(X_n) = u(X_\infty) \quad P_m\text{-a.s.}$$

Since $h'' \leqslant h$ everywhere, we may write $h = h'' + k$, where k is harmonic, and from the relation $P^h_{rm} = P^{h''}_{rm} + P^k_{rm}$, we derive that $\sigma^k = \sigma^h_s$. To complete the proof of the theorem it suffices to show that if σ^h and σ are mutually singular, then $\{h(X_n)\}$ converges to zero P_m-a.s.

Let h be such that σ and σ^h are mutually singular, and call h' the harmonic part in the Riesz decomposition of $h \wedge 1$. Since $h' \leqslant h$ and $h' \leqslant 1$, we have $\sigma^{h'} \leqslant \sigma^h$ and $\sigma^{h'} \leqslant \sigma$, which entails that $\sigma^{h'} = 0$, hence $h' = 0$ m-a.e. It follows by a simple application of Fatou's lemma that

$$\lim_n h'(X_n) = 0 \quad P_m\text{-a.e.},$$

and by the first part of the theorem,

$$\lim_n (h \wedge 1)(X_n) = 0 \quad P_m\text{-a.e.}$$

As a result, $\lim_n h(X_n) = 0$, and the proof is complete.

Corollary 2.10. *If r is continuous and strictly positive, the map $u \to u(X_\infty)$ is a positive isometry from $L^1(M, \sigma)$ onto $L^1(\Omega, \mathscr{I}, P_{rm})$.*

Proof. Clearly, if u is in $L^1(M, \sigma)$, then $u(X_\infty)$ is in $L^1(\Omega, \mathscr{I}, P_{rm})$ and

$$\|u\|_{L^1(\sigma)} = \|u(X_\infty)\|_{L^1(P_{rm})}.$$

Now let Z be in $L^1_+(\Omega, \mathscr{I}, P_{rm})$; the harmonic function $h = E_.[Z]$ is in $L^1(rm)$, hence is locally integrable, and therefore there exists an element u of $L^1(M, \sigma)$ such that $\sigma^h = u\sigma + \sigma^h_s$. By the previous results, we know that $h(X_n)$ converges to $u(X_\infty)$ P_{rm}-a.s., and on the other hand

$$h(X_n) = E_{X_n}[Z] = E_{rm}[Z \mid \mathscr{F}_n],$$

which by the martingale theorem converges to Z P_{rm}-a.s. It follows that $Z = u(X_\infty)$ P_{rm}-a.s., and the proof is complete.

The following result is important. It asserts that the Banach space of bounded harmonic functions is isomorphic to the space L^∞ of a probability measure defined on a Borel subset of a compact metrizable space. This provides a generalization of what is known for spread-out random walks on groups. (See Exercise 1.18.)

Theorem 2.11. *If $P_x \ll P_m$ on $\mathscr{I}$ for every x in E and, if r is continuous and strictly positive, the Banach space* b$\mathscr{H}$ *of bounded harmonic functions is isomorphic to the Banach space $L^\infty(M, \sigma)$.*

Proof. It was seen in ch. 2 §3 that the vector space b$\mathscr{H}$ is isomorphic to the vector space of a.s. bounded invariant random variables. Since $P_x \ll P_{rm}$ on $\mathscr{I}$ for every x in E, the equivalence a.s. for invariant random variables is the same as equivalence P_{rm}-a.s. As a result, b$\mathscr{H}$ is isomorphic to $L^\infty(\Omega, \mathscr{I}, P_{rm})$ under the mapping $h \to R(h)$, whose inverse mapping is $Z \to E_.[Z]$. If $|Z| < c$ P_{rm}-a.s., it follows that $|Z| < c$ P_x-a.s. for every x in E, and clearly these two mappings are norm-decreasing; they are therefore Banach isomorphisms.

It remains to prove that the Banach space $L^\infty(\Omega, \mathscr{I}, P_{rm})$ is isomorphic to $L^\infty(M, \sigma)$. This is done by using the map $u \to u(X_\infty)$ in the same way as in the above corollary.

Remark. If we compare this result with Theorem 1.5, we see that there is a one-to-one correspondence between the Borel subsets of M, up to σ-equivalence, and the equivalence classes of regular sets.

We close this section with a few comments on what can be said if 1 is no longer a harmonic function. In that case, the chain may "disappear" at a finite time and is then placed in the cemetery Δ, which we shall consider as an isolated point. If the other hypotheses of this section are in force we may construct E^* and M as before and prove Theorem 2.8; the measure σ^h will be carried by M. The boundary may be viewed as the union of M and the isolated point Δ, and the image of P_{rm} by X_∞ will be equal to $P_{rm}[\zeta < \infty]\,\varepsilon_\Delta + \sigma^h$, where h is the harmonic part of the Riesz decomposition of 1.

Exercise 2.12. With the notation of Theorem 2.9, prove that the following two conditions are equivalent:

(i) $h(X_n)$ converges in $L^1(P_{rm})$;

(ii) the measure σ_s^h vanishes.

These two conditions are satisfied if h is bounded.

Exercise 2.13. Prove that, if X is a spread-out random walk, every a.e. finite superharmonic function is locally integrable.

[*Hint*: Choose n such that $\mu^n > cm$ on an open set V. For any x in G, there is a point y such that $f(y) < \infty$ and $x \in yV$. Prove that $\int_{yV} f\,dm < \infty$.]

Exercise 2.14. Let X be a right random walk of law μ on a group G.

(1) Prove that the following two conditions are equivalent:

(i) there exist reference functions in C_K^+;

(ii) there exists a compact set K such that $G = K\,T_\mu$.

(2) Assume henceforth that the conditions of (1) are satisfied. Prove that the space E^* obtained for different reference functions in C_K^+ are homeomorphic G-spaces. Prove that in Theorem 2.7 the convergence holds a.s. even if X is not spread out.

(3) Let ν be a probability measure on G equivalent to m. Call σ_ν the image of P_ν by X_∞ and N the support of σ_ν. Prove that N does not depend on ν and is a compact G-space.

(4) From now onwards X is assumed to be spread out. Prove that there is an isomorphism $h \to \tilde{h}$ between the Banach spaces $b\mathscr{H}$ of bounded harmonic functions and $L^\infty(N, \sigma_\nu)$ such that

$$\tilde{h}(X_\infty) = \lim h(X_n) \quad \text{a.s.} \quad \text{and } h(g) = \langle g\gamma, \tilde{h}\rangle,$$

where γ is the image of P_e by X_∞. Compare with Exercise 1.18.

(5) Assume further that $\mu \ll m$, and prove that there exists a function k on $G \times N$ such that $k(\cdot, x)$ is invariant for σ_ν-almost every x in N, and if

$h \in \mathrm{b}\mathscr{H}$, then

$$h(g) = \int_N k(g, x)\, h(x)\, \mathrm{d}\sigma_\nu(x).$$

Exercise 2.15. (1) Prove that for any pair of points x, y in E, the equality $\hat{K}(x, \cdot\,) = \hat{K}(y, \cdot\,)$ holds if and only if $x = y$.

(2) Prove that there exists a unique (up to homeomorphism) compact metrizable space F such that: (i) E is a dense subset of F and the relative topology of E is coarser than its original topology; (ii) the functions $\hat{K}f, f \in C_K$, extend to continuous functions on F which separate F.

(3) State and prove a theorem of convergence to the boundary in F.

Exercise 2.16. Denote by (a, b), $a > 0$, b real, the generic point in the group of positive affine motions of the real line. Let X be *a right* random walk of law μ and suppose that

$$\int |\log a|\, \mathrm{d}\mu(a, b) < \infty, \quad \int \log a\, \mathrm{d}\mu(a, b) < 0, \quad \int \log(1 + b)\, \mathrm{d}\mu(a, b) < \infty.$$

(1) Prove that $\{X_n\}$ converges P_e-a.s. to $(0, Z)$ where Z is a real random variable. The group being identified with the part $\{a > 0\}$ of the plane, the line $\{a = 0\}$ may obviously be seen as a boundary and the foregoing is thus a theorem of convergence to the boundary.

[*Hint*: One may use the fact that if $\{U_n\}$ is a sequence of positive, independent and equally distributed random variables such that $E[\log(1 + U_1)] < \infty$, then $\lim U_n^{1/n} = 1$ P_e-a.s.]

(2) Prove further that if ν is the law of Z, then $\mu * \nu = \nu$ and deduce therefrom that there are non-constant bounded continuous μ-harmonic functions.

[*Hint*: For the last point, see ch. 2, Exercise 4.18.]

Exercise 2.17. Suppose that Gg is a strictly positive potential and that $m(rGg) \leqslant 1$. Define P^{Gg} as P^h was defined in the text and prove a theorem of convergence similar to Theorem 2.8. What will be the distribution of X_∞?

3. Integral representation of harmonic functions

In the preceding section, we have associated with each positive locally integrable harmonic function h of $\mathscr{H}_r$ a measure σ^h on the boundary. As a result, the co-invariant Radon measure hm had an integral representation as the image of σ^h by the kernel $\hat{K}$. Conversely, if η is a bounded measure on the boundary, $\eta\hat{K}$ is a co-excessive Radon measure such that $\langle \eta\hat{K}, r\rangle < \infty$. We wish to go deeper into this to get an integral representation of all co-excessive Radon measures—subject to some integrability condition as above—and study the question of the uniqueness of such a representation; then we want to turn this into an integral representation of harmonic functions recalling the classical Poisson representation in the unit disk.

In what follows r is a fixed reference function. We call $\mathscr{S}_r$ the set of co-excessive measures ξ such that $\xi(r) \leqslant 1$. We will have to add another hypothesis to the set of conditions in §2, namely

Condition 3.1. For every x in E, the measures $P(x, \cdot\,)$ and $\hat{P}(x, \cdot\,)$ are absolutely continuous with respect to m.

This condition is obviously in force for discrete chains and for random walks with an absolutely continuous law. It entails the following proposition for which we need some further notation: we denote by M_h the set of points s in M such that $\hat{K}(s, \cdot\,)$ is co-invariant. If M_h is a proper subset of M we will see below that it is a Borel subset of M.

Proposition 3.2. *If ξ is co-invariant then $\xi \ll m$ and there is a unique harmonic function h such that $\xi = hm$. Moreover there is a unique function k on $E \times M_h$ such that*

(i) *k is $\mathscr{E} \otimes \mathscr{E}^*$-measurable;*

(ii) *for every $s \in M_h$, the function $k(\cdot\,, s)$ is in $\mathscr{H}_r$ and $\hat{K}(s, \cdot\,) = k(\cdot\,, s)\, m$;*

(iii) *for any $h \in \mathscr{H}_r$,*

$$h = \int_M k(\cdot\,, s)\, \sigma^h(\mathrm{d}s).$$

Proof. That $\xi \ll m$ is obvious from Condition 3.1. Furthermore if g is any function within the class of $\mathrm{d}\xi/\mathrm{d}m$, it is also clear by duality that $Pg = g$ m-a.e. and that Pg does not depend on g. If we set $h = Pg$, then h is harmonic and $\xi = hm$; the uniqueness of h follows also from Condition 3.1. Finally if ξ is in $\mathscr{S}_r$, then h is in $\mathscr{H}_r$ as is seen by duality.

To prove the second part of the statement, we observe that by Proposition 5.3 in ch. 1, we can choose $g(\cdot\,, s)$ in the equivalence class of $\mathrm{d}\hat{K}(s, \cdot\,)/\mathrm{d}m$ in such a way that the resulting bivariate function is $\mathscr{E} \otimes \mathscr{E}^*$ measurable.

Then using a standard argument, $k(\cdot, s) = Pg(\cdot, s)$ is also seen to be in $\mathscr{E} \otimes \mathscr{E}^*$ and is the desired function.

The third part follows then easily by Theorem 2.8.

We may now turn to the study of the set $\mathscr{S}_r$ and state

Theorem 3.3. *The set $\mathscr{S}_r$ is a compact and convex set for the topology of vague convergence. Moreover for any $\xi \in \mathscr{S}_r$ there is at least one measure σ on $E \cup M$ such that*

$$\xi = \int_{E \cup M} \hat{K}(s, \cdot)\, d\sigma(s).$$

Proof. The set $\mathscr{S}_r$ is obviously convex; it is proved to be compact by the same reasoning as in Proposition 2.6. Let now ξ be in $\mathscr{S}_r$; by the Riesz decomposition theorem in ch. 2 §4, $\xi = \eta\hat{G} + \lambda$ where λ is co-invariant. By Proposition 3.2, there is therefore a harmonic function h in $\mathscr{H}_r$ such that $\xi = \eta\hat{G} + hm$. The measure $\hat{G}r \cdot \eta + \sigma^h$ is the desired measure σ.

Counterexamples (see Exercise 3.17) show that the representation just obtained is not unique. We shall now look for a Borel subset M_e of M such that we get uniqueness provided that we restrict the measures σ to those carried by $E \cup M_e$. Clearly we must exclude the set $M \backslash M_h$; indeed if $\hat{K}(s, \cdot) = \eta\hat{G} + hm$ with $\eta \neq 0$, then $\hat{K}(s, \cdot)$ corresponds to both ε_s and $\hat{G}r \cdot \eta + \sigma^h$.

Proposition 3.4. *The set M_h is a Borel subset of M and $\sigma^h(M \backslash M_h) = 0$ for every $h \in \mathscr{H}_r$.*

Proof. Let $\{f_n\}$ be a countable dense subset of $C_K^+(E)$; the set $M \backslash M_h$ is the union of the sets $\{s: Kf_n(s) > KPf_n(s)\}$ which are clearly Borel subsets of M.

The second half of the statement is obvious.

Before we proceed, we must introduce the following terminology.

Definition 3.5. We call *extremal measures* the extremal points of the compact convex set $\mathscr{S}_r$.

The zero measure is extremal and a non-zero measure ξ is extremal if, whenever $\xi = \xi_1 + \xi_2$ with ξ_1 and ξ_2 in $\mathscr{S}_r$, there is a real $t \in [0, 1]$ such that

$\xi_1 = t\xi$ and $\xi_2 = (1-t)\,\xi$. This is equivalent to saying that if $\xi' \in \mathscr{S}_r$ and $\xi - \xi' \in \mathscr{S}_r$, then there is a real $t \in [0, 1]$ such that $\xi' = t\xi$. In particular if ξ is co-invariant and extremal and if ξ' is co-invariant and $\xi' \leqslant \xi$, then $\xi' = t\xi$ with $t \in [0, 1]$. Finally we observe that if ξ is non-zero and extremal then $\xi(r) = 1$. In the sequel we shall use the term "*extremal*" to mean "*extremal and non-zero*".

From the Riesz decomposition theorem we deduce that an extremal measure is either a co-invariant or a co-potential measure. Moreover we have

Lemma 3.6. *A co-potential measure is extremal if and only if it is equal to* $\hat{K}(y, \cdot\,)$ *with* y *in* E.

Proof. The sufficiency is easy to prove. Let conversely ξ be in $\mathscr{S}_r$ and such that $\hat{K}(y, \cdot\,) - \xi$ be in $\mathscr{S}_r$. The co-invariant part of ξ being bounded above by the co-potential $\hat{K}(y, \cdot\,)$, vanishes identically and $\xi = \eta\hat{G}$ for a positive measure η. It follows that $\hat{K}(y, \cdot\,) - \xi$ is a co-potential, hence $\varepsilon_y - \eta$ a positive measure which implies that η is a multiple of ε_y.

If there is a point s in M_h such that $\hat{K}(s, \cdot\,)$ is not extremal, then the uniqueness property that we are looking for cannot hold. Indeed, we would have $K(s, \cdot\,) = \xi_1 + \xi_2$ with $\xi_1 \neq t\hat{K}(s, \cdot\,)$ and $\hat{K}(s, \cdot\,)$ would correspond to ε_s as well as to $\sigma_1 + \sigma_2 \neq \varepsilon_s$, where σ_i is associated with ξ_i by Theorem 3.3. We shall therefore try to restrict the boundary still further, namely to the points s for which $\hat{K}(s, \cdot\,)$ is co-invariant extremal.

For $s \in M_h$, we can, following §2, associate with the harmonic function $k(\,\cdot\,, s)$ the probability measures $P_\nu^{k(\cdot,s)}$ and the measure $\sigma^{k(\cdot,s)}$ which will be denoted for short by P_ν^s and σ^s. We will also write E^s for $E^{k(\cdot,s)}$.

Lemma 3.7. *For any* $\Lambda \in \mathscr{F}$, *the map* $s \to P^s[\Lambda]$ *is* $\mathscr{E}^*$*-measurable and for* $h \in \mathscr{H}_r$,

$$P_\nu^h[\Lambda] = \int_M P_\nu^s[\Lambda]\, \sigma^h(ds).$$

For any Borel subset A *of* M, *the map* $s \to \sigma^s(A)$ *is* $\mathscr{E}^*$*-measurable and*

$$\sigma^h(A) = \int_M \sigma^s(A)\, \sigma^h(ds).$$

Proof. By §2, if $\Lambda \in \mathscr{F}_n$, we have

$$P_\nu^s[\Lambda] = E_\nu[1_\Lambda k(X_n, s)];$$

the result is thus clear for $\Lambda \in \mathscr{F}_n$, and the proof is completed by means of the monotone class theorem.

If now x is in E^h and Λ still in $\mathscr{F}_n$, we have for σ^h almost every s,

$$E_x[1_\Lambda k(X_n, s)] = P_x^s[\Lambda];$$

this is true if $x \in E^s$ by definition of P_x^s; if $k(x, s) = 0$, then since $k(X_n, s)$ is a supermartingale it vanishes P_x-a.s. and finally $\sigma^h\{s: k(x, s) = \infty\} = 0$ since x is in E^h. We can thus write

$$P_x^h[\Lambda] = E_x\left[1_\Lambda \int_M k(X_n, s)\, \sigma^h(\mathrm{d}s)\right]$$

$$= \int_M E_x[1_\Lambda k(X_n, s)]\, \sigma^h(\mathrm{d}s) = \int_M P_x^s[\Lambda]\, \sigma^h(\mathrm{d}s).$$

For ν carried by E^h we get the announced result by integrating with respect to ν, then by using the monotone class theorem.

The second half of the statement is an easy consequence of the first and of the definitions of σ^s and σ^h as the images of P_{rm}^s and P_{rm}^h by X_∞.

Proposition 3.8. *For $s \in M_h$, the measure $\hat{K}(s, \cdot)$ is extremal if and only if $\sigma^s = \varepsilon_s$ and then $P_{rm}^s[X_\infty = s] = 1$.*

Proof. Let us write for short h instead of $k(\cdot\,, s)$ and suppose that g is a non-constant, bounded, P^h-harmonic function. We may suppose that $\|g\| < 1$ and by Condition 3.1 that $m(\{g < 1\}) > 0$. The measure $(gh) \cdot m$ is co-invariant and strictly less than hm. Indeed, remembering that P^h and $\hat{P}$ are in duality with respect to hm we have, for any $f \in \mathscr{E}_+$,

$$\langle ghm, \hat{P}f\rangle = \langle g, \hat{P}f\rangle_{hm} = \langle P^h g, f\rangle_{hm} = \langle ghm, f\rangle.$$

Consequently, if $\hat{K}(s, \cdot)$ is extremal there are no non-constant bounded harmonic functions, and the σ-algebra $\mathscr{I}$ is P_{rm}^h-a.s.trivial. The random variable X_∞ is therefore P_{rm}^h-a.s.constant, hence σ^h is the unit mass at some point in M. As a result of the definition of the topology of E^* and of Theorem 2.8 this unit mass must be ε_s and $P_{rm}^s[X_\infty = s] = 1$.

Conversely, if $\hat{K}(s, \cdot) = \xi_1 + \xi_2$ where ξ_i is co-invariant and not proportional to $\hat{K}(s, \cdot)$, then $\xi_i = h_i m$ where h_i is not proportional to $k(\cdot, s)$ and therefore $\sigma^s = \sigma^{h_1} + \sigma^{h_2}$ cannot be equal to ε_s.

As a consequence we get

Proposition 3.9. *The set M_e of points in M_h such that $\hat{K}(s, \cdot)$ is extremal is a Borel subset of M, and for any $h \in \mathscr{H}_r$ we have $\sigma^h(M_e^c) = 0$.*

Proof. By Proposition 3.8, the set M_e is the set of points s in M such that $\sigma^s = \varepsilon_s$. Let $\{B_n\}_{n\geq 0}$ be a countable algebra of sets generating the σ-algebra of Borel subsets of M. Then by the monotone class theorem,

$$M_e = \bigcap_{n=0}^{\infty} \{s : \sigma^s(B_n) = 1_{B_n}(s)\},$$

and by Lemma 3.7, M_e is a Borel set.

Next, from the formula $\sigma^h(A) = \int \sigma^s(A)\, d\sigma^h(s)$ of Lemma 3.7, it follows that

$$\sigma^s(A) = 1_A(s) \quad \sigma^h\text{-a.s.}$$

for every Borel subset A of M. Consequently

$$\sigma^s(B_n) = 1_{B_n}(s) \quad \sigma^h\text{-a.s.}$$

for every $n \geq 0$; hence by the monotone class theorem, the set of points s for which $\sigma^s \neq \varepsilon_s$ has σ^h-measure zero.

We may now state the main result of this section, which is the uniqueness result that we were looking for.

Theorem 3.10. *Let $\xi = \eta\hat{G} + h \cdot m$ be a co-excessive measure in $\mathscr{S}_r$; the measure $\sigma = \hat{G}r \cdot \eta + \sigma^h$ is the unique measure carried by the set $E \cup M_e$ such that*

$$\xi = \int_{E\cup M_e} \hat{K}(x, \cdot)\, d\sigma(x).$$

Proof. By the previous discussions, it suffices to show that if θ is a probability measure on M_e and $h \cdot m = \int_{M_e} \hat{K}(s, \cdot)\, \eta(ds)$ then $\sigma^h = \theta$.

Let A be a Borel subset of M, by the same argument as in Lemma 3.7 we get

$$P^h_{rm}[X_\infty \in A] = \int P^s_{rm}[X_\infty \in A]\ \theta(ds);$$

but since for σ-almost all s, $P^s_{rm}[X_\infty \in A] = 1_A(s)$, we have $P^h_{rm}[X_\infty \in A] = \theta(A)$ and therefore $\sigma^h = \theta$.

From this uniqueness property and the fact that X_∞ is in M_e P^h_x-a.s. for every $h \in \mathscr{H}_\Pi$ and x in E, we see that the set M_e is the *useful part* of the boundary of E. Furthermore, it is often easier to describe the set M_e, which is isomorphic to a set of extremal points of a compact convex set, than the whole set M. In applications, the set M will therefore be forgotten, and the enlarged state space will be $E \cup M_e$. It may happen that it is no longer a compact space, but that causes no inconvenience in general.

Using Proposition 3.2 the above representation theorem may be turned into a representation theorem for harmonic functions.

Corollary 3.11. *For every harmonic function h in $\mathscr{H}_r$ there is one and only one measure σ^h carried by M_e such that*

$$h = \int_{M_e} k(\cdot\,, s)\ \sigma^h(ds),$$

and σ^h is the image of P^h_{rm} by X_∞. Conversely, for any probability measure σ on M_e the integral above defines a function of $\mathscr{H}_r$ and this function is extremal if and only if it is equal to $k(\cdot\,, s)$ for some s in M_e.

This corollary allows us to return to the exit distributions of the chain, namely the distributions of X_∞ under the different probability measures P_x.

Proposition 3.12. *For any harmonic function h in $\mathscr{H}_r$, any x in E^h, and any Borel subset A of M_e,*

$$P^h_x[X_\infty \in A] = \int_A k(x, s)\ \sigma^h(ds).$$

In particular, if σ is the measure on M_e of the representation of the harmonic function 1, then for every $x \in E$,

$$P_x[X_\infty \in A] = \int_A k(x, s)\ \sigma(ds).$$

Proof. Let us start with the case where $h = k(\cdot\,, s)$ for $s \in M_e$. We know from Proposition 3.8 that $\varphi = P_{\cdot}^{s}[X_\infty \in A] = 1_A(s)$ rm-a.e. Moreover $\varphi/k(\cdot\,, s)$ is $P^{k(\cdot,s)}$-harmonic and since this T.P. is easily seen to be absolutely continuous with respect to m, we have the equality everywhere. Consequently the proposition is true when $h = k(\cdot\,, s)$ and the general result follows from Lemma 3.7 and its proof.

It is finally probably worth recording that in the case of discrete chains the function k which appears in the above results may be computed without any reference to a dual chain. From now on we assume that X is discrete. Let ρ be a probability measure such that the potential ρG is bounded and strictly positive on E. By setting

$$m = \rho G \quad \text{and} \quad \hat{P}(x, y) = P(y, x)\, m(y)/m(x)$$

we get a chain $\hat{X}$ in duality with X with respect to m and we have the hypothesis of this and the last sections. If we set $r(x) = \rho(x)/\rho G(x)$ it is easily checked that $\hat{G}r = 1$ and thus r may be used as a reference function. Furthermore a function is integrable with respect to rm if and only it it is integrable with respect to ρ and all finite functions are locally integrable.

We may define the boundary M and the space E^* by using this particular reference function r and for $s \in M_e$, $x \in E$, we get

$$k(x, s) = \hat{K}(s, x)/\rho G(x).$$

We can extend k to $E \times E^*$ by putting

$$k(x, s) = G(x, s)/\rho G(x), \quad x \in E, \quad s \in E^*.$$

It is easily seen that the uniformity relative to which the functions $\hat{K}f$, $f \in C_K$ are uniformly continuous is the same as the uniformity relative to which the functions $k(x, \cdot)$ are uniformly continuous. As a result, the space E^* could have been defined by using the functions $k(x, \cdot)$ instead of the functions $\hat{K}f$. A sequence $\{s_n\}$ of points in E converges to a point s of E if and only if either s is in E and $s_n = s$ for n sufficiently large, or s is in M and $k(x, s_n) \to k(x, s)$ for every x in E. It thus turns out that the ancillary dual chain used above disappears completely and may therefore be forgotten.

Exercise 3.13. Let X be discrete and retain the notation at the end of the section.

(1) Prove that the set $\mathscr{S}\rho$ of superharmonic functions f such that $\rho(f) \leqslant 1$

is a compact convex set for the topology of pointwise convergence.

(2) Prove that the extremal elements of $\mathscr{S}\rho$ are either harmonic functions or potentials and that a potential is extremal if and only if it has the form $G(\cdot, y)$ for some y in E.

(3) Prove that the extremal harmonic functions are the functions $k(\cdot, s)$ for $s \in M_e$.

Exercise 3.14. With the hypothesis of this section show that

$$k(x, X_\infty) = (\mathrm{d}P_x/\mathrm{d}P_{rm})_{\mathscr{I}}.$$

Exercise 3.15. Prove that for a Borel subset A of M,

$$P_{rm}[X_\infty \in A \mid \mathscr{F}_n] = \int_A k(X_n, s)\, \sigma(\mathrm{d}s).$$

Exercise 3.16 (fine topology). (1) If $s \in M_e$ and A is a Borel subset of E^* the probability $P^s_{rm}[\underline{\lim}\{X_n \in A\}]$ is either 0 or 1.

(2) One can define a topology on E in the following way: if $s \notin M_e$ then $\{s\}$ is open; if $s \in M_e$, the collection of sets A which are closed for the $\delta + d$ metric on E^* and such that the probability in (1) is equal to 1, is a basis of neighbourhoods of s. Prove that this topology is finer than the $\delta + d$-topology.

Exercise 3.17. Let X be the random walk of ch. 3, Exercise 3.13(2).

(1) Notice that, with the notation of the end of this section, we may take $\rho = \varepsilon_0$ and then all finite harmonic functions are ρ-integrable. Compute k and prove that E^* is obtained by adjoining two points to the space $E = \mathbf{Z}$. Prove that both points correspond to a non-zero extremal function; compute these functions. Give all the harmonic functions and compare with ch. 5, Exercise 1.9.

(2) Take now $\rho(x) = 0$ for $x < 0$ and $\rho(x) = (p/q)^x((p-q)/p)$ for $x \geqslant 0$. Do the same work as in (1). One of the points of the boundary does not correspond to an extremal function, thus M_e is different from M in this case.

(3) More generally, if X is any random walk on $\mathbf{Z}$ and ρ any probability measure on $\mathbf{Z}$, there is at most one ρ-integrable extremal function beside the function 1.

Exercise 3.18. Let E be the set of lattice points (n, i) in the plane with $0 \leqslant i \leqslant n$, and define P by

$$P((n, i), (n+1, i)) = P((n, i), (n+1, i+1)) = 2^{-1}.$$

A point in E is thus identified with i heads in n tosses of a fair coin.

(1) Observe that the unit mass at (0, 0) is a reference measure, and compute the corresponding kernel K.

(2) Prove that a sequence (n_k, j_k) is Cauchy in the corresponding space E^* if and only if there is a number $s \in [0, 1]$ such that $\lim_k j_k/n_k = s$. Prove that M is homeomorphic to $[0, 1]$ endowed with its usual topology and that for $s \in M$,

$$K((m, i), s) = 2^m s^i (1 - s)^{m-i}.$$

Prove that $M_e = M$.

(3) What is the measure σ corresponding to the harmonic function 1? Describe h and P^h for the case in which σ^h is the Lebesgue measure on M, and that in which $\sigma^h = \varepsilon_s$ for $s \in M$.

Exercise 3.19. Let E be the set of all finite sequences of letters a and b. The empty sequence is taken as the starting state. A transition probability on E is defined in the following way: if x is a finite sequence, and we denote by xa and xb the sequences obtained by adjoining a or b on the right of x, then

$$P(x, xa) = P(x, xb) = \tfrac{1}{2}.$$

(1) Prove that the boundary M is the set of all infinite sequences of letters a and b. Find the topology of M.

(2) Compute the measure σ corresponding to the harmonic function 1.

CHAPTER 8

POTENTIAL THEORY FOR HARRIS CHAINS

In this chapter, we proceed with the study of a potential theory suitable for Harris chains, which was initiated in ch. 6. Special functions will still play an important role, but we shall consider only the bounded ones. Thus, in all statements of this chapter, "special" must be understood as "special and bounded".

1. Harris chains and duality

As was seen in ch. 4 §4, most Harris chains, and actually all the usual ones, have a dual chain which was shown to be equally Harris up to modification.

We shall give below a proof of the latter result that does not rely on the notion of essential irreducibility. Afterwards we shall study the relationship between special and co-special (special for the dual chain) functions.

Theorem 1.1. *If P is Harris and in duality with $\hat{P}$, there is a modification of $\hat{P}$ which is Harris. Furthermore if $h \in \mathscr{U}_+$, $h > 0$ and $U_h > 1 \otimes m$, this modification may be chosen so that $\hat{U}_h(h) = 1$ and $\hat{U}_h > 1 \otimes m$.*

Proof. As m is the only invariant measure for P, the duality between P and $\hat{P}$ holds with respect to m. For $f, g \in \mathscr{E}_+$,

$$\langle \hat{U}_h f, g \rangle = \langle f, U_h g \rangle > m(f)\, m(g),$$

which implies that, for all $f \in b\mathscr{E}_+$, $\hat{U}_h f > m(f)$ m-a.e. By letting f run through a denumerable algebra of sets generating $\mathscr{E}$ we get that there is an m-null set N_1 such that $\hat{U}_h(x, \cdot\,) > m$ for x outside N_1. Moreover, for every $g \in b\mathscr{E}_+$,

$$\langle \hat{U}_h(h), g \rangle = \langle h, U_h(g) \rangle = (hm)\, U_h(g) = m(g)$$

according to ch. 3, Proposition 2.9. Therefore $\hat{U}_h(h) = 1$ outside an m-null set N_2.

Since m is $\hat{P}$-excessive, one can find, as in ch. 3, Corollary 2.12, an m-null set N containing $N_1 \cup N_2$ and such that N^c is an absorbing set. We then pick a point x_0 in N^c and set

$$\hat{P}'(x, \cdot) = \hat{P}(x, \cdot) \quad \text{for } x \in N^c,$$

$$\hat{P}'(x, \cdot) = \hat{P}(x_0, \cdot) \quad \text{for } x \in N;$$

the T.P. $\hat{P}'$ is a modification of $\hat{P}$ for which the conditions $\hat{U}_h(h) = 1$ and $\hat{U}_h > 1 \otimes m$ hold. It is then clear from ch. 3, Proposition 2.7 that $\hat{P}'$ is Harris, and the proof is complete.

In the following, we do assume that there is at least one T.P. $\hat{P}$ in duality with P.

On account of ch. 6, Corollary 4.6, the function h in the preceding result is both special and co-special. It is natural to ask whether every special function is, as a rule, also co-special (up to a modification of $\hat{P}$). On account of ch. 6, Proposition 4.9, this would imply that, for every quasi-compact T.P., there is at least one quasi-compact modification of $\hat{P}$. This turns out to be false, as is shown in Exercise 1.9, and we shall below describe the set of functions both special and co-special. For this purpose, we need the following classical

Lemma 1.2. *Let $\{a_n\}_{n \in \mathbf{N}}$ be a bounded sequence of real numbers converging to zero in the sense of Cesaro and set $f(s) = \sum_0^\infty a_n s^n$ $(0 \leqslant s < 1)$; then*

$$\lim_{s \to 1} (1 - s) f(s) = 0.$$

Proof. The sequence $U_n = n^{-1} \sum_0^n a_k$ converges to zero, and an easy computation yields

$$(1 - s) f(s) = (1 - s)^2 \sum_0^\infty U_n \frac{s^n}{n} \cdot$$

For $\varepsilon > 0$, there is an integer N such that $|U_n| \leqslant \varepsilon$ for $n \geqslant N$, hence

$$|(1 - s) f(s)| \leqslant (1 - s)^2 \sum_0^{N-1} |U_n| \frac{s^n}{n} + \varepsilon (1 - s)^2 \sum_N^\infty \frac{s^n}{n} \cdot$$

Now, the first term on the right of this inequality can be made arbitrarily small by choosing s near one, and the second term is less than ε, which completes the proof.

Before we proceed, let us remark that if f is special for X and g special for $\hat{X}$, then $f \wedge g$ is special and co-special.

Proposition 1.3. *A function h in $b\mathscr{E}_+$ is special and co-special if and only if there is a real number c such that*

$$U_{ch} > 1 \otimes m \quad \textit{and} \; \hat{U}_{ch} > 1 \otimes m.$$

Proof. The "if" part is obvious. To prove the converse, we start with P and $\hat{P}$ quasi-compact. Since the constant functions are special there is, by virtue of ch. 6, Lemma 4.7, a function $f > 0$ and a constant $a < 1$ such that $U_a > 1 \otimes fm$. For $a' < a$, we have $U_{a'} > (a - a')\, U_a U_{a'}$, and therefore, for all $g \in b\mathscr{E}_+$ and $x \in E$,

$$\begin{aligned} U_{a'}\, g(x) &> (a - a')\, U_a U_{a'}\, g(x) \\ &> (a - a')\, \langle f, U_{a'} g \rangle = (a - a')\, \langle \hat{U}_{a'} f, g \rangle. \end{aligned}$$

Since $\hat{P}$ is quasi-compact, there is, by Lemma 1.2, a number a_0 such that $a' < a_0$ implies $a' \hat{U}_{a'} f \geqslant k > 0$; thus

$$U_{a'} g(x) > \frac{k(a - a')}{a'}\, m(g).$$

This proves that, for a' sufficiently small, $U_{a'} > 1 \otimes m$. Symmetrically, one can find a'' such that $\hat{U}_{a''} > 1 \otimes m$. Then, taking $c = a' \wedge a''$, we get the desired result for constant functions.

Let us now shift to the general case and suppose $h > 0$. The T.P.'s $Q = U_h I_h$ and $\hat{Q} = \hat{U}_h I_h$ are quasi-compact and in duality with respect to the measure hm. Since $U_c^Q = U_{ch} I_h$, the first part of the proof implies that for c sufficiently small,

$$U_{ch} I_h > 1 \otimes hm \quad \text{and} \; \hat{U}_{ch} I_h > 1 \otimes hm;$$

hence

$$U_{ch} > 1 \otimes m \quad \text{and} \; \hat{U}_{ch} > 1 \otimes m.$$

Now, any special and co-special function h is majorized by a strictly positive special function k and by a strictly positive co-special function $\hat{k}$. The function $k \wedge \hat{k}$ is strictly positive, special and co-special, and majorizes h. Since U_h increases when h decreases, the proof is complete.

The importance of functions h such that $U_h > 1 \otimes m$ was shown in ch. 6 §5. The above results lead one to think that it is exactly those functions which are both special and co-special. The following results lead in that direction.

Proposition 1.4. *Let h be special; there is a real c such that $U_{ch} > 1 \otimes m$ if and only if there is a modification of $\hat{P}$ such that h is co-special.*

Proof. The sufficiency follows from Proposition 1.3. The necessity may be shown in a way very similar to Theorem 1.1, and is left to the reader as an exercise.

Corollary 1.5. *Let P be quasi-compact; then there is a quasi-compact modification of $\hat{P}$ if and only if there is real c such that $U_c > 1 \otimes m$.*

Proof. Obvious.

In Proposition 1.4, the modification of $\hat{P}$ may depend on the function h. However, in important cases, one can find a modification of $\hat{P}$ which works for all functions h. To state the corresponding result we shall use the notational device of ch. 1 §5.3 and recall (ch. 3 §2.14) that if P_n^1 is the singular part of P_n with respect to m, then $\lim_n P_n^1(\cdot\,, E) = 0$ pointwise. The following statement covers the case where $\hat{P} \ll m$, hence in particular the case of discrete chains, the case of Harris random walks on groups and the case of quasi-compact $\hat{P}$.

Proposition 1.6. *If* $\lim_n \|\hat{P}_n(\cdot\,, E)\| = 0$ *and if h is special, then there is a real number c such that $U_{ch} > 1 \otimes m$ if and only if h is co-special.*

Proof. Only the necessity requires a proof. Without losing any generality we may assume that $\|h\| = a < 1$ and $U_h > 1 \otimes m$. As in Theorem 1.1, we may prove that for every g in $b\mathscr{E}_+$, we have

$$\hat{U}_h(g) > m(g) \quad m\text{-a.e.}$$

Pick c such that $a < c < 1$. The kernel $\hat{U}_c = \sum_0^\infty (1-c)^n \hat{P}_{n+1}$ may be written $\hat{U}_c = \hat{U}_c^0 + \hat{U}_c^1$, where $\hat{U}_c^1$ is the singular part. The hypothesis implies that there is a number $\alpha > 0$ such that $\hat{U}_c^0(\cdot\,, E) \geqslant \alpha$. By the resolvent equation, we thus have

$$\hat{U}_h g \geqslant \hat{U}_c(c-h)\,(\hat{U}_h g) \geqslant \hat{U}_c^0(c-a)\,\hat{U}_h g,$$

and since $\hat{U}_h g > m(g)$ m-a.e., for every $g \in b\mathscr{E}_+$,

$$\hat{U}_h g > \alpha(c - a)\, m(g).$$

It follows that h is co-special.

Exercise 1.7. (1) Let X and $\hat{X}$ be two Harris chains in duality and h a strictly positive, special and co-special function. If h is sufficiently small, prove that the associated kernels W_h and $\hat{W}_h$ are in duality relative to m. If φ and ψ are bounded functions, the kernels

$$\Gamma = I + W_h + 1 \otimes \varphi m + \psi \otimes m, \qquad \hat{\Gamma} = I + \hat{W}_h + \varphi \otimes m + 1 \otimes \psi m$$

are potential kernels in duality.

(2) If f is in $\mathscr{L}^1(m)$, prove that for any potential kernel Γ the functions $\Gamma|f|$ and $P\Gamma|f|$ are m-almost everywhere finite. If moreover $m(f) = 0$, then

$$(I - P)\,\Gamma f = f \quad m\text{-a.e.}$$

Exercise 1.8. Prove that in the results of this section one can avoid the use of modifications of $\hat{P}$ whenever E is an LCCB space, $\hat{P}$ a Feller kernel and h a continuous function.

[*Hint*: Use the results given in §4.]

Exercise 1.9. Let E be the unit interval $[0, 1]$ of the real line, $\mathscr{E}$ the σ-algebra of Borel subsets of E and m the Lebesgue measure on $\mathscr{E}$. Set $i(x) = x$ for $x \in E$, and define a pointwise transformation θ of E by

$$\theta = 2i 1_{[0,\frac{1}{4}]} + (i + \tfrac{1}{4})\, 1_{]\frac{1}{4},\frac{1}{2}]} + \tfrac{1}{2}(i + 1)\, 1_{]\frac{1}{2},1]}.$$

(1) Compute the Lebesgue derivative of $\theta^n(m)$ with respect to m.

(2) Calling φ the derivative $d\theta(m)/dm$, set, for $f \in \mathscr{E}$,

$$Sf = f \circ \theta^{-1}, \qquad \hat{S}f = (f \circ \theta)\, \varphi$$

and

$$P = \tfrac{1}{2}S + (1 \otimes \tfrac{1}{2}\varphi)\, m, \qquad \hat{P} = \tfrac{1}{2}\hat{S} + (1 - \tfrac{1}{2}\varphi) \otimes m.$$

Prove that P and $\hat{P}$ are two T.P.'s in duality relative to m, that P is quasi-compact and that no modification of $\hat{P}$ is quasi-compact.

(3) The same example provides a counter-example to the following sharpening of ch. 6, Exercise 2.13: if P is quasi-compact, P_n converges to $1 \otimes m$ in the norm of linear operators on $L^1(m)$.

2. Equilibrium, balayage and maximum principles

In this section we show that for each one of the potential operators Γ defined in ch. 6, Proposition 5.4 one can develop a potential theory which parallels the theory in ch. 2. The function φ in Γ is supposed to be bounded.

Theorem 2.1 (equilibrium principle). *For every set A in $\mathscr{E}$ such that $m(A) > 0$ and every potential kernel Γ, there is one and only one triple (u_A, C_A, v_A) where*

(i) *$u_A(v_A)$ is a function (measure) vanishing outside A and C_A a constant;*

(ii) $\langle m, u_A\rangle = \langle v_A, 1\rangle = 1$;

(iii) *$\Gamma u_A = C_A$ on A and $v_A\Gamma = C_A m$ on A.*

Moreover, $C_A = v_A\Gamma u_A$, and the following identities are satisfied:

(i) $P_A = \Gamma(I_A - \Pi_A) + 1 \otimes v_A$; (2.1)

(ii) $H_A = (I_A - \Pi_A)\,\Gamma + u_A \otimes m$; (2.2)

(iii) $\Gamma - G^A = P_A\Gamma + (\Gamma u_A - C_A) \otimes m = \Gamma H_A + 1 \otimes (v_A\Gamma - C_A m)$. (2.3)

Proof. We start with the kernel $\Gamma = I + W$ of ch. 6, Theorem 5.2. By the formulae therein, we have

$$U_A h + W I_A U_A h = Wh + m(h)^{-1}\,(hm)\,U_A h.$$

Restricting this equality to A yields

$$\Gamma I_A U_A h = Wh + m(h)^{-1}\,(hm)\,U_A h$$

on A. Since Wh is a constant function and $\langle m, I_A U_A h\rangle = m(h)$ by virtue of ch. 3, Proposition 2.9, the function $u_A = I_A U_A h/m(h)$ has the desired property, with

$$C_A = m(h)^{-2}\,(1 - m(h) + (hm)\,U_A h).$$

The process is similar for v_A. Integrating the equality

$$U_A I_A + U_A I_A W I_A = W I_A + m(h)^{-1}\,U_A(h) \otimes m I_A$$

with respect to the measure (hm) yields

$$(hm)\,U_A I_A \Gamma I_A = (hm)\,W I_A + m(h)^{-1}\,(hm)\,U_A h\, m I_A,$$

and since $(hm)\,W = ((1 - m(h))/m(h))\,m$, one gets

$$(hm)\,U_A I_A \Gamma I_A = m(h)^{-1}\,(1 - m(h) + (hm)\,U_A h)\,m I_A.$$

As $U_A I_A = 1$, we have $\langle (hm)\, U_A I_A, 1\rangle = m(h)$; the measure $m(h)^{-1}\,(hm)\, U_A I_A$ has thus the desired property, with the same value for C_A. Thus we have shown the existence of a triple (u_A, v_A, C_A). We turn to showing that it is unique.

From ch. 6, Theorem 5.2 it is easily seen that

$$U_A I_A + W I_A U_A I_A = W I_A + 1 \otimes v_A, \tag{2.4}$$

$$I_A U_A + I_A U_A I_A W = I_A W + u_A \otimes m. \tag{2.5}$$

Moreover, u_A and v_A are positive, and $C_A = v_A \Gamma u_A$.

Let u be a function with the same properties as u_A. Multiplying to the right by u on both sides of eq. (2.5) yields

$$u_A + I_A W u = (I_A U_A I_A)\, \Gamma u.$$

Since $\Gamma u = k$ on A and $U_A I_A = 1$, this may also be written

$$u_A + I_A W u = I_A \Gamma u = I_A u + I_A W u = u + I_A W u,$$

and since Wu is finite on A it follows that $u = u_A$. The uniqueness of v_A may be shown in the same way.

The identities of eqs. (2.1) and (2.2) are straightforward consequences from eqs. (2.4) and (2.5) and from the formulae giving P_A, Π_A, H_A as functions of U_A. To prove eq. (2.3), let us recall that $P_A + G^A = I + I_{A^c} U_A$; using this and ch. 6, Theorem 5.2, we find

$$P_A = \Gamma - G^A + m(h)^{-1}\, I_{A^c}\, U_A(h) \otimes m;$$

multiplying on the right by u_A the two members of this equality yields, thanks to the properties of u_A,

$$C_A = \Gamma u_A + m(h)^{-1}\, I_{A^c}\, U_A(h),$$

which completes the proof, the other equality being shown in a very similar way.

It remains to show that all is still true for a kernel Γ' equal to $\Gamma + 1 \otimes \nu + \varphi \otimes m$. For instance, it is easy to show that if the measure v'_A occurs in eq. (1.1) with Γ' instead of Γ, then $v'_A = v_A - \nu(I_A - \Pi_A)$. By means of eq. (2.5) and of the equality $\nu H_A I_A = \nu I_A$, one may compute $(v_A - \nu(I_A - \Pi_A))\, \Gamma'$ and find that it is a multiple of m on A; moreover, it is clear that $\langle v'_A, 1\rangle = 1$. In the same way one can show the existence of u'_A.

Remark. The result may be extended to kernels $\Gamma + 1 \otimes \nu + \varphi \otimes m$, where φ is merely finite, if we restrict ourselves to sets A such that $\Pi_A \varphi$ is finite. This will be of significance for Harris random walks.

Definition 2.2. We call $v_A(u_A)$ *the equilibrium measure (function)* of A with respect to Γ. We call C_A the *Robin's constant* of A with respect to Γ.

A probabilistic interpretation of v_A will be seen in the next section under an additional hypothesis. We may already note that v_A is absolutely continuous with respect to m.

Proposition 2.3. *Let Γ and $\hat{\Gamma}$ be potential kernels for X and $\hat{X}$ in duality with respect to m. Then for every A in $\mathscr{E}$ such that $m(A) > 0$, we have $C_A = \hat{C}_A$ and $v_A = \hat{u}_A m$.*

Proof. For every $g \in \mathscr{E}_+$, vanishing outside A,

$$C_A\, m(g) = \langle g, \Gamma u_A \rangle = \langle \hat{\Gamma} g, u_A \rangle.$$

Thus, $(u_A m)\, \hat{\Gamma} = C_A m$ on A, and since $m(u_A) = 1$ we have $u_A m = \hat{v}_A$ and $C_A = \hat{C}_A$.

In Exercise 1.7, the reader will find how to get potential kernels in duality.

We are now going to use the identities in Theorem 2.1 to prove Potential Theory principles, in a way parallel to ch. 2, Corollary 1.12 and ch. 2, Theorem 4.5 for the transient case.

Theorem 2.4 (balayage principle). *For every A in $\mathscr{E}$ such that $m(A) > 0$, and for every probability measure ν, the probability measure $\nu_A = \nu P_A$ is the only probability measure such that*

$$\nu_A \Gamma = \nu \Gamma + km$$

on A for a constant k.

Proof. In view of eq. (2.3), together with the fact that the measures $G^A(x, \cdot)$ vanish on A, it is clear that ν_A satisfies the required property with $k = \langle \nu, \Gamma u_A - C_A \rangle$.

Now let ν' be another probability measure such that $\nu' \Gamma = \nu \Gamma + k'm$ on A. The measures $H_A(x, \cdot)$ vanish for $x \notin A$, hence

$$\nu' \Gamma H_A = \nu \Gamma H_A + k' m H_A,$$

$$\nu_A \Gamma H_A = \nu \Gamma H_A + k m H_A,$$

and since by ch. 3, Proposition 2.9, $mH_A = m$, we get

$$\nu' \Gamma H_A = \nu_A \Gamma H_A + (k' - k)\, m.$$

Now eq. (2.3), along with the fact that $G^A(x, \cdot)$ vanishes for $x \in A$, implies that

$$\nu' \Gamma = \nu_A \Gamma + (k' - k)\, m.$$

Multiplying to the right the two sides of the latter equation by $(I - P)$ yields

$$\nu' - m(h)^{-1}.\,(hm)\,P + \mu(I - P) = \nu_A - m(h)^{-1}.\,(hm)\,P + \mu(I - P),$$

provided $\Gamma = I + W + 1 \otimes \mu + \varphi \otimes m$, and consequently $\nu' = \nu_A$.

Theorem 2.5 (maximum principle). *Let f be in $\mathcal{N}$ and such that $m(|f|) > 0$. If there is a number k such that $\Gamma f \leqslant k$ on $\{f > 0\}$, then $\Gamma f \leqslant k - f^-$ everywhere.*

Proof. Set $A = \{f > 0\}$; we have clearly $m(A) > 0$, and eq. (2.3) implies

$$\Gamma f \leqslant G^A f + P_A \Gamma f;$$

and since $G^A f \leqslant -f^-$ and $P_A \Gamma f \leqslant k$ everywhere, the theorem is established.

Definition 2.6. The kernel Γ is said to satisfy the *reinforced semi-complete maximum principle.*

The maximum principle may be stated in another fashion.

Corollary 2.7. *Let f_1 and f_2 be special and such that $m(f_1) = m(f_2) > 0$. If $\Gamma f_1 \leqslant \Gamma f_2 + k$ on $\{f_1 > 0\}$, then $\Gamma f_1 \leqslant \Gamma f_2 + k$ everywhere.*

Proof. We apply Theorem 2.5 to the function $f = f_1 - f_2$ after noticing that $\{f > 0\} \subset \{f_1 > 0\}$.

Exercise 2.8. If $\Gamma = I + W + 1 \otimes \nu + \varphi \otimes m$, with ν a positive measure, show that in Theorem 2.5 the assumption $m(|f|) > 0$ may be dropped.

[*Hint*: If $m(|f|) = 0$, then $\Gamma f = Gf + \nu(f)$ and G satisfies the reinforced maximum principle.]

Exercise 2.9. Show that the kernel $\Gamma = I + W$ satisfies the following maximum principle: if f and g are special and such that $m(f) \geqslant m(g)$, then the inequality $\Gamma f \leqslant \Gamma g + k$ holds everywhere if it holds on $\{f > 0\}$.

[*Hint*: This may be proved either by using the methods of this section or by calling upon ch. 2 §2.10.]

Exercise 2.10. If f is in $\mathscr{N}$ and vanishes outside A, then, as in the transient case, $P_A \Gamma f = \Gamma f$ for every potential kernel Γ.

Exercise 2.11. (1) Assume that there is a point x in E such that the set $\{x\}$ is recurrent (e.g., X is discrete and irreducible recurrent), then the kernel $G^{\{x\}}$ is a potential kernel. A function f is then special if and only if $G^{\{x\}}f$ is bounded.

(2) Let X be the chain with state space $E = \mathbf{N}$ and $P(0, \cdot) = \mu(\cdot)$, $P(n, \cdot) = \varepsilon_{n-1}(\cdot)$, where μ is a probability measure on $\mathbf{N}$. Prove that X is irreducible recurrent; find a necessary and sufficient condition on μ for X to be null, and characterize the special functions.

Exercise 2.12 (capacities). Let $\Gamma = I + W$; then for every A such that $m(A) > 0$ we have $C_A > 0$. Define a set function γ by $\gamma(A) = -C_A$ if $m(A) > 0$, $\gamma(A) = -\infty$ if $m(A) = 0$, and prove that γ is increasing and strongly subadditive.

Exercise 2.13. Prove that for every set A in $\mathscr{E}$ such that $m(A) > 0$, and every special function f,

$$\lim_N \sum_0^N [P_n f - P_A P_n f - P_n H_A f + P_A P_n H_A f] = G^A f.$$

[*Hint*: Use ch. 6, Exercise 5.17.]

Exercise 2.14. Prove the following domination principle which is dual to the maximum principle of Theorem 2.5: If μ and ν are two mutually singular probability measures and A is a set such that $\mu(A) = 0$, $\nu(A^c) = 0$ and $m(A) > 0$, if $(\mu - \nu)\,\Gamma + cm$ is positive on A for some constant $c > -\infty$, then $(\mu - \nu)\,\Gamma - \mu + cm$ is positive everywhere.

[*Hint*: Use the operator H_A.]

Exercise 2.15. Prove that the potential kernels of Harris resolvents satisfy the following *semi-complete maximum principle*: if $f \in \mathscr{N}$ and $k \in \mathbf{R}$, $\{Wf \leqslant k\}$

on $\{f > 0\}$ implies $Wf \leqslant k$ everywhere. Conversely, for every $\alpha > 0$, $I + \alpha W$ satisfies the reinforced semi-complete maximum principle.

3. Normal chains

In ch. 6 §6, we saw that for any pair (f, g) of special functions and any probability measure ν, the ratio of the numbers

$$m(f)^{-1} \sum_{0}^{n} \nu P_m f \quad \text{and } m(g)^{-1} \sum_{0}^{n} \nu P_m g$$

converges to 1 as n tends to infinity. One could think of sharpening this result by showing that their difference tends to a finite limit. In view of ch. 6, Proposition 5.7, this amounts to showing that for $f \in \mathcal{N}$, the sums $\sum_1^n P_m f$ converge pointwise.

This is also of interest from the potential theoretic point of view, since the above convergence would imply that f has a potential in the sense of ch. 2. As we will see below, such a result permits one to single out from all the kernels exhibited in ch. 6, Proposition 5.4 a subclass of canonical character. Notably, we have as yet not provided recurrent random walks with a potential kernel which is a convolution kernel. The following study is a first step in the search for such a kernel.

Unfortunately the above limit may fail to exist, as is shown in Exercise 3.11, and we shall proceed to what may be said if it exists.

Theorem 3.1. *For a Harris chain the three following statements are equivalent:*

(i) *for every charge f, $\lim_n \sum_1^n P_m f$ exists;*

(ii) *for every non-negligible special function h of $\mathcal{U}_+$ there is a probability measure λ_h such that, for every g in $b\mathcal{E}_+$ and x in E,*

$$\lim_n P_n U_h I_h g(x) = \lambda_h(g);$$

(iii) *for every non-negligible special function h of $\mathcal{U}_+$, every x in E and B in $\mathcal{E}$, the sequence $\{P_n U_h I_h(x, B)\}$ has a finite limit as n tends to infinity.*

Whenever $h = 1_A$, we write λ_A in place of λ_h; we then have $\lambda_A = \lim_n P_n P_A$, and λ_A vanishes outside A. Let us also remark that by ch. 6, Proposition 5.7, the convergence in (i) is bounded.

Proof. Obviously (ii) implies (iii). Conversely, for every $B \in \mathcal{E}$, the function

$$\lambda_h(\cdot, B) = \lim_n P_n U_h I_h(\cdot, B)$$

is bounded and harmonic, hence constant, and we can set

$$\lambda_h(B) = \lim_n P_n U_h I_h(\cdot, B).$$

The Vitali–Hahn–Saks theorem then implies that λ_h is a probability measure for which (ii) is satisfied.

We proceed to show that (ii) implies (i). Let $f \in \mathcal{N}$ and h a function in $\mathcal{U}_+$, special, strictly positive and such that $|f|/h$ is bounded (for instance a multiple of $|f| + g$ where g is special and strictly positive). For any kernel W, we have

$$Wf = U_h f + U_h I_h W f = U_h I_h(f/h + Wf),$$

and since $Wf = \sum_1^n P_m f + P_{n+1} W f$, this may be written

$$\sum_1^n P_m f = Wf - P_{n+1} U_h I_h(f/h + Wf).$$

The function $f/h + Wf$ is bounded and consequently (ii) implies (i).

Conversely, assume h special and non-negligible, and let g be in $b\mathscr{E}_+$. The function $f = hg - hU_h(hg)$ is a charge because $|f| \leqslant 2h\|g\|$ and

$$m(f) = \langle hm, g\rangle - \langle hm, U_h I_h g\rangle = 0$$

by ch. 3, Proposition 2.9. By the resolvent equation it is easily seen that $Pf = (I - P)\, U_h I_h g$, which leads to

$$\sum_1^n P_m f = U_h(hg) - P_{n+1} U_h I_h g,$$

and it follows that (i) implies (iii) and therefore (ii).

Definition 3.2. A Harris chain satisfying the equivalent conditions of Theorem 3.1 will be called *normal.*

Proposition 3.3. *A chain X is normal if and only if the pointwise limit* $\lim_n P_n W f$ *exists for every function f such that* $|f|$ *is special and any kernel W. It suffices that the limit exists for f in* $\mathcal{N}$.

Proof. The identity

$$Wf = \sum_1^n P_m f + P_{n+1} W f,$$

which holds for $f \in \mathcal{N}$, shows that if the limit in the statement exists for $f \in \mathcal{N}$, then X is normal.

Conversely if X is normal, then $P_{n+1}Wf$ converges for f in $\mathcal{N}$. Let f be special and h be the function used to construct W; the function $m(h)\,f - m(f)\,h$ is in $\mathcal{N}$, hence

$$P_n W(m(h)\,f - m(f)\,h) = m(h)\,P_n Wf - m(f)\,(1 - m(h))/m(h)$$

has a limit when n tends to infinity, which completes the proof.

Theorem 3.4. *If X is normal, there exists a potential kernel Γ such that for every $f \in \mathcal{N}$,*

$$\lim_n \sum_0^n P_m f = \Gamma f \quad \textit{and} \lim_n P_n \Gamma f = 0.$$

Moreover, the probability measure λ_A of Theorem 3.1(ii) is the equilibrium measure of A with respect to Γ.

Proof. By the preceding proof, it suffices to show that there is a measure ν such that, for every special function f, $\nu(f) = \lim_n P_n Wf$. The kernel $\Gamma = I + W - 1 \otimes \nu$ will then satisfy the first part of the statement.

The function $\lim_n P_n Wf$ is bounded and harmonic and therefore constant. Let h_1 be special and > 0; the sequence of bounded measures $P_n W I_{h_1}$ converges on every bounded function by Proposition 3.3, so that, by the Vitali–Hahn–Saks theorem, there is a bounded measure ν_1 such that for f with $|f| < h_1$,

$$\lim_n P_n Wf = \nu_1(f).$$

Let h_2 be another function with the same properties and ν_2 the corresponding measure. The measures ν_1 and ν_2 agree on the positive functions less than $h_1 \wedge h_2$, hence are equal. Since every special function is majorized by a strictly positive special function, there is indeed a measure ν such that

$$\lim_n P_n Wf = \nu(f)$$

for every special function f.

Now let v_A be the equilibrium measure of A with respect to Γ. By eq. (2.1) we have, for A special and non-negligible,

$$P_A = \Gamma(I_A - \Pi_A) + 1 \otimes v_A.$$

For f in $b\mathscr{E}$ the function $(I_A - \Pi_A) f$ is in $\mathscr{N}$, and passing to the limit in

$$P_n P_A f = P_n \Gamma(I_A - \Pi_A) f + v_A(f)$$

yields $v_A(f) = \lambda_A(f)$.

By ch. 6, Proposition 5.4 if Γ and Γ' are two kernels with the properties stated in Theorem 3.4, then $\Gamma - \Gamma' = \varphi \otimes m$, where φ is a finite function. The equilibrium measures are the same for the two kernels but the Robin's constants are different. If we assume that $\hat{X}$ is also normal we may narrow still further the class of potential kernels with a canonical character.

Proposition 3.5. *If X and $\hat{X}$ are normal, there is a pair of kernels Γ and $\hat{\Gamma}$ in duality with respect to m and such that $\Gamma(\hat{\Gamma})$ is a potential kernel for $X(\hat{X})$ satisfying the conditions of Theorem* 3.4.

Proof. We first show that the measure ν in the proof of Theorem 3.4 is absolutely continuous with respect to m. Indeed if f is special and $m(f) = 0$ we have for every $g \in \mathscr{L}^1_+$

$$\nu(f) = \lim_n m(g)^{-1} \langle P_n W f, g \rangle = \lim_n m(g)^{-1} \langle f, \hat{W}\hat{P}_n g \rangle = 0.$$

We can thus choose a finite measurable function $\hat{\gamma}$ such that $\nu = \hat{\gamma} m$.

By the first section h may be chosen such that also $\hat{U}_h > 1 \otimes m$. Then $\hat{W}$ is in duality with W. If $\hat{X}$ is normal, then $\hat{P}_n \hat{W}$ converges to $1 \otimes \hat{\nu}$, and by the above discussion we can choose a finite function γ such that $\hat{\nu} = \gamma m$. The kernels

$$\Gamma = I + W - 1 \otimes \hat{\gamma} m - \gamma \otimes m,$$

$$\hat{\Gamma} = I + \hat{W} - 1 \otimes \gamma m - \hat{\gamma} \otimes m$$

thus have the desired properties. We notice that they map special functions into finite ones but not always into bounded ones.

The functions γ and $\hat{\gamma}$ are defined up to equivalence. In the next chapter we shall see that for random walks there is a canonical choice for these functions. They are also uniquely determined for discrete chains, where for instance

$$m(y)\, m(x)\, \hat{\gamma}(x) = \lim_n \hat{W}\hat{P}_n(x, y)$$

for any $y \in E$. We recall (ch. 6, Exercise 4.11) that all functions with finite support are special. In these cases, the pair $(\Gamma, \hat{\Gamma})$ is unique up to addition of the same multiple of $1 \otimes m$, so that the Robin's constant, which is the same for Γ and $\hat{\Gamma}$, is determined up to addition of a constant which is the same for all sets. For a further study of Robin's constant see Exercises 3.12 and 4.18.

Normality is a rather stringent condition, and one could think of demanding the convergence of $\{P_n W f\}$ only for functions f in subclasses of $\mathscr{S}$. Actually if these subclasses are rich enough, one can develop the same theory as above with the obvious changes. One can even solve the Poisson equation with second member the opposite of a special function. This will be sketched for discrete chains in the next section, and studied for random walks in ch. 9.

Unfortunately, Exercise 3.11 shows that there are chains for which the convergence property fails to be true for any reasonable subclass. It is therefore a natural task to find out whether all "classical" chains are normal in a more or less stringent sense. Here is a first easy result.

Proposition 3.6. *All aperiodic positive chains are normal.*

Proof. The proposition is an obvious consequence of ch. 6 §2.

In the following chapter, we shall prove results of this kind for Harris random walks on abelian groups, and the reader will also find examples of normal chains in Exercise 4.17.

Exercise 3.7. If $f \in \mathscr{L}^1(m)$ and $\sum_1^n P_m f$ has a finite limit when n tends to infinity, then $m(f) = 0$. If $g \in b\mathscr{E} \cap \mathscr{L}^1(m)$ and $f = (I - P)\, g$, then $\sum_1^n P_m f$ tends to a finite limit when n tends to infinity. Prove that the functions of this form are not always charges.

Exercise 3.8. If X is normal, prove that for any $f \in \mathscr{U}_+$ and g special, the limit $\lim_n P_n U_f g$ exists.

Exercise 3.9. A set $A \in \mathscr{E}$ is said to be *small* if there is a probability measure λ_A such that for all g in $b\mathscr{E}$,

$$\lim_n P_n P_A g = \lambda_A(g).$$

(1) Prove that a subset B of a small set A is small and that $\lambda_B = \lambda_A P_B$. Then show that for a positive chain all sets are small and that $\lambda_A = m P_A$.

(2) If X is normal, special sets are small sets, but the converse is not true.

Show that the union of a small set and a special set is small.

Exercise 3.10. Assume that $\hat{X}$ is normal and show that in ch. 6, Theorem 6.5 one may assume that f and g in $\mathscr{L}^1(m)$ provided that ν_1 and ν_2 are absolutely continuous with respect to m, the Radon–Nikodym derivatives being co-special.

Exercise 3.11. Let X be discrete and irreducible recurrent.

(1) Show that if $\hat{X}$ is normal, then for every pair (x, y) of points in E the limit

$$\lim_N \sum_{n=0}^{N} (P_n(y, y) - P_n(x, y))$$

exists and is finite.

(2) Pick two points a and b in E and set $F = \{a, b\}$ and for $t \in [0, 1[$

$$A(t) = \sum_{n=1}^{\infty} t^n P_a[S_F = n], \qquad B(t) = \sum_{n=1}^{\infty} t^n P_b[S_F = n].$$

Show that if $\hat{X}$ is normal, the following limit exists:

$$\lim_{t\to 1} R(t) = \lim_{t\to 1} (1 - A(t))/(1 - B(t)).$$

[*Hint*: Express the sums $\sum_0^\infty P_n(x, y)\, t^n$ as functions of

$$\sum_1^{\infty} t^n P.[\{S_F = n\} \cap \{X_n = \cdot\ \}],$$

and then use (1).]

(3) Assume now that the states are labelled

$$a_0 = a, a_1, a_2, \ldots, a_n, \ldots$$

$$b_0 = b, b_1, b_2, \ldots, b_n, \ldots$$

and that the transition probability is given by

$$P(a, a_i) = p_i = P(a_i, b), P(a_i, a_i) = q_i = 1 - p_i,$$

$$P(b, b_i) = p_i' = P(b_i, a), P(b_i, b_i) = q_i' = 1 - p_i',$$

with $\sum_i p_i = \sum_i p_i' = 1$, all other entries being 0. The chain X is irreducible recurrent, as is easily checked.

Let $\{n_k\}$ and $\{n_k'\}$ be two increasing sequences of integers such that $n_k < n_k' < n_{k+1}$. The numbers p_i are chosen in the following way: the first

n_1 of them are equal to $\varepsilon_1 = (2n_1)^{-1}$, the n_2 following ones are equal to $\varepsilon_2 = (2^2 n_2)^{-1}$, and so forth; there are therefore n_k consecutive p_i's equal to $\varepsilon_k = (2^k n_k)^{-1}$. The same pattern is followed for the numbers p_i', but with the use of $\{n_k'\}$ in place of $\{n_k\}$.

Show that one can choose the sequences $\{n_k\}$ and $\{n_k'\}$ in such a way that

$$\lim_{n\to\infty} R(1-\varepsilon_n) \neq \lim_{n\to\infty} R(1-\varepsilon_n'),$$

and deduce that the corresponding chain $\hat{X}$ is not normal.

(4) By a slight modification of the above example, prove that the limit in (1) may exist for $\hat{X}$ and fail to exist for X.

Exercise 3.12 (capacities). (1) Let X and $\hat{X}$ be two normal chains in duality, Γ and $\hat{\Gamma}$ two potential kernels as in Proposition 3.5. For every non-negligible special set A, show that there is a measure ν^A absolutely continuous with respect to m and such that

$$\lim_n P_n G^A = 1 \otimes \nu^A, \qquad \lambda_A \Gamma = C_A m + \nu^A.$$

(2) Let A, B be special and L be any special set containing $A \cup B$, then

$$C_B - C_A = \langle \nu^A - \nu^B, u_L \rangle.$$

(3) Show that $C_A \leqslant C_B$ if $A \subset B$.

For a negligible set A, we shall set $C_A = -\infty$. Then show that for any special sets $A_i, \ldots, A_n$,

$$C_{\bigcap_{i=1}^n A_i} \leqslant \sum_{i=1}^n C_{A_i} - \sum_{i \neq j} C_{A_i \cup A_j} + \cdots + (-1)^{n+1} C_{A_1 \cup A_2 \cdots \cup A_n}.$$

Exercise 3.13. Extend the definitions and results in this section to Harris resolvents.

4. Feller chains and recurrent boundary theory

Throughout this section, E is an LCCB and the transition probabilities P and $\hat{P}$ are Feller kernels. Since $P1 = 1$, in order that P be Feller it suffices that P maps C_K into C_b.

Since the topology of E has a countable basis, it is readily checked that there is a largest open set O such that $m(O) = 0$. We leave to the reader as an exercise the task of showing that O^c is an absorbing set. The set O

is thus of no significance in the study of X, and in the sequel we shall always assume that the following condition is satisfied.

Condition 4.1. For every open set U we have $m(U) > 0$. This condition is naturally satisfied for all the usual chains such as discrete chains and Harris random walks.

Proposition 4.2. *There exists a special function which is strictly positive and continuous.*

Proof. Let h be a strictly positive special function. There exists a number a such that $\infty > m(\{h \geqslant a\}) > 0$. The set $\{h \geqslant a\}$ contains a compact set K with $m(K) > 0$, and K is special.

For $0 < c < 1$, the function $U_c(\cdot, K)$ is upper-semi-continuous, strictly positive and special. The space E is the union of the closed sets $\{U_c(\cdot, K) \geqslant n^{-1}\}$. By Baire's theorem one of these sets has a non-empty interior; there exists a non-zero function φ in C_K^+ that has support in this set and so is special. By Condition 4.1 we have $m(\varphi) > 0$, and therefore $U_c\varphi$ is special, continuous and strictly positive.

This proposition has the following obvious

Corollary 4.3. *Every function in* $b\mathscr{E}_+$ *with compact support is special. The measure* m *is a Radon measure.*

Let $f(\hat{f})$ be continuous, special (co-special) and strictly positive. Then by Proposition 1.3, there exists a multiple, say h, of $f \wedge \hat{f}$ such that $U_h > 1 \otimes m$ and $\hat{U}_h > 1 \otimes m$. The function h is continuous and strictly positive. For any non-zero function in C_K^+ there is a multiple, say g, such that $U_g > 1 \otimes m$ and $\hat{U}_g > 1 \otimes m$. In the sequel h and g will always stand for functions enjoying the above properties.

Proposition 4.4. *If* $f \in C_b$ *and* $|f| < ch$ *for a positive number* c *(in particular if* $f \in C_K$*) then* $U_h(f)$ *and* $W_h(f)$ *are bounded and continuous. Furthermore* $W_g(f)$ *is bounded and continuous for every* g.

Proof. If suffices to consider a function f such that $0 \leqslant f \leqslant h$. Then $U_h(f)$ is lower semi-continuous because it is defined by a series of positive continuous functions. Since $h - f \leqslant h$, $U_h(h - f) = 1 - U_h(f)$ is also lower semi-continuous; hence $U_h(f)$ is continuous. The same argument works for $W_h(f)$, because $W_h(h)$ is constant and therefore continuous. Finally, by ch. 6 §5, we have

$$W_g(f) = W_h(f) - \frac{m(f)}{m(g)} W_h(g) + \frac{1}{m(h)} \langle h, W_g f \rangle,$$

and the latter part of the statement follows from the former.

In §3, we saw that aperiodic positive chains are normal. This was due to the fact that the powers P_n of the T.P. converge to the bounded invariant measure m. We are now going to give for null chains a result relating normality to some kind of convergence of the powers P_n.

Let $\Gamma = I + W$ be a potential kernel, where W is associated with one of the functions h or g above. Choose a denumerable set $\{f_n\}_{n \geqslant 1}$ of functions dense in C_K for the topology of uniform convergence on compact sets. By the above proposition the functions Γf_n are bounded and continuous. As in ch. 7 §2, we choose a metric δ on E, compatible with the topology of E and such that the Cauchy completion of (E, δ) is obtained by adjoining the Alexandrof's point Δ. Furthermore, for x, y in E we set

$$d(x, y) = \sum_{n=1}^{\infty} |\Gamma f_n(x) - \Gamma f_n(y)|/2^n \|\Gamma f_n\|.$$

Then $d + \delta$ is a metric on E, and we lay down

Definition 4.5. We define E^* to be the Cauchy completion of the metric space $(E, d + \delta)$ and set $M = E^* \setminus E$. The set M is called the *recurrent boundary* of E for X.

For every compact subset K of E, there is a number M_K such that if $f, g \in b\mathscr{E}$ and vanish outside K, then

$$|\Gamma(f - g)| \leqslant M_K \|f - g\|.$$

As a result, the sequence $\{\Gamma f_n\}$ is dense in the set $\{\Gamma f, f \in C_K\}$ for the topology of uniform convergence, and Γf is uniformly continuous on $(E, d + \delta)$ for every $f \in C_K$. These functions thus extend continuously to E^*; the extended functions will still be denoted by Γf. If x is a point in M, the map $f \to \Gamma f(x)$ defines clearly a Radon measure on E which will be denoted by $\Gamma(x, \cdot)$.

Let us observe that a sequence $\{x_n\}$ of points in E converges to a point x in M, if and only if $\{x_n\}$ converges to Δ in the original topology of E and if the measures $\Gamma(x_n, \cdot)$ converge vaguely to $\Gamma(x, \cdot)$. The metric $d + \delta$ is also extended to E^* by using the extended kernel Γ.

The space E^* may depend on the particular function h or g used in defining Γ. In the following proposition, we accordingly write E_g^* and E_h^*.

Proposition 4.6. *Up to homeomorphism, the space E_g^* does not depend on the particular function g, and there is a continuous map from E_h^* onto E_g^*.*

Proof. From the identity

$$W_g(x, \cdot) + \frac{1}{m(g)} W_h g(x) \otimes m = W_h(x, \cdot) + \frac{1}{m(h)} 1 \otimes (hm) W_g,$$

it follows that if $\{x_n\}$ is a sequence of points in E converging to Δ and such that $W_h(x_n, \cdot)$ converges vaguely, then $W_g(x_n, \cdot)$ converges vaguely. We define the map j from E_h^* to E_g^* by setting $W_g(j(x), \cdot) = \lim W_g(x_n, \cdot)$ if $W_h(x, \cdot) = \lim W_h(x_n, \cdot)$. If $\{x_n\}$ converges to x in E_g^*, one may extract a subsequence $\{x_m\}$ from $\{x_n\}$ such that $W_h(x_m, \cdot)$ converges; consequently the map j is onto. It is also clearly continuous, and that E_g^* does not depend on g is now evident.

Remark. If x and y are two points in E_h^* such that $j(x) = j(y)$, the kernels $W_h(x, \cdot)$ and $W_h(y, \cdot)$ are equal up to addition of a multiple of $1 \otimes m$. This will be useful in the following chapter, where E^* will be computed in the case of random walks.

From now on, E^* will denote the Cauchy completion of E obtained for a function $g \in C_K \cap \mathscr{U}_+$ and such that $U_g > 1 \otimes m$. We observe that if x and y are two points in E such that $\Gamma(x, \cdot) = \Gamma(y, \cdot)$, then, by ch. 6, Theorem 5.2, we have

$$W(x, \cdot) + m(g)^{-1} \otimes (gm) P = W(y, \cdot) + m(g)^{-1} \otimes (gm) P,$$

and since $W = \Gamma - I$, it follows that $x = y$. If x and y are in M and $\Gamma(x, \cdot) = \Gamma(y, \cdot)$, the sequence whose terms are alternately x and y is a Cauchy sequence and thus $x = y$. But it may happen that for a point x in E and a point s in M we have $\Gamma(x, \cdot) = \Gamma(s, \cdot)$. Also $\Gamma(s, \cdot)$ may be a barycenter of measures $\Gamma(s', \cdot)$ for other points s' in M. This recalls the situation of ch. 7 which explains the additional condition under which we are going to work.

If s is in M, the measure $\Gamma(s, \cdot)$ is the limit of the measures $\Gamma(x_n, \cdot)$ for a suitable sequence $\{x_n\}$. By passing to the limit, for the vague topology on E, in the identity

$$\Gamma P f = \Gamma f - f + (gm)\, Pf/m(g)$$

where $f \in C_K^+$, we get

$$\Gamma(s, \cdot\,)\, P \leqslant \Gamma(s, \cdot\,) + (gm)\, P/m(g).$$

Let us call ϑ the set of positive Radon measures λ such that $\lambda(g) = (1 - m(g))/m(g)$ and

$$\lambda P \leqslant \lambda + (gm)\, P/m(g).$$

The significance of the following condition will be made clearer later on in the case of discrete chains.

Condition 4.7. For any $\lambda \in \vartheta$, there is a unique measure β_λ on E^* such that

$$\lambda = \int_{E^*} \Gamma(s, \cdot\,)\, \beta_\lambda(\mathrm{d}s).$$

We may now turn to the result we were aiming at.

Definition 4.8. The chain X is said to be *weakly normal* if the sequences $\sum_1^n P_k f$ have a finite limit whenever n converges to $+\infty$, for every $f \in \mathcal{N} \cap C_K$.

Proposition 4.9. *The chain X is weakly normal if and only if $\{P_n \Gamma f\}$ converges pointwise for any $f \in C_K$. The limit is then equal to $\nu(f)$ where $\nu \in \vartheta$.*

Proof. The reasonings are the same as in Proposition 3.3 and Theorem 3.4. The fact that ν belongs to ϑ follows upon passing to the limit in the equation

$$P_n \Gamma P f = P_n \Gamma f - P_n f + (gm)\, Pf/m(g).$$

Theorem 4.10. *If X is null and $\{P_n(x, \cdot\,)\}$ converges vaguely on E^* for every x in E, then X is weakly normal. If Condition 4.7 is in force and X is weakly normal, then $\{P_n(x, \cdot\,)\}$ converges vaguely on E^* for any $x \in E$. The limit β is independent of x and carried by M.*

Proof. Since the functions Γf, $f \in C_K$, have a continuous extension to E^*, the first sentence is obvious.

Conversely, since E^* is compact, for any x, we may find a subsequence $\{n_k\}$ such that $\{P_{n_k}(x, \cdot)\}$ converges vaguely on E^* to a limit β. By Theorem 2.6 in ch. 6, the measure β is carried by M. Since for any $f \in C_K$, the function Γf extends continuously to E^* we have

$$\beta\Gamma f = \lim_k P_{n_k}\Gamma f(x) = \nu(f),$$

in other words

$$\nu = \int \Gamma(s, \cdot)\, \beta(ds).$$

By the uniqueness in Condition 4.7 we see that β does not depend on x nor on the particular subsequence. Thus $\{P_n\}$ converges vaguely on E^* which completes the proof.

Remarks. (1) If the equivalent conditions in the above theorem hold, the sequences $\{P_n\Gamma f\}$ converge for every function f such that Γf extends to a function in $C(E^*)$. We will see in the next chapter that, even for random walks, the class of such functions does not always include all special and continuous functions.

(2) The measure β has the following intuitive interpretation. Assume E to be discrete and let K be a finite set; then $P_nP_K(x, y)$ may be viewed as the probability that the chain started at x at time $-n$ enters K at point y after time zero. If λ_K is the probability measure defined in Theorem 3.1, then $\lambda_K(y)$ appears as the probability that the chain started at time $-\infty$ enters K at point y after time zero. If we let K swell to E, then λ_K must converge to an entrance distribution for the chain. It will be seen in Exercise 4.16 that the limit of λ_K is precisely the measure β. That β is an entrance distribution is also well seen from the discussion below.

In the remainder of this section we shall assume that X is a recurrent irreducible discrete chain and we shall try to illustrate the results of this and the previous sections. In the following chapter a similar study will be performed for Harris random walks on abelian groups.

In the sequel, e will stand for an arbitrary but fixed point in E. By the above discussions, we may define $\Gamma = I + W$ by taking as function g a suitable multiple of $1_{\{e\}}$. To lighten the notation we shall drop the brackets and write 1_e for the characteristic function of the set $\{e\}$ and $G^e(x, \cdot)$ instead of $G^{\{e\}}(x, \cdot)$.

From ch. 3, Theorem 1.9 it is known that $H_e(x, y) = 0$ if $x \neq e$ and $H_e(e, y) = m(y)/m(e)$. It follows then from Theorem 2.1 that

$$\Gamma = G^e + \frac{\Gamma(\cdot, e)}{m(e)} \otimes m + 1 \otimes (v_e \Gamma - C_e m);$$

but $\Gamma(\cdot, e) = 1_e + W(1_e)$, and $W(1_e)$ is a constant function; hence

$$\Gamma = G^e + \frac{1_e}{m(e)} \otimes m + 1 \otimes \nu,$$

where $\nu = km + v_e\Gamma - C_e m$ for a constant k. This will be used to prove

Proposition 4.11. *The space E^* is isomorphic to the Martin spacc of the transient chain $\hat{Y}$ whose T.P. is defined by $\hat{Q}(x, y) = \hat{P}(x, y)$ if $y \neq e$ and $\hat{Q}(x, e) = 0$, started with distribution $\Pi = \varepsilon_e$.*

Proof. The chain $\hat{Y}$ is clearly transient, and is in duality relative to m with the chain Y whose T.P. is given by $Q(x, y) = P(x, y)$ if $x \neq e$ and $Q(e, y) = 0$. The potential of Y is easily computed and is equal to

$$G^Y(x, \cdot) = G^e(x, \cdot) + 1 \otimes \varepsilon_e.$$

Furthermore $G^Y 1_e = 1$ so that $1_e/m(e)$ may be used as a reference function for $\hat{Y}$. Now the space E^* is homeomorphic to the closure of the set $\{\Gamma(x, \cdot), x \in E\}$ for the vague topology, but this is also homeomorphic to the closure of the set $\{G^e(x, \cdot), x \in E\}$, which in turn is homeomorphic to the closure of the set $\{G^Y(x, \cdot), x \in E\}$, and the proof is completed.

We can now see the significance of Condition 4.7; in the present case it means that, with the notation of ch. 7 §3, the boundary M is equal to its extremal or useful part M_e. We may also use the representation theorem for $\hat{Q}$-harmonic functions whose significance for the chain $\hat{X}$ is given by the following

Proposition 4.12. *A finite positive function f with $f(e) = 1$ is $\hat{Q}$-harmonic if and only if it is a solution of the Poisson equation*

$$(I - \hat{P}) f = - \hat{P}(\cdot, e).$$

Proof. By definition of $\hat{Q}$, the equation $\hat{Q}f = f$ reads

$$\sum_{y \neq e} \hat{P}(x, y) f(y) = f(x);$$

in other words

$$\hat{P}f(x) - \hat{P}(x, e) = f(x),$$

and the result follows.

We thus see that $\hat{Q}$-harmonic functions are co-subharmonic functions and provide solutions of the Poisson equation with second member the opposite of a multiple of $\hat{P}(\cdot, e)$. The function $\hat{P}(\cdot, e)$ is co-special and if k is another co-special function, then $\varphi = m(k)\,\hat{P}(\cdot, e) - k\,m(e)$ is a co-charge and the Poisson equation $(I - P)\,f = \varphi$ has the solution $\hat{\Gamma}\varphi$. If f is $\hat{Q}$-harmonic, such that $f(e) = m(k)$, then $\hat{\Gamma}\varphi + f$ is a solution of the Poisson equation $(I - \hat{P})\,f = -k$. As a result, if there exists a $\hat{Q}$-harmonic function, one can solve the above Poisson equation whatever the co-special function k is.

It is not obvious that there are always $\hat{Q}$-harmonic functions such that $f(e) = 1$. The above discussion shows that weak normality entails the existence of such functions for which we have an integral representation. Indeed if s is in M_e, the density $\hat{\gamma}^s = \mathrm{d}\Gamma(s, \cdot)/\mathrm{d}m$ is such that $\hat{P}\hat{\gamma}^s = \hat{\gamma}^s + \hat{P}(\cdot, e)$ and $\hat{\gamma}^s(e) = m(e)\,(1 - (m(g))/m(g) = \alpha$. In the present setting the integral representation theorem becomes

Theorem 4.13. *If X is null, for every positive solution f of the equation $(I - \hat{P})\,f = -\hat{P}(\cdot, e)$ such that $f(e) = \alpha$, there is a unique probability measure β^f, carried by M_e and such that*

$$f = \int_M \hat{\gamma}^s\,\mathrm{d}\beta^f(s).$$

This raises the problem of describing M_e for classical recurrent chains; that will be done in the next chapter for Harris random walks on abelian groups.

Exercise 4.15. Call F the Cauchy completion of E with respect to the metric d in the text. Prove that X is weakly normal if and only if the sequences $P_n(x, \cdot)$ converge vaguely on F. Is the corresponding measure β still carried by the boundary?

Exercise 4.16. Let X be weakly normal and $\{f_k\}$ a sequence of functions in C_K^+ increasing to 1_E; let $\lambda_k = \lim_n P_n U_{f_k} I_{f_k}$ (cf. Theorem 3.1), and prove that, under Condition 4.7, the sequence $\{\lambda_k\}$ converges vaguely on E^* to the measure β.

Exercise 4.17. Assume that X is an irreducible null recurrent discrete chain on $E = \mathbf{N}$ such that $P(k, l) = 0$ if $k \geqslant l + 2$. (See, for instance, ch. 2, Exercise 1.24.)

(1) Prove that m is the only invariant measure for P even if we allow the measures to take negative values.

(2) Prove that from every sequence of integers converging to infinity one may extract a sub-sequence such that, for all pairs (x, y) of points in E,

$$\lim_j P_{n_j} W_h(x, y) = \nu(y)$$

exists. Notice that Pf has finite support if f has, and prove that

$$\nu P = \nu + \frac{(hm)\, P}{m(h)}.$$

Then derive that X is weakly normal.

Exercise 4.18 (degenerate discrete chains). We use this term for discrete, weakly normal chains such that the value C_A of the Robin's constant is the same for all finite non-empty sets A.

(1) Assume that $\hat{X}$ is also normal in the same sense and show that the kernels Γ and $\hat{\Gamma}$ in Proposition 3.5 may be chosen in such a way that $C_A = 0$ for every finite non-empty set A, and that in that case $\Gamma(x, x) = \hat{\Gamma}(x, x) = 0$ for all x in E.

(2) Let (x, y) be any pair of points in E and set $A = \{x, y\}$. Prove that either

$$\Gamma(y, x) = 0, \qquad \Gamma(x, y) > 0, \qquad \lambda_A(y) = 1, \qquad \lambda_A(x) = 0$$

or

$$\Gamma(y, x) > 0, \qquad \Gamma(x, y) = 0, \qquad \lambda_A(y) = 0, \qquad \lambda_A(x) = 1.$$

(3) Let $y < x$ stand for $\Gamma(y, x) = 0$, and show that $<$ is an ordering on E. Show that if $z < y < x$, then $P_z[T_y < T_x] = 1$, and conclude that in moving to the right, the chain can move at most at one step at a time.

(4) Prove that the ordering of states must be that of the integers, the positive integers, or the negative integers.

(5) Let X be the renewal chain of Exercise 1.16 in ch. 3. Find the condition under which it is null recurrent, and prove that it is then weakly normal and degenerate. Other important examples of degenerate chains will be given in the next chapter.

CHAPTER 9

RECURRENT RANDOM WALKS

This chapter is devoted to the study of Harris random walks. Although everything may be carried over to non-abelian groups, we shall deal here only with abelian groups, in order to avoid some technicalities and keep within reasonable bounds. Throughout this chapter, we thus study a Harris random walk with law μ on an abelian locally compact metrizable group G. To save space, we shall use the multiplicative notation in G. When G is compact, all the results are either trivial or meaningless, and we shall therefore always assume G to be non-compact. Finally, a Harris random walk being a Feller chain, we shall use freely the results of ch. 8, items 4.1–4.4.

1. Preliminaries

In the sequel, τ will stand for any translation operator in G, and we shall write τ_a when we want to emphasize the relevant element a in G, namely $\tau_a f(x) = f(xa)$. We write τ^{-1} for the inverse of τ.

Lemma 1.1. *For any function $h \in \mathscr{U}_+$ and such that $U_h > 1 \otimes m$, the function τh has the same properties and*

$$W_{\tau h} = \tau W_h \tau^{-1}.$$

Proof. Left to the reader as an exercise.

We shall first construct a function h with particularly nice properties.

Lemma 1.2. *Let $f \in C_0$ and K be a compact set in G; the function $g: x \to \inf_{b \in K} f(xb)$ is in C_0.*

Proof. Since the function f is uniformly continuous, for every $\varepsilon > 0$ there is a neighbourhood V of e such that if $xy^{-1} \in V$; then for all b in G,

$$|f(xb) - f(yb)| < \varepsilon.$$

Thus

$$f(xb) < f(yb) + \varepsilon,$$

and taking the infimum over $b \in K$, first in the left member then in the right one, yields

$$g(x) < g(y) + \varepsilon.$$

The inequality $g(y) < g(x) + \varepsilon$ may be shown in the same way, and thus the proof is complete.

Proposition 1.3. *There is a strictly positive symmetric function h in $\mathscr{U}_+ \cap C_0$ and such that*

(i) $U_h > 1 \otimes m$ *and* $\hat{U}_h > 1 \otimes m$;

(ii) $\lim_{b\to e} \|W_h(\tau_b h - h)\| = \lim_{b \to e} \|\hat{W}_h(\tau_b h - h)\| = 0$.

Proof. Let u be a symmetric non-negligible function in C_K^+. Since X is Harris, the function

$$h_1 = \alpha \sum_0^\infty 2^{-n} P_n u = \alpha(u + 2^{-1} U_{2^{-1}} u),$$

where $\alpha \in \mathbf{R}_+$, is strictly positive, in C_0, and special.

In the same way put

$$\hat{h}_1 = \alpha \sum_0^\infty 2^{-n} \hat{P}_n u;$$

by ch. 8, Proposition 1.3, if we take α small enough, then $h_2 = h_1 \wedge \hat{h}_1$ is symmetric, in $C_0 \cap \mathscr{U}_+$ and $U_{h_2} > 1 \otimes m$, $\hat{U}_{h_2} > 1 \otimes m$.

Let V be a symmetric compact neighbourhood of e and set

$$h(x) = \inf_{b \in V} h_2^2(xb).$$

We will show that h satisfies (ii), and since it clearly satisfies all the other requirements it is the desired function.

Pick $\varepsilon > 0$ and call K the compact set $\{h_2 \geq \varepsilon\}$. Since V is symmetric, for every x in $\boldsymbol{G}$ and b in V we have $h(xb) < h_2^2(x)$; hence $|h(x) - h(xb)| < h_2^2(x)$, and it follows that, for x in K^c and b in V,

$$|h(x) - h(xb)| < \varepsilon h_2(x).$$

On the other hand, since h is uniformly continuous, there is a neighbourhood $V' \subset V$ such that, for $b \in V'$ and $x \in K$,

$$|h(x) - h(xb)| \leqslant \varepsilon^2 \leqslant \varepsilon h_2(x).$$

Combining the two inequalities yields, for every $b \in V'$,

$$|\tau_b h - h| < \varepsilon h_2.$$

Consequently $\|W_{h_2}(\tau_b h - h)\| < \varepsilon m(h_2)/((1 - m(h_2))$, and since, by ch. 6, Corollary 5.5,

$$W_h(\tau_b h - h) = W_{h_2}(\tau_b h - h) - m(h)^{-1} \langle h, W_{h_2}(\tau_b h - h)\rangle,$$

it follows that $\lim_{b \to e}\|W_h(\tau_b h - h)\| = 0$. The dual result is obtained in the same way.

Throughout the sequel we shall use the kernels W_h and $\hat{W}_h$ (and we often write simply W and $\hat{W}$) associated with this particular function h. For every τ, the function τh has similar properties, and we shall also use the kernels $W_{\tau h}$ and $\hat{W}_{\tau h}$. They satisfy the following

Lemma 1.4. *For every special function φ and $a \in \mathbf{G}$,*

$$m(h)\, \|W_h\varphi - W_{\tau_a h}\varphi\| \leqslant m(\varphi)\, (\|\hat{W}_h(\tau_a h - h)\| + \|W_h(\tau_a^{-1} h - h)\|).$$

Proof. By ch. 6, Corollary 5.5,

$$W_h\varphi - W_{\tau h}\varphi = m(h)^{-1} [\langle \tau h, W_h\varphi\rangle - m(\varphi)\, W_{\tau h}h],$$

and since by ch. 6, Theorem 5.2 we have $W_h(h) = W_{\tau h}(\tau h)$, it follows that

$$W_h\varphi - W_{\tau h}\varphi = m(h)^{-1} [\langle \hat{W}_h(\tau h - h), \varphi\rangle - m(\varphi)\, W_{\tau h}(h - \tau h)].$$

The desired result is then a consequence of Lemma 1.1.

We now proceed to the key technical results.

Lemma 1.5. *For any x in $\mathbf{G}$ and $f \in C_K^+$, the set of numbers $\{W_h(\tau_a f)\,(x),\ a \in \mathbf{G}\}$ is bounded.*

Proof. Let V be a compact neighbourhood of e, and choose $f' \in C_K^+$ such that $\tau_b f' \geqslant f$ for every $b \in V$. Then

$$W_h(\tau_a f')\,(xb) = W_{\tau_b h}(\tau_{ab} f')\,(x) \geqslant W_{\tau_b h}(\tau_a f)\,(x).$$

Suppose there exists a sequence $\{a_n\}$ converging to Δ, and such that $W_h(\tau_{a_n}f)(x)$ converges to $+\infty$; the formula

$$W_{\tau_b h}(\tau_{a_n}f)(x) + \frac{m(f)}{m(h)} W_h(\tau_b h) \geqslant W_h(\tau_{a_n}f)(x)$$

then implies that

$$\lim_n W_{\tau_b h}(\tau_{a_n}f)(x) = +\infty,$$

hence that

$$\lim_n W_h(\tau_{a_n}f')(y) = +\infty,$$

the convergence being uniform in $y \in xV$. It follows if $g \in C_K^+$ and $g \geqslant 1_{xV}$ that

$$\lim_n \int g\, W_h(\tau_{a_n}f')\, dm = +\infty;$$

but since

$$\langle g, W_h(\tau_{a_n}f')\rangle \leqslant \|\hat{W}_h(g)\|\, \|f'\|_1 < \infty,$$

we have a contradiction.

Proposition 1.6. *For $f \in C_K$, the sets of functions*

$$\{W(\tau_a f),\ a \in \mathbf{G}\} \quad \text{and} \quad \{WP_n f,\ n \in \mathbf{N}\}$$

are equi-uniformly continuous on every compact set in $\mathbf{G}$.

Proof. Pick x in $\mathbf{G}$. For any $b \in \mathbf{G}$, we have

$$\begin{aligned} A(b) &= W_h(\tau_a f)(xb) - W_h(\tau_a f)(x) = W_{\tau_b h}(\tau_{ab} f)(x) - W_h(\tau_a f)(x) \\ &= W_{\tau_b h}(\tau_{ab} f)(x) - W_h(\tau_{ab} f)(x) + W_h(\tau_a(\tau_b f - f))(x) \\ &= \frac{1}{m(h)} (hm)\, W_{\tau_b h}(\tau_{ab} f) - \frac{m(f)}{m(h)} W_h(\tau_b h)(x) + W_h(\tau_a(\tau_b f - f))(x) \\ &= \frac{1}{m(h)} \langle \hat{W}_h(\tau_b^{-1} h), \tau_a f\rangle - \frac{m(f)}{m(h)} W_h(\tau_b h)(x) + W_h(\tau_a(\tau_b f - f))(x). \end{aligned}$$

Let K be the compact support of f, H a symmetric compact neighbourhood of e and f' a function in C_K^+ equal to 1 on HK. For $\varepsilon > 0$, one can find a neighbourhood V of e such that, for $b \in V$,

$$|\tau_b f - f| < \varepsilon f'.$$

By the choice of h, we may now assert that there is a neighbourhood V' of e such that, for $b \in V'$,

$$|A(b)| \leqslant \varepsilon + \varepsilon W_h(\tau_a f')\,(x).$$

By the preceding lemma, the functions $W_h(\tau_a f)$ are thus equi-uniformly continuous on a neighbourhood of x, and this implies the result by the Heine–Borel property.

For the set $\{WP_n f,\ n \in \mathbf{N}\}$, the conclusion follows readily from the first one and the fact that

$$WP_n f(x) = \int W(\tau_a f)\,(x)\,\mu^n(\mathrm{d}a).$$

Symmetrically, we have

Proposition 1.7. *For $f \in C_K$, the set of functions $\{P_n Wf,\ n \in \mathbf{N}\}$ is equi-uniformly continuous on every compact set in $\boldsymbol{G}$.*

Proof. This is much simpler owing to the fact that $P_n Wf \leqslant \|Wf\|$ for every n, and it is left to the reader as an exercise.

The next result is a first step towards normality theorems.

Proposition 1.8. *Let f be special; for every pair (x, y) of points in $\boldsymbol{G}$,*

$$\lim_{n\to\infty} (P_n W f(x) - P_n W f(y)) = 0.$$

Proof. Let $\mathfrak{U}$ be an ultrafilter finer than Frechet's filter. Since Wf is bounded for every x in $\boldsymbol{G}$, $L(x) = \lim_{\mathfrak{U}} P_n W f(x)$ exists. Moreover, for k fixed,

$$L(x) = \lim_{\mathfrak{U}} P_{n+k} W f(x).$$

Indeed

$$P_n W f(x) - P_{n+k} W f(x) = \sum_{i=1}^{k} P_{n+i}[f - (m(f)\,h/m(h))]$$

and since $f - (m(f)\,h/m(h))$ is bounded and m-integrable, by ch. 6, Theorem 2.6 the right member converges to zero as n tends to infinity.

To simplify the notation, we write φ_n below in place of $P_n Wf$. The measure μ being spread out, for k sufficiently large we have $\mu^k = \alpha + \beta$, where $\alpha = gm$. But

$$\varphi_{n+k}(x) = \langle \alpha * \varepsilon_x, \varphi_n \rangle + \langle \beta * \varepsilon_x, \varphi_n \rangle,$$

and therefore

$$\left|\varphi_{n+k}(x) - \int \varphi_n(y)\, g(yx^{-1})\, \mathrm{d}y\right| \leqslant \|Wf\|\, \|\beta\|.$$

The sequence $\{\varphi_n\}$ is bounded; hence it admits a limit ψ along $\mathfrak{U}$ in $\sigma(L^\infty, L^1)$. (As usual we shall not distinguish between functions and equivalence classes.) Passing to the limit in the last inequality, we thus get

$$\left|L(x) - \int \psi(y)\, g(yx^{-1})\, \mathrm{d}y\right| \leqslant \|Wf\|\, \|\beta\|,$$

which implies

$$|L(x) - P_k\, \psi(x)| \leqslant 2\|Wf\|\, \|\beta\|.$$

As k tends to infinity, $P_k\psi$ converges uniformly to L, which is therefore measurable and invariant, hence constant. The proof is then easily completed.

The following result is the key to the whole of the sequel.

Proposition 1.9. *Every sequence of integers converging to infinity contains a subsequence $s = \{n_j\}$ with the following property: for every f in C_K, the four sequences $\{WP_{n_j}f\}$, $\{P_{n_j}Wf\}$, $\{\hat{W}\hat{P}_{n_j}f\}$, $\{\hat{P}_{n_j}\hat{W}f\}$ converge uniformly on every compact subset as j tends to infinity.*

Moreover, the sequence of measures $\varepsilon_x WP_{n_j}$ converges vaguely to a multiple $\gamma^s(x)\, m$ of the Haar measure. The function γ^s is positive, uniformly continuous, and such that

$$P\gamma^s = \gamma^s + m(h)^{-1}\, Ph,$$

and for every τ, the function $(I - \tau)\, \gamma^s$ is bounded.

Of course, the same result holds for $\varepsilon_x \hat{W}\hat{P}_{n_j}$.

Proof. The group G being metrizable, the space C_K contains a dense countable subset. It suffices to prove the first statement for the functions of this subset; but this follows from Propositions 1.6–1.7 and from repeated applications of Cantor's diagonal process.

We turn to the second part of the proposition. From the first part, it is plain that the sequence $\{\varepsilon_x WP_{n_j}\}$ converges vaguely to a Radon measure ν^x and that the map $x \to \nu^x(f)$ is continuous. For $g \in C_K$, we have

$$\int \nu^x(f)\, g(x)\, m(\mathrm{d}x) = \lim_j \langle WP_{n_j}f, g\rangle = \lim_j \langle f, \hat{P}_{n_j}\hat{W}g\rangle,$$

so that if $m(f) = 0$, then $\nu^x(f) = 0$ which shows that $\nu^x = \gamma^s(x)\, m$. We thus have $\gamma^s(x) = \lim_j \hat{W}\hat{P}_{n_j} f(x)$ if f is any function in C_K^+ with $m(f) = 1$. Let f be such a function; then $\tau^{-1}f$ is also such a function, and we have

$$\tau\gamma^s = \lim_j \tau W P_{n_j} \tau^{-1} f = \lim_j W_{\tau h} P_{n_j} f.$$

By Lemma 1.4, we thus have

$$\|(I - \tau)\, \gamma^s\| \leqslant m(h)^{-1} \left(\|\hat{W}(\tau h - h)\| + \|W_h(\tau^{-1}h - h)\|\right);$$

by the choice of h the function γ^s is therefore uniformly continuous, and we also see that $(I - \tau)\, \gamma^s$ is the bounded limit of $(I - \tau)\, W P_{n_j} f$, hence is bounded for every τ.

We proceed to show that γ^s satisfies the equality in the statement. First passing to the limit in the relation

$$P_{n+j} f + P W P_{n_j} f = W P_{n_j} f + m(h)^{-1}\, Ph$$

yields, by mean of Fatou's lemma, that

$$P\gamma^s \leqslant \gamma^s + m(h)^{-1}\, Ph.$$

Next, by the bounded convergence above, $P(I - \tau)\, W P_{n_j} f$ converges to $P(I - \tau)\, \gamma^s$. It follows that, for every τ,

$$P(I - \tau)\, \gamma^s = (I - \tau)\, \gamma^s + m(h)^{-1}\, (I - \tau)\, Ph,$$

and since all the functions herein are finite, we may rewrite this as

$$\gamma^s - P\gamma^s + m(h)^{-1}\, Ph = \tau(\gamma^s - P\gamma^s + m(h)^{-1}\, Ph).$$

There is consequently a number $k \geqslant 0$ such that

$$P\gamma^s = \gamma^s + m(h)^{-1}\, Ph - k,$$

which implies

$$P_n\gamma^s = \gamma^s + m(h)^{-1} \sum_1^n P_k h - nk.$$

The function h being integrable, the term on the right converges to $-\infty$ if $k > 0$, and this contradicts the fact γ^s is positive. Hence $k = 0$ and

$$P\gamma^s = \gamma^s + m(h)^{-1}\, Ph.$$

The following result will be important in the sequel.

Proposition 1.10. *Every co-special function is integrable by the measure $\gamma^s m$. In particular,*

$$\int h\gamma^s \, dm = m(h)^{-1}(1 - m(h)) = q.$$

Proof. Let g be co-special and f in C_K^+ with $m(f) = 1$. By Fatou's lemma

$$\int g\gamma^s \, dm \leqslant \varliminf_j \langle g, WP_{n_j}f\rangle = \varliminf_j \langle \hat{P}_{n_j}\hat{W}g, f\rangle \leqslant \|\hat{W}_h(g)\|,$$

which proves the first conclusion. Since $\hat{W}_h(h) = q$ we also have

$$\int h\gamma^s \, dm \leqslant q. \tag{1.1}$$

If g is in C_K^+, since the convergence of $WP_{n_j}f$ is uniform on compact sets, we actually have

$$\int g\gamma^s \, dm = \lim_j \langle \hat{P}_{n_j}\hat{W}_h g, f\rangle = \lim_j \hat{P}_{n_j}\hat{W}_h g,$$

since this last limit is a constant function. The same equality is true for all functions $\hat{P}_k g$. Indeed

$$\int \hat{P}g\gamma^s \, dm = \int gP\gamma^s \, dm = \int g\gamma^s \, dm + m(h)^{-1}\int gPh \, dm$$

$$= \lim_j \hat{P}_{n_j}\hat{W}g + m(h)^{-1}\int \hat{P}gh \, dm,$$

which by ch. 6, Theorem 5.2 may be written

$$\int \hat{P}g\gamma^s \, dm = \lim_j \hat{P}_{n_j}\hat{W}(\hat{P}g) + \lim_j \hat{P}_{n_j+1}g,$$

and the last term is zero, by ch. 6, Theorem 2.6. Inductively we get the result for all k's. Now set $\varphi = \sum_0^\infty 2^{-k}\hat{P}_k g$; we have

$$\int \varphi\gamma^s \, dm = \sum_0^\infty 2^{-k}\left(\lim_j \hat{P}_{n_j}\hat{W}(\hat{P}_k g)\right),$$

but for every j, ch. 6, Theorem 5.2 implies that

$$\hat{P}_{n_j}\hat{W}(\hat{P}_k g) \leqslant \|\hat{W}g\| + k\|g\|,$$

so that we can interchange summation and passage to the limit to get

$$\int \varphi \gamma^s \, dm = \lim_j \hat{P}_{n_j} \hat{W} \varphi.$$

By the way h was chosen, we can manage to have $\varphi \geqslant h$. Yet another application of Fatou's lemma then gives

$$\int (\varphi - h) \, \gamma^s \, dm \leqslant \lim_j \hat{P}_{n_j} \hat{W} (\varphi - h) = \int \varphi \gamma^s \, dm - q.$$

Combining this with eq. (1.1) yields the desired result.

Clearly if γ^s were the only solution of the equation $Pf = f + m(h)^{-1} Ph$, the measures $\varepsilon_x W P_n$ would converge vaguely to a limit, and using the duality it would not be difficult to derive a theorem of normality for random walks. That is why we shall now study the uniqueness of the solutions of the above equation. In what follows a *real character* of $\mathbf{G}$ will be a homomorphism of $\mathbf{G}$ into $\mathbf{R}$.

Proposition 1.11. *Let g be a special function and f_1, f_2 two functions in $\mathscr{E}_+$ such that*

$$Pf_1 - f_1 = Pf_2 - f_2 = g.$$

If for every $\tau \in \mathbf{G}$, the function $(I - \tau)(f_1 - f_2)$ is bounded, then $f_1 - f_2 = k + \chi$ where k is a constant and χ a continuous real character of $\mathbf{G}$ such that $P|\chi| - |\chi|$ is m-integrable.

Proof. Set $f = f_1 - f_2$; plainly

$$P(I - \tau) f - (I - \tau) f = 0,$$

and since $(I - \tau) f$ is bounded it follows that for every $a \in \mathbf{G}$, there is a constant $C(a)$ such that for all x in $\mathbf{G}$,

$$f(x) - f(xa) = C(a).$$

Making $x = e$ in this equation yields $C(a) = f(a) - f(e)$; hence

$$f(xa) = f(x) + f(a) - f(e).$$

It remains to set $k = f(e)$ and $\chi(x) = f(x) - f(e)$ to get the desired constant and character. Since the character χ is measurable, it is also continuous.

It is easily checked that $P\chi = \chi$, which implies $P|\chi| \geqslant |\chi|$. If the functions $v = P|\chi| - |\chi|$ were not m-integrable, by the Chacon–Ornstein theorem (ch. 4, Theorems 3.2 and 3.5) the sequence $\sum_1^n P_k v / \sum_1^n P_k g$ would converge to infinity with n. But since $|\chi| \leqslant k + f_1 + f_2$ for every n,

$$P_n|\chi| \leqslant k + f_1 + f_2 + 2\sum_1^n P_k g,$$

and we would have a contradiction.

The existence of a non-trivial character demands that the group $\boldsymbol{G}$ as well as the law μ satisfy some stringent conditions that we describe below.

Proposition 1.12. *If there is a non-trivial real continuous character χ such that $P|\chi| - |\chi|$ is m-integrable, the group $\boldsymbol{G}$ is isomorphic to $\mathbf{R} \oplus \boldsymbol{K}$ or $\mathbf{Z} \oplus \boldsymbol{K}$, where $\boldsymbol{K}$ is a compact abelian group. Moreover χ is unique up to multiplication.*

Proof. The kernel Ker χ of the character χ is a closed subgroup of $\boldsymbol{G}$. Let us call m_1 and m_2 Haar measures on Ker χ and $\boldsymbol{G}/\text{Ker}\,\chi$. For any $f \in \mathscr{E}_+$, the integral

$$I(g) = \int_{\text{Ker}\,\chi} f(gx_1)\, dm_1(x_1)$$

is constant on the co-sets of Ker χ, and therefore we may write

$$\int_G (P|\chi| - |\chi|)\, dm = \int_{G/\text{Ker}\,\chi} dm_2(x_2) \int_{\text{Ker}\,\chi} (P|\chi| - |\chi|)\,(x_2 x_1)\, dm_1(x_1),$$

if m_1 and m_2 are suitably chosen. Now the function $P|\chi| - |\chi|$ is constant on the co-sets of Ker χ, because if $g \in \text{Ker}\,\chi$,

$$P|\chi|\,(xg) = \int |\chi|\,(yxg)\,\mu(dy) = P|\chi|\,(x),$$

and of course $|\chi|(xg) = |\chi|(x)$. Consequently, we have

$$\int_G (P|\chi| - |\chi|)\, dm = m_1(\text{Ker}\,\chi) \int_{G/\text{Ker}\,\chi} (P|\chi| - |\chi|)\,(x_2)\, dm_2(x_2).$$

If the last integral vanishes, then $P|\chi| = |\chi|$. But this implies that $|\chi|$ is harmonic and hence constant, which is impossible since χ is non-trivial. It follows then from the hypothesis that $m_1(\text{Ker}\,\chi) < \infty$, and therefore that Ker χ is a compact sub-group of $\boldsymbol{G}$.

Now the group $G/\text{Ker}\ \chi$ is isomorphic to a sub-group of $\mathbf{R}$, and is locally compact for a topology finer than the topology of $\mathbf{R}$. By the structure theorem, $G/\text{Ker}\ \chi$ has an open sub-group isomorphic to $\mathbf{R}^n \oplus K$, where K is compact; since χ is one-to-one on $G/\text{Ker}\ \chi$, it follows that $G/\text{Ker}\ \chi$ is either isomorphic to $\mathbf{R}$ or discrete. In the latter case, a moment's thought proves that $P|\chi| - |\chi|$ is the restriction to $G/\text{Ker}\ \chi$ of a continuous function on $\mathbf{R}$; hence is greater than a number $k > 0$ on a neighbourhood of e in $\mathbf{R}$. Consequently the integral of this function with respect to the Haar measure of the sub-group is not finite unless there is only a finite number of elements of the sub-group in this neighbourhood of e. But this implies that the sub-group is isomorphic to $\mathbf{Z}$.

The group G is therefore compactly generated, hence isomorphic to $\mathbf{R}^n \oplus \mathbf{Z} \oplus K$, with K a compact group. By the above argument it is either $\mathbf{R} \oplus K$ or $\mathbf{Z} \oplus K$. The uniqueness property in the statement is now clear.

From now on, when we deal with such a group, it will always be assumed that its Haar measure m is the product of the Lebesgue measure by the normalized Haar measure on the compact factor. Moreover, a point in G will be a pair (x, k), where x is in $\mathbf{R}(\mathbf{Z})$ and k is in K, and χ will denote the character such that $\chi(x,k) = x$. It is then known, by ch. 3 §3.12, that $\int_G \chi \, \mathrm{d}\mu = 0$ if the integral is meaningful, and we define

$$\sigma^2 = \int_G \chi^2 \, \mathrm{d}\mu.$$

We have

Theorem 1.13. *There is a non-trivial real continuous character such that $P|\chi| - |\chi|$ is m-integrable if and only if the group G is of the above type and $\sigma^2 < +\infty$.*

Proof. Call $\tilde{\mu}$ the image of μ by the mapping χ and $\tilde{P}$ the T.P. on $G/\text{Ker}\ \chi$ associated with $\tilde{\mu}$. Since χ^2 and $P|\chi| - |\chi|$ are constant on the co-sets of $\text{Ker}\ \chi$, it is easily seen that

$$\sigma^2 = \int_G \chi^2 \, \mathrm{d}\mu = \int_{G/\text{Ker}} x^2 \, \mathrm{d}\tilde{\mu}(x),$$

$$I = \int_G (P|\chi| - |\chi|) \, \mathrm{d}m = \int_{G/\text{Ker}\,\chi} (\tilde{P}|\chi| - |\chi|) \, \mathrm{d}m_2.$$

It suffices therefore to prove that $I = \sigma^2$ when $\mathbf{G} = \mathbf{R}$ or $\mathbf{Z}$. We do it for $\mathbf{R}$, the proof for $\mathbf{Z}$ being identical.

Let χ^+ and χ^- be the positive and negative parts of χ. From the equality $P\chi = \chi$, we deduce that $P\chi^+ - \chi^+ = P\chi^- - \chi^-$, whereupon

$$P|\chi| - |\chi| = 2(P\chi^+ - \chi^+) = 2(P\chi^- - \chi^-).$$

Since χ^+ vanishes on $\mathbf{R}_-$ and χ^- on $\mathbf{R}_+$ it follows that

$$I = 2\int_{-\infty}^{0} P\chi^+(x)\,dx + 2\int_{0}^{+\infty} P\chi^-(x)\,dx,$$

and using the dual random walk,

$$\begin{aligned} I &= 2\int_{-\infty}^{+\infty} \chi^+(x)\,\hat{P}1_{]-\infty,0]}(x)\,dx + 2\int_{-\infty}^{+\infty} \chi^-(x)\,\hat{P}1_{]0,+\infty[}(x)\,dx \\ &= 2\int_{0}^{+\infty} x\,\hat{P}1_{]-\infty,0]}(x)\,dx - 2\int_{-\infty}^{0} x\,\hat{P}1_{]0,+\infty]}(x)\,dx \\ &= 2\int_{0}^{+\infty} x\,\mu([x,\infty[)\,dx - 2\int_{-\infty}^{0} x\,\mu(]-\infty,x[)\,dx. \end{aligned}$$

We may apply Fubini's theorem in each of the two integrals, since the functions therein have constant signs, and we get the desired result.

Definition 1.14. The random walk is of *type II* if the conditions of the preceding theorem hold. Otherwise it is of *type I*.

In the sequel, when we deal simultaneously with type I and type II random walks, the character χ occurring in the formulae will be understood as identically zero in the former case.

If the random walk is of type II we shall set

$$\mathbf{G}^+ = \{x \in \mathbf{G} : \chi(x) \geqslant 0\} \quad \text{and} \quad \mathbf{G}^- = (\mathbf{G}^+)^{-1}.$$

We shall need the following

Lemma 1.15. *If the random walk is of type II, for every $a \in \mathbf{G}$ the sequence $(I - \tau_a)\,P_n|\chi|$ converges boundedly to zero.*

Proof. Pick x and a in $\mathbf{G}$; we may write

$$|\chi|\,(yxa) - |\chi|\,(yx) = \alpha 1_{\mathbf{G}^+}(y) - \alpha 1_{\mathbf{G}^-}(y) + \psi(yx),$$

where α is equal to $|\chi|\,(a)$ and ψ is a function with compact support which depends on x but is majorized by a constant which depends only on a. (The reader is invited to draw the graph of $(I - \tau_a)\,|\chi|$ when $\boldsymbol{G} = \mathbf{R}$.) It follows that

$$(I - \tau_a)\,P_n|\chi|\,(x) = \alpha\,\mu^n(\boldsymbol{G}^+) - \alpha\,\mu^n(\boldsymbol{G}^-) + P_n\,\psi(x).$$

Since $\sigma^2 < \infty$, we may use the central limit theorem to the effect that

$$\lim_n \mu^n(\boldsymbol{G}^+) = \lim_n \mu^n(\boldsymbol{G}^-) = \tfrac{1}{2},$$

and by ch. 6, Theorem 2.11 we have $\lim_n P_n\,\psi(x) = 0$, which completes the proof.

Exercise 1.16. Extend all the results of this section to non-abelian groups.
[*Hint*: In Proposition 1.6 use left translations instead of right translations.]

2. Normality and potential kernels

In this section we prove a first result of normality, from which we derive the construction of convolution kernels that are potential kernels in the sense of ch. 8. We start by showing that the function γ^s in Proposition 1.9 does not depend on s.

Theorem 2.1. *There is a uniformly continuous positive function γ such that $P\gamma - \gamma = m(h)^{-1}\,Ph$, and such that, for every f in C_K,*

$$\lim_n WP_n\,f(x) = m(f)\,\gamma(x)$$

uniformly on compact sets.

If we want to emphasize the dependence of γ on h we shall write γ_h.

Proof. Let $s = \{n_j\}$ and $s' = \{n'_j\}$ be two sequences such that the sequences $\{WP_{n_j}\}$ and $\{WP_{n'_j}\}$ converge vaguely to $\gamma^s \otimes m$ and $\gamma^{s'} \otimes m$ with $\gamma^s \neq \gamma^{s'}$. By Propositions 1.9–1.11, there are two constants α and β such that $\gamma^s - \gamma^{s'} = \alpha + \beta\chi$, where χ vanishes if the random walk is of type I. By Proposition 1.10, we have

$$\int h\gamma^s\,dm = \int h\gamma^{s'}\,dm = (1 - m(h)/m(h));$$

hence

$$\alpha \int h \, dm + \beta \int h\chi \, dm = 0;$$

since h is symmetric, $\int \chi h \, dm = 0$ and therefore $\alpha = 0$, and the proof is completed for type I random walks.

Now let X be type II and g a function in C_K^+ such that $m(g) = 1$; then

$$\beta\chi = \lim_j (WP_{n_j}g - WP_{n'_j}g).$$

By taking sub-sequences, we may assume that $n_j \leqslant n'_j$. Let a be in $\mathbf{G}$ outside $\operatorname{Ker} \chi$; from the equality

$$WP_n g = Wg + m(h)^{-1} \langle h, Pg + \cdots + P_n g \rangle - (Pg + \cdots + P_n g),$$

we derive that

$$-\beta\chi(a) = \beta(I - \tau_a)\chi = \lim_j \sum_{n_j}^{n'_j} (I - \tau_a) P_k g.$$

By ch. 6, Proposition 5.7, the sequence on the right is bounded, and since $\hat{P}|\chi| - |\chi|$ is m-integrable, we have

$$\begin{aligned} -\beta\chi(a)\,\sigma^2 &= \beta \int ((I - \tau_a)\chi)(\hat{P}|\chi| - |\chi|)\, dm \\ &= \lim_j \int \left(\sum_{n_j}^{n'_j} (I - \tau_a) P_k g \right) (\hat{P}|\chi| - |\chi|)\, dm \\ &= \lim_j \langle g, (I - \tau_{a^{-1}})(\hat{P}_{n'_j+1}|\chi| - \hat{P}_{n_j}|\chi|) \rangle, \end{aligned}$$

and this limit is zero, by Lemma 1.15. It follows that $\beta = 0$, which completes the proof.

Remark. Of course, the symmetric result holds for the dual chain, and we call $\hat{\gamma}$ the corresponding function. It is easily seen that $\hat{\gamma}(x) = \gamma(x^{-1})$.

The following result of normality will be sharpened at the end of the section.

Theorem 2.2. *For every positive function f majorized by a multiple of h, we have*

$$\lim_n P_n W f = \int \hat{\gamma} f \, dm, \qquad \lim_n \hat{P}_n \hat{W} f = \int \gamma f \, dm.$$

Proof. It suffices to prove the result for the dual chain and for $f \leqslant h$.

By Proposition 1.8, we may find a sequence n_j such that

$$\lim_{n_j} \hat{P}_{n_j} \hat{W} f = \underline{\lim}_n \hat{P}_n \hat{W} f,$$

and this limit is a constant function. Let g be in C_K^+, with $m(g) = 1$; passing to the limit in the equality

$$\langle \hat{P}_{n_j} \hat{W} f, g \rangle = \langle f, W P_{n_j} g \rangle,$$

we get, using Fatou's lemma, that

$$\underline{\lim}_n \hat{P}_n \hat{W} f \geqslant \int \gamma f \, dm.$$

But $h - f$ is also positive and less than h, and therefore

$$\underline{\lim}_n \hat{P}_n \hat{W}(h - f) \geqslant \int \gamma (h - f) \, dm.$$

Since $\hat{P}_n \hat{W} h = \int \gamma h \, dm$, it follows that

$$\overline{\lim}_n \hat{P}_n \hat{W} f \leqslant \int \gamma f \, dm,$$

which completes the proof.

We turn to the construction of a potential kernel.

Theorem 2.3. *The kernel*

$$A = I + W - 1 \otimes \hat{\gamma} m - \gamma \otimes m$$

is a convolution kernel which has the following properties:

(i) *it maps special functions into functions finite and bounded above, and maps charges into bounded functions;*

(ii) *if f is a charge such that $|f| \leqslant h$, then*

$$\lim_n \sum_0^n P_k f = A f \quad \textit{and} \lim_n P_n A f = 0;$$

(iii) *for every special function f, we have*

$$(I - P)\, Af = f.$$

Proof. If f is a charge, it is easily seen that

$$\sum_1^n P_k f = Wf - P_{n+1} Wf.$$

If, in addition, $|f| \leqslant h$, the preceding result yields

$$\lim_n \sum_0^n P_k f = f + Wf - \int \hat{\gamma} f \, dm = Af,$$

and it is obvious that $\lim_n P_n Af = 0$.

We proceed to show that A is a convolution kernel. Let f be a positive function majorized by a multiple of h and such that $m(f) = 1$. The function $\varphi = f - m(h)^{-1}\, h$ satisfies the above condition, and thus, using Proposition 1.10, we have

$$\lim_n \sum_1^n P_k \varphi = W\varphi - \int \hat{\gamma} f \, dm + m(h)^{-2}\, (1 - m(h)).$$

On the other hand,

$$WP_n f = Wf + m(h)^{-1} \left\langle h, \sum_1^n P_k \varphi \right\rangle + m(h)^{-2} \left\langle h, \sum_1^n P_k h \right\rangle - \sum_1^n P_k f.$$

Letting n tend to infinity, it follows that

$$\lim_n \left[\sum_1^n P_k f - m(h)^{-2} \left\langle h, \sum_1^n P_k h \right\rangle \right] = Wf - \gamma\, m(f) + m(h)^{-1}\, \langle h, W\varphi \rangle$$
$$- \int \hat{\gamma} f \, dm + m(h)^{-2}\, (1 - m(h)).$$

Since

$$\langle h, W\varphi \rangle = \langle \hat{W} h, \varphi \rangle = 0,$$

by setting

$$C_n = m(h)^{-2}\, (1 - m(h)) + m(h)^{-2} \left\langle h, \sum_1^n P_k h \right\rangle,$$

we get

$$Af = (I + W - \gamma \otimes m - 1 \otimes \hat{\gamma} m)\, f = \lim_n \left[\sum_0^n P_k f - C_n\, m(f) \right]$$
$$= \lim_n \left(\sum_0^n P_k - C_n (1 \otimes m) \right) f,$$

which proves that A is a convolution kernel.

Finally, A satisfies (iii) because, for special f, since γ and $P\gamma$ are finite, Af and PAf are also finite. Thus

$$(I - P)\,Af = f - Pf + (I - P)\,Wf - \gamma\, m(f) + P\gamma\, m(f),$$

and since

$$(I - P)\,Wf = Pf - m(h)^{-1}\,Ph\, m(f) \quad \text{and } P\gamma = \gamma + m(h)^{-1}\,Ph,$$

we get

$$(I - P)\,Af = f.$$

Of course the same results are true for the dual chain, and we have the following

Corollary 2.4. *The kernel*

$$\hat{A} = I + \hat{W} - \hat{\gamma} \otimes m - 1 \otimes \gamma m$$

is a convolution kernel in duality with A, and has the same property with respect to $\hat{X}$ as A has with respect to X.

Proof. Straightforward.

Definition 2.5. If f is special we say that Af is the ***potential*** of f.

We shall denote by α the Radon measure such that $A(x, \cdot\,) = \alpha * \varepsilon_x$. If $\hat{\alpha}$ is the image of α by the mapping $x \to x^{-1}$, we also have $\hat{A}(x, \cdot\,) = \hat{\alpha} * \varepsilon_x$ and

$$\alpha - \alpha * \mu = \hat{\alpha} - \hat{\alpha} * \hat{\mu} = \varepsilon_e.$$

We have shown that the kernel A provides us with a solution to the Poisson equation with special second member. We now turn to the problem of knowing to what extent it is the only kernel with such a property.

Theorem 2.6. *Let A' be a convolution kernel satisfying properties* (i) *through* (iii) *in Theorem* 2.3; *then there is a constant c such that $A' = A + c \otimes m$. Let A' be a convolution kernel satisfying only* (i) *and* (iii) *in Theorem* 2.3; *then if X is type I, there is a constant c such that $A' = A + c \otimes m$, and if X is type II, there are two constants c and d such that*

$$A' = A + c \otimes m + d(1 \otimes \chi m - \chi \otimes m).$$

Remarks. The kernel $1 \otimes \chi m - \chi \otimes m$ is a convolution kernel, since it may be written $(\chi m) * \varepsilon_x$.

In order that (i) be satisfied, the constant d must lie in the interval $[-\sigma^{-2}, \sigma^{-2}]$, as will be shown in the next section.

Proof. If A' is a convolution kernel satisfying (i), then for every τ in $\boldsymbol{G}$ and f special, $(I - \tau) A'f = A'(I - \tau) f$ is bounded. If A' also satisfies (iii), it follows from Proposition 1.11 that

$$Af - A'f = a(f) + \chi\, b(f),$$

where $a(f)$ and $b(f)$ are constants. If we set $x = e$ in this equality we get

$$a(f) = \int A(e, \mathrm{d}y)\, f(y) - \int A'(e, \mathrm{d}y)\, f(y),$$

thus $a(f) = \int f \,\mathrm{d}\nu_1$ for a measure ν_1. If, in the same way, we set $x = x_0$, where x_0 is not in Ker χ, we get that $b(f) = \int f\, \mathrm{d}\nu_2$ for a measure ν_2. Thus $A' = A + 1 \otimes \nu_1 + \chi \otimes \nu_2$. But since A and A' are convolution kernels, for every a we must have

$$1 \otimes \nu_1 \tau_a + \chi \otimes \nu_2 \tau_a = 1 \otimes \nu_1 + \tau_a \chi \otimes \nu_2,$$

that is

$$1 \otimes \nu_1 \tau_a + \chi \otimes \nu_2 \tau_a = 1 \otimes \nu_1 + \chi(a) \otimes \nu_2 + \chi \otimes \nu_2;$$

this implies that for every a in $\boldsymbol{G}$, $\nu_2 \tau_a = \nu_2$, so that $\nu_2 = dm$ for some constant d. It follows that for every a,

$$\nu_1 \tau_a = \nu_1 + d\chi(a)\, m.$$

Setting $\nu_1' = \nu_1 + d\chi m$ we get at once that $\nu_1' \tau_a = \nu_1'$, and therefore that $\nu_1 = cm - d\chi m$. We thus have

$$A' = A + c(1 \otimes m) + d(\chi \otimes m - 1 \otimes \chi m),$$

and the proof is now easily completed.

As a result, we see that if we take another function h, the corresponding kernel will differ from A only by a multiple of $1 \otimes m$.

We end this section by strengthening the normality theorem (2.2). One might hope that random walks were normal in the sense of ch. 8 §3; actually this is not always true, and we shall prove a slightly less general result. We need some preliminary results and notation. We define

$$\gamma'(a) = m(h)^{-1} \int h \tau_a \gamma \, dm;$$

the function γ' is positive, uniformly continuous and unbounded. Furthermore, from $\hat{\gamma}(x) = \gamma(x^{-1})$ it is easily derived that $\hat{\gamma}'(x) = \gamma'(x^{-1})$.

Lemma 2.7. *For every x and a in* **G**,

$$m(h)^{-1} W(\tau_a h)(x) = \gamma(x) - \gamma(xa) + \gamma'(a).$$

Proof. Choose f in C_K^+ such that $m(f) = 1$; then

$$\tau_a \gamma = \lim_n \tau_a W P_n f = \lim_n W_{\tau_a h} P_n(\tau_a f),$$

so that by ch. 6, Corollary 5.5,

$$\tau_a \gamma = \gamma - m(h)^{-1} W(\tau_a h) + m(h)^{-1} \lim_n \langle h, W_{\tau_a h} P_n(\tau_a f)\rangle.$$

But

$$\langle h, W_{\tau_a h} P_n(\tau_a f)\rangle = \langle \hat{P}_n \hat{W}(\tau_a^{-1} h), f\rangle,$$

and by the argument in Proposition 1.10 we could, by choosing $\varphi \geqslant \tau_a^{-1} h$, find a subsequence $\{n_j\}$ such that $\{\hat{P}_{n_j} \hat{W}(\tau_a^{-1} h)\}$ converges boundedly to $\int \gamma \tau_a^{-1} h \, dm$. Consequently

$$\lim_n \langle h, W_{\tau_a h} P_n(\tau_a f)\rangle = \int \tau_a \gamma h \, dm,$$

which completes the proof.

Proposition 2.8. *For every x and a in* **G**,

$$\gamma(xa) \leqslant \gamma(x) + \gamma'(a) \quad \textit{and } \gamma'(xa) \leqslant \gamma'(x) + \gamma'(a).$$

Proof. The first part follows immediately from the preceding lemma, and implies in turn that

$$m(h)^{-1} \int h(y)\, \gamma(yxa) \, dm(y) \leqslant m(h)^{-1} \int h(y)\, \gamma(yx) \, dm(y) + \gamma'(a),$$

which is the second conclusion in the statement.

Lemma 2.9. *The function $\gamma - \gamma'$ is bounded.*

Proof. Choose a function f in C_K^+ such that $m(f) = 1$ and integrate the equality in Lemma 2.7 with respect to the measure fm. We get

$$\langle f, m(h)^{-1} W(\tau_a h)\rangle = \int f\gamma \, dm - \int \tau_a \gamma f \, dm + \gamma'(a);$$

since the left-hand side is a bounded function of a, it remains to prove that $\gamma(a) - \int \tau_a \gamma f \, dm$ is bounded; but this is equal to

$$\int f(x) \, (I - \tau_x) \, \gamma(a) \, dm(x),$$

so that the result follows from the fact that $(I - \tau_x) \, \gamma(a)$ is bounded uniformly for x in a compact set.

The following result is the key technical one.

Proposition 2.10. *There exists a constant C such that, for every a in* $\mathbf{G}$,

$$m(h)^{-1} W(\tau_a h) \leqslant \gamma + \hat{\gamma} + C.$$

Proof. In view of the two preceding results, there is a constant C' such that

$$m(h)^{-1} W(\tau_a h) \, (x) \leqslant \gamma(x) + \gamma(a) - \gamma(ax) + C'.$$

In addition,

$$\gamma(a) = \gamma(axx^{-1}) \leqslant \gamma(ax) + \gamma'(x^{-1}),$$

in view of Proposition 2.8; hence

$$m(h)^{-1} W(\tau_a h) \, (x) \leqslant \gamma(x) + \gamma'(x^{-1}) + C' = \gamma(x) + \hat{\gamma}'(x) + C',$$

and another application of the preceding lemma yields the desired conclusion.

We turn to our main result on normality.

Theorem 2.11. *For every special and co-special function f,*

$$\lim_n P_n W f = \int \hat{\gamma} f \, dm.$$

Proof. We first observe that by integrating the inequality of the preceding proposition, written for $\hat{X}$, with respect to $\hat{\mu}^n(\mathrm{d}a)$, we get

$$m(h)^{-1}\,\hat{W}\hat{P}_n h \leqslant \gamma + \hat{\gamma} + C$$

for every integer n. The left-hand side tends to $\hat{\gamma}$ as n tends to infinity and a special and co-special function f is integrable by the measure $(\gamma + \hat{\gamma} + C)\,m$; we may thus apply the Lebesgue theorem to obtain

$$\lim\langle P_n W f, m(h)^{-1}\,h\rangle = \lim\langle f, m(h)^{-1}\,\hat{W}\hat{P}_n h\rangle = \int \hat{\gamma} f\,\mathrm{d}m.$$

We now conclude the proof by using Proposition 1.8.

If the special functions are also co-special (in particular if $\mu = \hat{\mu}$), then X is normal in the sense of ch. 8 §3. It will also be shown in §4 that all type II random walks are normal in the sense of ch. 8 §3.

With the obvious changes we may now apply the results of ch. 8 §§2 and 3. For instance, if K is a special and co-special set, then $P_n P_K$ has a limit λ_K which is the equilibrium measure of K with respect to A. In the same way, we may define the Robin's constant. The reader will have no difficulty in translating the results of ch. 8 to the present situation.

Exercise 2.12. Prove that $\lim WP_n f(x) = \gamma(x)\,m(f)$ for every special function f.

[*Hint*: Use ch. 6, Exercise 5.17.]

Exercise 2.13. Prove that $\gamma(x^{-1}) = \hat{\gamma}(x)$.

Exercise 2.14. If f is in C_K^+, then Af is continuous.

Exercise 2.15. (1) Let G be discrete, and prove that the kernel

$$A(x, y) = \lim_{N\to\infty} \sum_0^N (P_n(x, y) - P_n(e, e))$$

exists and is finite and that $A(x, y) \leqslant 0$ for every pair (x, y) of points in G. Furthermore, if f is a charge with finite support, then

$$\lim_N \sum_0^N P_n f(x) = A\,f(x);$$

hence A is a potential kernel.

(2) Prove that

$$G^{\{e\}}(x, y) = A(x, y) - A(x, e) - A(e, y).$$

(3) Set, according to Definitions 2.5, $\alpha(x) = A(e, x)$. Prove that $\alpha(x) < 0$ for every $x \neq e$ and that $\alpha(x + y) \geqslant \alpha(x) + \alpha(y)$.

(4) Prove that $f \in b\mathscr{E}_+$ is special and co-special if and only if

$$\sum (|\alpha(x)| + |\hat{\alpha}(x)|) f(x) < \infty.$$

Prove that if f is special and symmetric it is also co-special.

(5) For any pair (x, y) of points of G, the function α satisfies the following two inequalities:

$$\alpha(x)\,\alpha(x^{-1}) \leqslant \alpha(xy)\,\alpha(x^{-1}) + \alpha(x)\,\alpha(y),$$

$$\alpha(x)\,\alpha(x^{-1}) \leqslant \alpha(yx)\,\alpha(x^{-1}) + \alpha(x)\,\alpha(y).$$

[*Hint*: Use the maximum principle of ch. 8 §2.9.]

Exercise 2.16. (1) Prove the following generalization of the non-positivity property in Exercise 2.15. There is a positive bounded measure α_1 such that $\alpha - \alpha_1$ is absolutely continuous with respect to m, and with a density p bounded above.

[*Hint*: See ch. 5 §5.1.]

(2) Denote by β_n the singular part of μ^n; then prove that the singular parts of the kernels A and W are the convolution kernels $\varepsilon_x * \sum_0^\infty \beta_n$ and $\varepsilon_x * \sum_1^\infty \beta_n$.

Exercise 2.17. For every special and co-special function h', there is a function $\gamma_{h'}$ such that $\lim_n W_{h'} P_n f = m(f)\,\gamma'$ for every f in C_K. Check the properties of $\gamma_{h'}$.

Exercise 2.18. If $g \in C_K^+ \cap \mathscr{U}$, there exists a function L_g such that $\lim_n U_g P_n f = L_g\, m(f)$ for every f in C_K and

$$\lim_n \hat{P}_n \hat{U}_g f = \int L_g f \, dm.$$

Exercise 2.19. Prove that the kernel A determines μ uniquely.

[*Hint*: Prove that A determines Π_K for any compact set K and prove that $P = \lim_n \Pi_{K_n}$, where K_n is an increasing sequence of compact sets whose interiors cover E. The reader should also refer to ch. 10 §3.]

Exercise 2.20 (Fourier criterion for recurrence). Let $\boldsymbol{G}$ be countable and retain the notation of ch. 3 §4. The random walk X is assumed to be recurrent.

(1) Prove that for every x in $\boldsymbol{G}$, the function

$$\gamma \to 2(1 - \operatorname{Re} \gamma(x))/\operatorname{Re}(1 - \varphi(\gamma))^{-1}$$

is integrable on Γ. We will call its integral $I(x)$.

[*Hint*: For every x in $\boldsymbol{G}$, there is an integer $n = n(x)$ such that $\mu^n(x) = c > 0$; then $1 - \operatorname{Re} \varphi(\gamma)^n \geqslant c(1 - \operatorname{Re} \varphi(\gamma))$.]

(2) With the notation of Exercise 2.15 we have

$$- I(x) = \alpha(x) + \alpha(-x) = G^{\{e\}}(x, x).$$

(3) Prove that

$$B = \int_\Gamma \operatorname{Re}(1 - \varphi(\gamma))^{-1} \, d\gamma = +\infty.$$

[*Hint*: If the integral were finite then we would have $-I(x) \leqslant B < \infty$, and this is impossible since it may be shown that

$$\lim_{x \to \Delta} G^{\{e\}}(x, x) = +\infty.$$

For the latter point see, for instance, §4 below.]

(4) Try to extend the above result to arbitrary abelian groups.

Exercise 2.21. (1) Compute the potential kernel of the Bernoulli random walk on $\mathbf{Z}$ ($\mu(1) = \mu(-1) = 2^{-1}$) by means of Exercise 2.20.

Exercise 2.22. Extend the results of this section to non-abelian groups.

3. Martin boundary

As has been shown in ch. 8 §4, the boundary points of $\boldsymbol{G}$ for X are in one-to-one correspondence with the cluster points for the vague topology of the sets $\{W(x, \cdot), x \in \boldsymbol{G}\}$. We are therefore going to study these cluster points, and we shall find that there are either one or two such points, according to whether the random walk is of type I or II.

Our first result parallels Proposition 1.8.

Proposition 3.1. *For every special function f such that $\lim_{x\to\Delta} Pf(x) = 0$ (in particular $f \in C_0$), and every x in* $\mathbf{G}$,

$$\lim_{a\to\Delta} (W\,f(xa) - W\,f(a)) = 0.$$

Proof. Set $\varphi_a = \tau_a Wf$ and $B = \|Wf\|$. Let $\mathfrak{U}$ be an ultrafilter finer than the filter of neighbourhoods of Δ, and define

$$L(x) = \lim_{\mathfrak{U}} \varphi_a(x).$$

Since

$$P_n Wf = Wf - \sum_0^n P_{k+1}f + m(h)^{-1} \sum_0^n P_{k+1}h\, m(f),$$

for fixed n we have

$$\lim_{a\to\Delta} (P_n\varphi_a - \varphi_a) = \lim_{a\to\Delta} \tau_a \left(- \sum_0^n P_{k+1}f + m(h)^{-1} \sum_0^n P_{k+1}h\, m(f)\right) = 0,$$

owing to the assumed property of f.

If n is sufficiently large, $\mu^n = g_n m + \beta_n$, where g_n is a non-zero function in $L^1_+(m)$, and

$$\left| P_n\varphi_a(x) - \int \varphi_a(yx)\, g_n(y)\, m(\mathrm{d}y) \right| \leqslant B\|\beta_n\|.$$

Pick a Borel function ψ within the equivalence class of the limit of φ_a along $\mathfrak{U}$ in $\sigma(L^\infty, L^1)$; clearly

$$\left| L(x) - \int \psi(yx)\, g_n(y)\, m(\mathrm{d}y) \right| \leqslant B\|\beta_n\|,$$

and consequently

$$|L(x) - P_n\,\psi(x)| \leqslant 2B\|\beta_n\|.$$

By letting n tend to infinity, it follows that $L = \lim_n P_n\psi$, hence that L is a constant function, and that $\lim_{\mathfrak{U}} (\varphi_a(x) - \varphi_a(e)) = 0$; the desired conclusion is then easily obtained.

Remark. The reader will notice without difficulty that the condition on f is also necessary.

Proposition 3.2. *Every sequence converging to Δ contains a sub-sequence $\{a_n\}$ such that, for every f in C_K, the sequence $\{W(\tau_{a_n} f)\}$ converges uniformly on* *compact sets to $\Gamma(x)\, m(f)$, where Γ is a continuous function.*

Proof. By Lemma 1.5, for any x in $\boldsymbol{G}$ the mappings

$$\mu_a^x : f \to W(\tau_a f)\,(x)$$

form a vaguely relatively compact set of positive Radon measures. Next, by Proposition 1.6, and by the Cantor diagonal process, every sequence converging to Δ contains a sub-sequence $\{a_n\}$ such that the sequence $\{W(\tau_{a_n} f_i)\}$ converges uniformly on compact sets for every function f_i in a denumerable dense subset of C_K. As a result, the measures $\mu_{a_n}^x$ converge vaguely to a measure ν^x, and for every $f \in C_K$ the function $x \to \nu^x(f)$ is continuous.

Let f, g be in C_K and b in $\boldsymbol{G}$. We have

$$\begin{aligned}\int \nu^x(f - \tau_b f)\, g(x)\, \mathrm{d}m(x) &= \lim_n \int W(\tau_{a_n}(f - \tau_b f))\, g\, \mathrm{d}m \\ &= \lim_n \int \tau_{a_n}(f - \tau_b f)\, \hat{W} g\, \mathrm{d}m \\ &= \lim_n \int f(x)\, (Wg(a_n^{-1}x) - Wg(b^{-1}a_n^{-1}x))\, \mathrm{d}m(x) = 0,\end{aligned}$$

in view of Proposition 3.1. It follows that ν^x is translation invariant, hence that $\nu^x = \Gamma(x)\, m$, where Γ is continuous.

As in the preceding sections, we observe that if Γ did not depend on the sub-sequence we would obtain a limit theorem; we therefore study the function Γ.

Proposition 3.3. *There are two constants α and β such that $\Gamma = \gamma + \alpha + \beta\chi$.*

Proof. It is easily seen from the definition of Γ that $P\Gamma \leqslant \Gamma + m(h)^{-1} Ph$. Define $C = \Gamma - \gamma$; if $f \in C_K^+$ and $m(f) = 1$, then

$$C(x) = \lim_n W(\tau_{a_n} - P_n)\, f(x),$$

so that for $b \in \boldsymbol{G}$,

$$\begin{aligned}C(x) - C(bx) &= \lim_n (I - \tau_b)\, W(\tau_{a_n} - P_n)\, f(x) \\ &= \lim_n (I - \tau_b)\, W_{\tau_x h}(\tau_{a_n} - P_n)\, \tau_x f(e).\end{aligned}$$

But $(\tau_{a_n} - P_n)\,\tau_x f$ is a charge, and if g is a charge, then $(I - \tau_b)\,(W_{\tau_x h} - W_h)\,g = 0$ by ch. 6, Corollary 5.5; consequently,

$$\lim_n (I - \tau_b)\, W_{\tau_x h}(\tau_{a_n} - P_n)\,\tau_x f(e) = \lim_n (I - \tau_b)\, W(\tau_{a_n} - P_n)\,\tau_x f(e),$$

and since $m(\tau_x f)$ is still equal to 1, this limit is equal to $C(e) - C(b)$. It follows that $C = \alpha + \tilde{\chi}$, where $\tilde{\chi}$ is a real character. As $|\tilde{\chi}| \leqslant \Gamma + \gamma + \alpha$, it follows that $P|\tilde{\chi}| < +\infty$ and

$$P_n|\tilde{\chi}| \leqslant \Gamma + \gamma + 2m(h)^{-1}\,(Ph + \cdots + P_n h),$$

hence that $\lim n^{-1}P_n|\tilde{\chi}| = 0$. Next, if

$$P\tilde{\chi} - \tilde{\chi} = \int \tilde{\chi}\, \mathrm{d}\mu = \beta,$$

then

$$P_n\tilde{\chi} = \tilde{\chi} + n\beta \quad \text{and } n^{-1}P_n\tilde{\chi} = n^{-1}\tilde{\chi} + \beta;$$

by letting n tend to infinity it follows that $\beta = 0$, hence $P\tilde{\chi} = \tilde{\chi}$ and

$$P\Gamma = \Gamma + m(h)^{-1}\,Ph;$$

as in Proposition 1.11 we may now prove that $P|\tilde{\chi}| - |\tilde{\chi}|$ is m-integrable. The proof is then complete.

Proposition 3.4. *From every sequence converging to Δ, we may select a sub-sequence $\{a_n\}$ such that the measures $W(a_n, \cdot\,)$ converge vaguely to $(\hat{\gamma} + \beta\chi)\,m$, where β is a constant.*

Proof. Since the set $\{W(x, \cdot\,),\, x \in G\}$ is vaguely relatively compact it is clear that we may find a sub-sequence $\{a_n\}$ such that the sequence $W(a_n, \cdot\,)$ converges. In view of Proposition 3.1, for every f in C_K, $Wf(a_n x)$ converges then to a constant C.

For g in C_K we have therefore

$$\begin{aligned} C\, m(g) &= \lim\langle Wf(a_n \cdot\,), g\rangle \\ &= \lim\langle f, \hat{W}(\tau_{a_n}^{-1} g)\rangle = m(g) \int (\hat{\gamma} + \alpha + \beta\chi)\, f\, \mathrm{d}m, \end{aligned}$$

and it follows that $C = \int (\hat{\gamma} + \alpha + \beta\chi)\, f\, \mathrm{d}m$.

Now if f is C_0 and if $Wf(a_n)$ converges to $\int (\hat{\gamma} + \alpha + \beta\chi)\, f\, \mathrm{d}m$, the same property is true for Pf, because by ch. 6, Theorem 5.2,

$$W\,Pf(a_n) = Wf(a_n) - Pf(a_n) + m(h)^{-1}\int hPf\,dm,$$

and thus the right-hand side has a limit which is equal to

$$\int (\hat{\gamma} + \alpha + \beta\chi)\,f\,dm + m(h)^{-1}\int \hat{P}hf\,dm$$

$$= \int (\hat{\gamma} + \alpha + \beta\chi)\,f\,dm + \int (\hat{P}\hat{\gamma} - \hat{\gamma})\,f\,dm$$

$$= \int \hat{P}\hat{\gamma}f\,dm + \int \alpha f\,dm + \int \beta\chi f\,dm = \int (\hat{\gamma} + \alpha + \beta\chi)\,Pf\,dm.$$

If $v \in C_K^+$, this property is still true for $\varphi = \sum 2^{-n}P_n v$, because the series $\sum 2^{-n}WP_n v$ converge uniformly, and we can manage to choose v so that $\varphi \geqslant h$. By an argument of lower semi-continuity it follows that

$$\underline{\lim}\; W(\varphi - h)\,(a_n) \geqslant \int (\hat{\gamma} + \alpha + \beta\chi)\,(\varphi - h)\,dm$$

and

$$\underline{\lim}\; Wh(a_n) \geqslant \int (\hat{\gamma} + \alpha + \beta\chi)\,h\,dm,$$

so that

$$\lim Wh(a_n) = \int (\hat{\gamma} + \alpha + \beta\chi)\,h\,dm.$$

Since $Wh = \int \hat{\gamma}h\,dm$ and $\int \chi h\,dm = 0$, it follows that $\alpha = 0$, and the proof is thus complete. Actually, using lower-semi-continuity we can show a bit more, namely that $Wf(a_n)$ converges to $\int (\hat{\gamma} + \beta\chi)\,f\,dm$ for every continuous function dominated by a multiple of h.

Definition 3.5. If the random walk is of type II, we say that x tends to $\pm\infty$ if $\chi(x)$ tends to $\pm\infty$.

The following theorem is the main result of this section. It states that for type I random walks, the Martin compactification is obtained by adjoining one point to the group $\boldsymbol{G}$, whereas for type II random walks one must adjoin two points, one at each "end" (in the obvious sense) of the group $\boldsymbol{G}$. In view of the remark below ch. 8, Proposition 4.6, the compactification thus obtained is the same as if W had been associated with a function $h \in C_K^+$.

Theorem 3.6. *If the random walk is of type I, the measures $W(x, \cdot)$ converge vaguely to $\hat{\gamma} m$ when x tends to Δ. If the random walk is of type II, the measures $W(x, \cdot)$ converge to $\hat{\gamma} m \pm \sigma^{-2}\chi m$ when x tends to $\pm\infty$.*

Proof. For type I random walks, there is nothing to prove, in view of the preceding result.

Consider a type II random walk, and let $\{a_n\}$ and $\{a_n'\}$ be two sequences such that $\{W(a_n, \cdot)\}$ and $\{W(a_n', \cdot)\}$ converge vaguely. By the above proposition, there is a constant β such that

$$\lim(W(a_n, \cdot) - W(a_n', \cdot)) = \beta\chi m.$$

Pick f in C_K and $b \notin \operatorname{Ker} \chi$; on account of Proposition 3.1,

$$W(I - \tau_b) f(xa_n) - W(I - \tau_b) f(xa_n')$$

converges boundedly to

$$\beta \int \chi(I - \tau_b) f \, dm = \beta\chi(b) \, m(f).$$

If $v = P|\chi| - |\chi|$, we have therefore

$$\lim\langle v, W(I - \tau_b) f(\cdot\, a_n) - W(I - \tau_b) f(\cdot\, a_n')\rangle = \beta\chi(b) \, m(f) \, m(v). \tag{3.1}$$

Now, $(I - \tau_b)f$ is a charge with compact support, so that by §2,

$$W(I - \tau_b) f - \int \hat{\gamma}(I - \tau_b) f \, dm = \lim_N \sum_1^N P_n(I - \tau_b) f,$$

and the convergence is bounded; consequently

$$\begin{aligned}
&\langle v, W(I - \tau_b) f(\cdot\, a_n) - W(I - \tau_b) f(\cdot\, a_n')\rangle \\
&= \lim_N \left\langle v, \sum_1^N P_n(I - \tau_b) f(\cdot\, a_n) - \sum_1^N P_n(I - \tau_b) f(\cdot\, a_n')\right\rangle \\
&= \lim_N \langle \hat{P}_{N+1}|\chi| - \hat{P}|\chi|, (I - \tau_b)(f(\cdot\, a_n) - f(\cdot\, a_n'))\rangle \\
&= \lim_N \langle \hat{P}_{N+1}(I - \tau_b^{-1})|\chi| - \hat{P}(I - \tau_b^{-1})|\chi|, f(\cdot\, a_n) - f(\cdot\, a_n')\rangle,
\end{aligned}$$

and, because of Proposition 1.15, this limit is equal to

$$- \langle \hat{P}(I - \tau_b^{-1})\, |\chi|, f(\cdot\, a_n) - f(\cdot\, a_n')\rangle$$

$$= - \langle Pf, (I - \tau_b^{-1})\, |\chi|(\cdot\, a_n^{-1}) - (I - \tau_b^{-1})\, |\chi|(\cdot\, a_n'^{-1})\rangle. \quad (3.2)$$

From the form of the function $(I - \tau_b^{-1})\, |\chi|$ which was described in §1, it follows that if $\{a_n\}$ and $\{a_n'\}$ either both converge to $+\infty$ or both converge to $-\infty$, the above limit is zero. As a result $\beta = 0$, and therefore the measures $W(x, \cdot)$ have a limit when x converges either to $+\infty$ or to $-\infty$.

The functions Wf thus have a limit at each end of the group G, and since $P_n W$ converges to $\hat{\gamma} m$ and

$$\lim P_n 1_{G+} = \lim P_n 1_{G-} = \tfrac{1}{2},$$

it follows that if $W(x, \cdot)$ converges to $\hat{\gamma} m + \lambda\chi m$ as $x \to +\infty$, then $W(x, \cdot)$ converges to $\hat{\gamma} m - \lambda\chi m$ as $x \to -\infty$. It remains to compute λ.

Let $\{a_n\}$ be a sequence converging to $+\infty$ and $\{a_n'\}$ a sequence converging to $-\infty$. If we resume the above calculations it is seen that $\beta = 2\lambda$ and that the limit in eq. (3.2) is equal to $2\, m(f)\, \chi(b)$. Comparing with eq. (3.1), it follows that $\lambda\, m(v) = 1$; that is, $\lambda = \sigma^{-2}$, and the proof is complete.

As remarked at the end of the proof of Proposition 3.4, the above convergences hold not only for functions in C_K but also for continuous functions majorized by a multiple of h. Actually, the scope of validity of the preceding theorem may be still further enlarged by the use of the same device as in Theorem 2.11.

Theorem 3.7. *If f is both special and co-special, and if $\lim_{a\to\Delta} Pf(a) = 0$, then*

$$\lim_{x\to\Delta} Wf(x) = \int \hat{\gamma} f\, dm$$

if X is type I, and

$$\lim_{x\to\pm\infty} Wf(x) = \int (\hat{\gamma} \pm \sigma^{-2}\chi)\, f\, dm$$

if X is type II.

Proof. Let us write down the proof for type I walks, leaving to the reader the obvious changes needed for type II. Let g be a function in C_K^+ with $m(g) = 1$; from Proposition 3.2 and Theorem 3.6, it is easily derived that $\lim_{a\to\Delta} W(\tau_a g) = \gamma$, and by Proposition 2.10, there is a constant D such that $W(\tau_a g) \leqslant D(\gamma + \hat{\gamma} + C)$.

Now on account of Proposition 3.1, from every sequence converging to Δ we may select a sub-sequence $\{a_n\}$ such that $Wf(a_n x)$ converges boundedly to a constant ρ. Then

$$\rho = \lim_n \langle \tau_{a_n} Wf, g \rangle = \lim_n \langle f, \hat{W}(\tau_{a_n^{-1}} g) \rangle = \int \hat{\gamma} f \, dm,$$

by applying the preceding remarks to the dual chain.

The above results may also be stated for the potential kernel A.

Corollary 3.8. *If f is a charge such that $|f|$ is co-special and $\lim_{a \to \Delta} P|f|(a) = 0$, then*

$$\lim_{x \to \Delta} Af(x) = 0$$

for type I random walks, and

$$\lim_{x \to \pm\infty} Af(x) = \pm \sigma^{-2} \int \chi f \, dm$$

for type II random walks.

Proof. Straightforward.

Exercise 3.9. If $g \in C_K^+ \cap \mathscr{U}$, then for type I random walks,

$$\lim_{x \to \Delta} U_g f(x) = \int L_g f \, dm$$

(L_g was defined in Exercise 2.18). For type II random walks there exist two functions $L_g^{\pm}$ such that $2L_g = L_g^+ + L_g^-$ and

$$\lim_{x \to \pm\infty} U_g f(x) = \int L_g^{\pm} f \, dm.$$

4. Renewal theory

In the preceding section, we have studied the asymptotic behaviour of potentials Af when f is a charge. We want now to fulfil the same task when $f \geqslant 0$. Clearly this amounts to studying the asymptotic behaviour of functions γ. Since these functions are positive and sub-harmonic, they are unbounded; therefore when x tends to Δ, $\gamma(x)$ has always $+\infty$ as a cluster point. The problem is to find out whether it is the only one, and in that case to compute the speed of convergence of γ to $+\infty$. We shall find that type I and type II random walks have quite different features with regard to this problem, and require different proofs. We start, however, with some applications of the preceding section which concern both types.

Proposition 4.1. *If f is in C_K^+ and $m(f) = 1$, then for any a in $\mathbf{G}$,*

$$\lim_{x\to\Delta}(I - \tau_a)\, Af(x) = \lim_{x}(I - \tau_a)\, \gamma(x) = 0$$

for type I random walks, and

$$\lim_{x\to\pm\infty}(I - \tau_a)\, Af(x) = -\lim_{x\to\pm\infty}(I - \tau_a)\, \gamma(x) = \pm\, \sigma^{-2}\, \chi(a)$$

for type II random walks. Moreover, the convergence is uniform for a in a compact set.

Proof. Since

$$(I - \tau_a)\, Af(x) = (I - \tau_a)\, f(x) + (I - \tau_a)\, Wf(x) - (I - \tau_a)\, \gamma(x),$$

it follows from Theorem 3.6 that the two limits are opposite. Indeed for type II random walks x and ax converge to the same end of the group whatever a is.

On the other hand, since A is a convolution kernel, we have $(I - \tau_a)\, Af = A(I - \tau_a)\, f$, and therefore by Corollary 3.8, the limit is zero in the former case, and

$$\pm\, \sigma^{-2} \int \chi(I - \tau_a)\, f\, dm = \pm\, \sigma^{-2}\, \chi(a)$$

in the latter. The last conclusion follows from the uniform continuity of functions γ.

Corollary 4.2. *Let a be a point in $\mathbf{G}$ such that a^n converges to Δ; then*

$$\lim_n n^{-1}\gamma(a^n) = 0$$

for type I random walks, and

$$\lim_n n^{-1}\gamma(a^n) = \sigma^{-2}|\chi|\,(a)$$

for type II random walks.

Proof. By the preceding result, the differences $\gamma(a^n) - \gamma(a^{n-1})$ converge to the assigned limits; since Cesaro convergence is weaker than ordinary convergence, the proof is complete.

For a while we shall deal only with type II random walks. The asymptotic behaviour of γ is then completely settled by the following

Theorem 4.3. *For type II random walks,*

$$\lim_{x \to \Delta} \gamma(x)/|\chi|(x) = \sigma^{-2}.$$

Proof. When a is in Ker χ, the functions $(I - \tau_a)\gamma$ are uniformly bounded; it suffices thus to prove the result when G is either $\mathbf{R}$ or $\mathbf{Z}$. Since the function γ' is subadditive (Proposition 2.8), it is a classical result (see ch. 4 §6.2) that $\gamma'(x)/|\chi|(x)$ has a limit whenever $|\chi|(x)$ tends to infinity, or equivalently x tends to Δ. As the difference $\gamma' - \gamma$ is bounded (Lemma 2.9), it follows that $\gamma(x)/|\chi|(x)$ also has a limit. This limit is σ^{-2}, since by Corollary 4.2, $\gamma(na)/|\chi|(na)$ tends to σ^{-2} as $n \to \infty$. This result may be restated in terms of potentials.

Corollary 4.4. *Let f be special, then*

$$\lim_{x \to \Delta} Af(x)/|\chi|(x) = -\sigma^{-2}\, m(f).$$

Proof. Obvious.

This result enables us to complete Theorem 2.6 by stating

Proposition 4.5. *The constant d which appears in Theorem 2.6 must lie in the interval $[-\sigma^{-2}, \sigma^{-2}]$.*

Proof. Consider the kernel $A' = A + c(1 \otimes m) + d(1 \otimes \chi m - \chi \otimes m)$ and let f be a special function; in the right-hand side of the equality

$$A'f = f + Wf + c\, m(f) - \int \hat{\gamma} f \, dm + d \int \chi f \, dm - (\gamma + d\chi)\, m(f),$$

all the summands are bounded but the last one. It is now plain that $A'f$ cannot be bounded from above unless d lies in the interval $[-\sigma^{-2}, \sigma^{-2}]$.

We now turn to type I random walks, for which the situation as well as the proofs are much more intricate.

Proposition 4.6.

$$\lim_{x\to\Delta}(\gamma(x) + \gamma(x^{-1})) = \lim_{x\to\Delta}(\gamma'(x) + \gamma'(x^{-1})) = +\infty.$$

Proof. Since $\gamma - \gamma'$ is bounded, it suffices to prove the result for γ'. If the result were false, there would exist a sequence $\{a_n\}$ and a constant C such that $\gamma'(a_n) + \gamma'(a_n^{-1}) \leqslant C < \infty$. But as γ' is subadditive for every x in $\boldsymbol{G}$, we have

$$\gamma'(x) = \gamma'(xa_na_n^{-1}) \leqslant \gamma'(xa_n) + \gamma'(a_n^{-1}),$$

and since $\gamma'(xa_n) - \gamma'(a_n)$ converges to zero as $n \to \infty$ (cf. Proposition 4.1) the function γ' would be bounded, which is a contradiction.

Remark. The above result may also be stated as

$$\lim(\gamma(x) + \hat{\gamma}(x)) = \lim(\gamma'(x) + \gamma'(x^{-1})) = +\infty,$$

and also if f is special as

$$\lim(Af(x) + Af(x^{-1})) = \lim(Af(x) + \hat{A}f(x)) = -\infty.$$

We now turn to the main result for type I random walks. The proof is somewhat similar to a proof given for transient random walks.

Theorem 4.7. *If the function γ does not converge to $+\infty$, there is one and only one finite number L for which there exist sequences $\{x_n\}$ such that* $\lim_n \gamma(x_n) = L$, *and then* $\lim_n \gamma(x_n^{-1}) = +\infty$.

Proof. The last conclusion follows at once from Proposition 4.6. To prove the first one, consider a function f in C_K^+ with support V, and set $\varphi = -Af$. For any a in $\boldsymbol{G}$, we have

$$(\tau_a - I)\,\varphi = (I + W)\,(I - \tau_a)\,f + \text{constant}.$$

Since $(I - \tau_a)f$ is a charge, we may apply the semi-complete maximum principle (ch. 8 §2) to infer that there exists $\xi \in V$ such that, for all x in $\boldsymbol{G}$,

$$\varphi(xa) - \varphi(x) \leqslant \varphi(\xi a) - \varphi(\xi),$$

or equivalently

$$\varphi(xa) + \varphi(\xi) \leqslant \varphi(\xi a) + \varphi(x).$$

Repeated replacements of x by xa yield

$$\varphi(xa^n) + \varphi(\xi) \leqslant \varphi(\xi a) + \varphi(xa^{n-1})$$

and by summing up and dividing by n,

$$n^{-1}\,\varphi(xa^n) + \varphi(\xi) \leqslant \varphi(\xi a) + n^{-1}\,\varphi(x).$$

Finally, $\varphi(\xi) \leqslant n^{-1}\,\varphi(x) + \varphi(\xi a)$, and since n is arbitrary, for every $a \in G$ there exists ξ in V such that $\varphi(\xi) \leqslant \varphi(\xi a)$, hence such that $\inf_{x\in V} \varphi(x) \leqslant \varphi(\xi a)$. By Proposition 4.1, we know that $\varphi(a) - \varphi(\xi a)$ converges uniformly in $\xi \in V$ to zero whenever $a \to \Delta$, and we may therefore conclude that

$$\varliminf_{a\to\Delta} \varphi(a) \geqslant \inf_{x\in V} \varphi(x). \tag{4.1}$$

Now let L be a finite cluster point for φ and $\{a_n'\}$ a sequence converging to Δ and such that $L = \lim_n \varphi(a_n')$. We may select from $\{a_n'\}$ a sub-sequence $\{a_n\}$ such that

$$\lim_n \varphi(\xi a_j^{-1} a_n) = L \qquad \text{uniformly if } \xi \in V \text{ and } j < n,$$

$$\lim_n \varphi(\xi a_n^{-1} a_j) = +\infty.$$

This may be done recursively in the following way. The points $a_1, \ldots, a_{j+1}$ being already chosen, we choose a_{j+1} in such a way that

$$\varphi(\xi a_k^{-1} a_{j+1}) - \varphi(a_{j+1}) \leqslant 2^{-j}$$

for every $\xi \in V$ and $k \leqslant j$; this is possible on account of Proposition 4.1. But we may also choose a_{j+1} in order that $\varphi(\xi a_{j+1}^{-1} a_k) \geqslant 2^k$ for every $k \leqslant j$ and $\xi \in V$; indeed, by Proposition 4.6, $\varphi(a_n'^{-1} a_k)$ converges to $+\infty$, and by Proposition 4.1, $(I - \tau_\xi)\,\varphi(a_n'^{-1} a_k)$ converges to zero uniformly for ξ in a compact set.

Let us next set

$$\Phi = n^{-1} \sum_{i=1}^{n} \varphi(\cdot\, a_i).$$

This function is also of the form $-Af$, and as such satisfies the inequality (4.1) with the appropriate set V. On the other hand it is plain from Proposition 4.1 that

$$\underline{\lim}_{x \to \Delta} \Phi(x) = \underline{\lim}_{x \to \Delta} \varphi(x).$$

It follows that

$$\inf_{\substack{\xi \in V \\ 1 \leqslant k \leqslant n}} \left(n^{-1} \sum_{i=1}^{n} \varphi(\xi a_k^{-1} a_i) \right) \leqslant \underline{\lim}_{x \to \Delta} \varphi(x),$$

and a fortiori

$$n^{-1} \sum_{i=1}^{n} \inf\{\varphi(\xi a_k^{-1} a_i) : \xi \in V, 1 \leqslant k \leqslant n\} \leqslant \underline{\lim}_{x \to \Delta} \varphi(x).$$

As soon as i is sufficiently large, the above infimum is assumed for a number $k \leqslant i$ and thus is near L; as a result, we obtain $L \leqslant \underline{\lim}_{x \to \Delta} \varphi(x)$, and consequently $L = \underline{\lim}_{x \to \Delta} \varphi(x)$. It follows that L is the only finite cluster point.

Definition 4.8. A type I random walk is said to be of *type Ib* if γ has a finite cluster point; it is otherwise said to be of *type Ia*.

There cannot be type Ib random walks on any recurrent abelian group, and our next goal is to give the relevant classification of groups. The process is somewhat parallel to the proof of the renewal theorem for transient random walks. In the following proofs we use the additive notation for the operation in $\boldsymbol{G}$.

Lemma 4.9. *If $\boldsymbol{G}$ has an open, non-compact, compactly generated sub-group $\boldsymbol{G}_1$ such that $\boldsymbol{G}/\boldsymbol{G}_1$ is infinite, then every random walk is of type Ia.*

Proof. We use the notation of the proof of Theorem 4.7, and assume that X is *type Ib*. By Theorem 4.7, and the same device as in the proof of ch. 5, Proposition 4.1, we may find a sequence $\{z_n\}$ of points in $\boldsymbol{G}$ such that the sets $z_n \boldsymbol{G}_1$ are pairwise disjoint and such that $\lim_n \varphi(z_n) = L < \infty$.

Since $\boldsymbol{G}_1$ is compactly generated but not compact, we can choose $x \in \boldsymbol{G}_1$ such that $nx \to \Delta$ as $n \to \infty$, and by Proposition 4.1 and Theorem 4.7, we can suppose that $\lim_n \varphi(nx) = +\infty$.

Arguing as in ch. 5, Proposition 4.1, it is easily seen that for n sufficiently large, there is an integer $k_n > 0$ such that

$$\varphi(z_n + k_n x) < L + 1, \qquad \varphi(z_n + (k_n + 1)\, x) \geqslant L + 1.$$

Now $z_n + kx \to \Delta$ uniformly in k as $n \to \infty$, so that by Proposition 4.1 we get

$$\lim_n \varphi(z_n + k_n x) = L + 1,$$

which contradicts Theorem 4.7.

Corollary 4.10. *If G has a closed sub-group isomorphic to $\mathbf{R}^{d_1} \oplus \mathbf{Z}^{d_2} \oplus \mathbf{K}$, where K is a compact group and $d_1 + d_2 \geqslant 2$, then all random walks are type Ia.*

Proof. We choose x and y in G such that $nx + ky \to \infty$ as $n + k \to \infty$, and use the proof of Lemma 4.9. One could also argue as in ch. 5, Corollary 4.2.

Lemma 4.11. *If every element of G is compact, then every Harris random walk on G is of type Ia.*

Proof. We suppose that X is of type Ib, and we denote by H an open compact sub-group containing the support of f (we recall that $\varphi = -Af$). Then G/H is infinite, and arguing as in Lemma 4.9, we may pick a sequence $\{z_n\}$ such that $z_0 = e$ and the co-sets $z_n H$ are disjoint, and $\lim_n \varphi(z_n) = L < +\infty$. By Theorem 4.7, $\lim_n \varphi(-z_n) = +\infty$. By the hypothesis, for every element z in G, the sequence $\{nz\}_{n \geqslant 0}$ intersects H, so that we can set for every n

$$h_n = \inf\{m \geqslant 1 : m z_n \in H\}.$$

For $n \geqslant 1$, $h_n > 1$, and since $(h_n - 1)\, z_n \in z_n^{-1} H$, we have $\lim_n \varphi((h_n - 1)\, z_n) = +\infty$. Thus $h_n > 2$ for n sufficiently large, and we can choose k_n to be the largest positive integer less than $h_n - 1$ such that $\varphi(k_n z_n) \leqslant 2L + 2$. Then for n sufficiently large,

$$\varphi(k_n z_n) \leqslant 2L + 2 \leqslant \varphi((k_n + 1)\, z_n).$$

Now, by using the identities of ch. 8, Theorem 2.1, we can write

$$\begin{aligned}
\varphi(x+y) - \varphi(x) - \varphi(y) &= -Af(x+y) + Af(x) + Af(y) \\
&= -A(\tau_y f)(x) + Af(x) + A(\tau_y f)(e) \\
&= -G^H(\tau_y f)(x) - P_H A(\tau_y f)(x) - (\Gamma u_H(x) - C_H)\, m(f) \\
&\quad + G^H f(x) + P_H A f(x) + (\Gamma u_H(x) - C_H)\, m(f) \\
&\quad + G^H(\tau_y f)(e) + P_H A(\tau_y f)(e) + (\Gamma u_H(e) - C_H)\, m(f).
\end{aligned}$$

Since the support of f is contained in $\boldsymbol{H}$, we have $G^{\boldsymbol{H}}f(x) = 0$; on the other hand, $\Gamma u_{\boldsymbol{H}} = C_{\boldsymbol{H}}$ in e, since $e \in \boldsymbol{H}$. Finally, owing to the definitions of $P_{\boldsymbol{H}}$ and $G^{\boldsymbol{H}}$, we get, after cancellation,

$$\varphi(x+y) - \varphi(x) - \varphi(y) = -G^{\boldsymbol{H}}(\tau_y f)(x) - P_{\boldsymbol{H}}A(\tau_y f)(x) + P_{\boldsymbol{H}}Af(x) + A(\tau_y f)(e).$$

We may suppose that Af is negative, since this amounts to adding to A a multiple of $1 \otimes m$. On the other hand, as a consequence of Proposition 4.1, we have

$$\lim_{y \to \Delta}[A(\tau_y f)(e) - P_{\boldsymbol{H}}A(\tau_y f)(x)] = 0$$

uniformly in x. We may thus conclude that, uniformly in $x \in \boldsymbol{G}$,

$$\varphi(x + y) - \varphi(x) - \varphi(y) \leqslant 0(y)$$

where $0(y) \to 0$ as $y \to \Delta$. Letting $y = z_n$ and $x = k_n z_n$, we get that

$$\varphi((k_n + 1) z_n) - \varphi(k_n z_n) \leqslant \varphi(z_n) + 0_n(1),$$

where $0_n(1) \to 0$ as $n \to \infty$. For n sufficiently large, we thus have

$$L + 1 \leqslant \varphi(k_n z_n) \leqslant 2L + 2 \leqslant \varphi((k_n + 1) z_n) \leqslant 3L + 3.$$

Since either $\{k_n z_n\}$ or $\{(k_n + 1) z_n\}$ has a sub-sequence which converges to infinity, we have a contradiction with Theorem 4.7.

Theorem 4.12. *If the random walk is of type Ib, then* $\mathbf{G}$ *is isomorphic either to* $\mathbf{R} \oplus \boldsymbol{K}$ *or* $\mathbf{Z} \oplus \boldsymbol{K}$, *where* $\boldsymbol{K}$ *is a compact group. Furthermore, there is a finite number* L *such that (with the obvious meaning for* $\pm \infty$)

$$\lim_{x \to +\infty} Af(x) = L\, m(f), \qquad \lim_{x \to -\infty} Af(x) = -\infty$$

or

$$\lim_{x \to +\infty} Af(x) = -\infty, \qquad \lim_{x \to -\infty} Af(x) = L\, m(f)$$

for every function $f \in C_K^+$.

Proof. The first part is shown in the same way as ch. 5, Theorem 4.4. The second part then follows easily from the above results and the arguments in ch. 5, Theorem 3.1.

Unfortunately we do not know of a characterization of the laws of type Ib walks. Exercise 4.16, however, gives some information on this subject.

Exercise 4.13. (1) Prove that type II random walks are normal in the sense of ch. 8 §3.

[*Hint*: Observe that in Proposition 2.10, $\hat{\gamma}$ may be replaced by a multiple of γ.]

(2) Derive that the resolvent of linear Brownian motion is normal in the sense of ch. 8 §3.

Exercise 4.14. Prove that with the notation of Exercise 2.15, if X is a recurrent random walk on $\mathbf{Z}$, then $\alpha(x) = \lambda x$ for all $x \leqslant 0$ if and only if X is left continuous (see ch. 3, Exercise 3.13) and type II with $\lambda = -\sigma^{-2}$. Compute the potential kernel of the Bernoulli random walk ($\mu(1) = \mu(-1) = 2^{-1}$).

Exercise 4.15. For a type I random walk on $\mathbf{R}$ or $\mathbf{Z}$, prove that $\lim_{x\to\Delta} x^{-1}\gamma(x) = 0$.

Exercise 4.16. (1) All recurrent random walks on $\mathbf{Z}$ have the property that $\alpha(x) < 0$ for $x \neq 0$ with one single exception: left- or right-continuous random walks of type I. In the left-continuous case, $\alpha(x) = 0$ for $x \leqslant 0$ and $\alpha(x) < 0$ for $x > 0$. The reader is invited to return to ch. 8, Exercise 4.18.

(2) Prove that if a random walk on $\mathbf{Z}$ is type I, and $\mu(x) = 0$ for all but a finite number of positive entries x, then there is a finite cluster point for γ.

(3) Use the random walks described in (1) to prove that there exist random walks which are not normal in the sense of ch. 8 §3.

CHAPTER 10

CONSTRUCTION OF MARKOV CHAINS AND RESOLVENTS

We have seen that the potential kernels of Markov chains and resolvents satisfy maximum principles in the recurrent as well as in the transient case. We will deal here with the converse problem: is any kernel satisfying an appropriate maximum principle the potential kernel of a Markov chain or of a resolvent? The answer is negative (cf. Exercise 1.8) in general, but can be made positive by means of additional conditions.

1. Preliminaries and bounded kernels

In this and the following section, we study positive and proper kernels denoted by G or V. We denote by $\mathscr{B}$ the set of functions h in $b\mathscr{E}$ such that $G|h|$ (or $V|h|$ according to the context) is bounded, and by $\mathscr{B}_+$ the set of positive functions in $\mathscr{B}$. Once for all, we remark that two measures which are equal on $\mathscr{B}_+$ are equal.

These kernels are supposed to satisfy some of the maximum principles defined in ch. 2. From now on we abbreviate "complete maximum principle" into C.M.P. and "reinforced complete maximum principle" into R.C.M.P., and shall write $G \in$ (R.C.M.P.) or $V \in$ (C.M.P.) to mean that $G(V)$ satisfies the corresponding principle.

It may happen that the kernels satisfy the maximum principles only for subclasses of $\mathscr{E}_+$, for instance $\mathscr{B}_+$; when this is the case it will be explicitly stated.

Proposition 1.1. *The kernel V satisfies the C.M.P. if and only if for every f in $\mathscr{E}$ such that Vf is meaningful and assumes positive values,*

$$\sup_{x \in E} Vf(x) = \sup_{x \in \{f > 0\}} Vf(x).$$

Proof. Let a be the right-hand side of the above equation. We have $Vf^+ \leqslant a \vee 0 + Vf^-$ on $\{f^+ > 0\}$, hence everywhere, and $Vf \leqslant a \vee 0$. Since Vf assumes positive values, we have $a \vee 0 = a > 0$, which proves the necessity.

Assume conversely that $f, g \in \mathscr{E}_+$ and $Vf \leqslant a + Vg$ $(a > 0)$ on $\{f > 0\}$; then $V(f - g) \leqslant a$ on $\{f - g > 0\} \subset \{f > 0\}$, and if V satisfies the above equation we get $V(f - g) \leqslant a$ everywhere, which completes the proof.

Proposition 1.2. *If $V \in (C.M.P.)$ and $f \in \mathscr{E}_+$, the kernel $VI_f \in (C.M.P.)$.*

Proof. Left to the reader as an exercise.

Proposition 1.3. *If $V \in (C.M.P.)$, then for every $\alpha > 0$, $I + \alpha V \in (R.C.M.P.)$. If $G \in (R.C.M.P.)$, then $G \in (C.M.P.)$ and G is a one-to-one operator and may be written $G = I + V$, where V is a proper kernel.*

Proof. Let $f, g \in \mathscr{E}_+$ and assume that

$$f + \alpha Vf \leqslant a + g + \alpha Vg - g = a + \alpha Vg$$

on $\{f > 0\}$. A fortiori $Vf \leqslant a/\alpha + Vg$ on $\{f > 0\}$, hence everywhere. On $\{f = 0\}$, we consequently have

$$f + \alpha Vf \leqslant a + g + \alpha Vg - g,$$

so that $I + \alpha V \in$ (R.C.M.P.).

Now let $G \in$ (R.C.M.P.) and assume that $f, g \in \mathscr{E}_+$ are such that $Gf \leqslant a + Gg$ on $\{f > 0\}$. It would be easy to show that this inequality holds everywhere if g vanished on $\{f > 0\}$; we reduce the problem to this case by setting $f' = f - f \wedge g$ and $g' = g - f \wedge g$. We then have

$$Gf' \leqslant a + Gg' \quad \text{on } \{f > 0\} \supset \{f' > 0\}.$$

Since $g' = 0$ on $\{f' > 0\}$, we get

$$Gf' \leqslant a + Gg' - g' \quad \text{on } \{f' > 0\},$$

hence everywhere, and a fortiori $Gf' \leqslant a + Gg'$ everywhere, which implies

$$Gf = Gf' + G(f \wedge g) \leqslant a + Gg' + G(f \wedge g) = a + Gg$$

everywhere.

Let now $f \in \mathscr{B}$ be such that $Gf \geqslant 0$. We have

$$Gf^+ - f^+ \geqslant Gf^- \quad \text{on } \{f^- > 0\};$$

hence $Gf \geqslant f^+ \geqslant f$. If $Gf = 0$, it follows that $f^+ = 0$, and symmetrically that $f^- = 0$, which proves that G is one-to-one.

On the other hand if we set $V = G - I$, it is clear that the mapping V from $\mathscr{B}$ to $b\mathscr{E}$ is positive and majorized by G, hence is a positive proper kernel.

Corollary 1.4 (Uniqueness theorem). *If $V \in (C.M.P.)$, there is at most one resolvent such that $V_0 = V$. More precisely, for every $\alpha \geqslant 0$, there is at most one kernel V_α such that*

$$\alpha V V_\alpha = \alpha V_\alpha V = V - V_\alpha.$$

If $G \in (R.C.M.P.)$, there is at most one T.P. such that $G = \sum_0^\infty P_n$.

Proof. Assume that there are two such kernels, V_α and $\tilde{V}_\alpha$: then for $f \in \mathscr{B}_+$,

$$(I + \alpha V)(V_\alpha f - \tilde{V}_\alpha f) = 0,$$

and since $I + \alpha V$ is one-to-one, $V_\alpha f = \tilde{V}_\alpha f$.

In the same way, if P and $\tilde{P}$ are two T.P.'s with potential kernel G, we have

$$GP = G\tilde{P} = G - I,$$

and the result follows once more from the fact that G is one-to-one.

As we have already pointed out, there does not always exist such a resolvent (transition probability). We begin with the special case where V is a bounded kernel; the space $\mathscr{B}$ is then equal to $b\mathscr{E}$, and V is an endomorphism of $b\mathscr{E}$ with norm $||V||$. The following result is basic for the sequel.

Theorem 1.5. *A bounded kernel V satisfies the complete maximum principle if and only if there exists a submarkovian resolvent $\{V_\alpha\}_{\alpha>0}$ such that $V_0 = V$.*

Proof. The sufficiency was shown in ch. 2 §6. We proceed to the necessity.

Suppose that there exists an operator V_α, $\alpha > 0$, on $b\mathscr{E}$ such that

$$\alpha V V_\alpha = \alpha V_\alpha V = V - V_\alpha. \tag{1.1}$$

This operator is positive; indeed, let g be a negative function in $b\mathscr{E}$ such that $V_\alpha g = V(g - \alpha V_\alpha g)$ assumes a positive maximum. By Proposition 1.1, this maximum is assumed on $\{g > \alpha V_\alpha g\}$, that is on the set where $V_\alpha g$ is strictly negative, which is a contradiction.

Furthermore, $\|\alpha V_\alpha\| \leqslant 1$; indeed, V_α being positive, it suffices to check that $\alpha V_\alpha 1 \leqslant 1$. First, we observe that it is not true that $1 \leqslant \alpha V_\alpha 1$ everywhere, since we would then have

$$0 \geqslant V[\alpha(1 - \alpha V_\alpha 1)] = \alpha V_\alpha 1 \geqslant 1.$$

Then $V[\alpha(1 - \alpha V_\alpha 1)] \leqslant 1$ on the non-empty set $\{\alpha(1 - \alpha V_\alpha 1) > 0\}$, and hence everywhere, by the complete maximum principle.

For $\alpha > 0$, we now set

$$V_\alpha = V(I - \alpha V + \alpha^2 V^2 - \cdots);$$

this series converges for $\alpha < \|V\|^{-1}$, and thus defines an operator V_α which satisfies eq. (1.1); it follows that V_α is positive and that $\|V_\alpha\| < \alpha^{-1}$. Now choose $\alpha < \|V\|^{-1}$ and $\beta < \alpha$, and set

$$V_{\alpha+\beta} = V_\alpha(I - \beta V_\alpha + \beta^2 V_\alpha^2 - \cdots); \tag{1.2}$$

the operator thus defined satisfies the relation

$$\beta V_{\alpha+\beta} V_\alpha = \beta V_\alpha V_{\alpha+\beta} = V_\alpha - V_{\alpha+\beta}. \tag{1.3}$$

The operator V commutes with V_α, hence with $V_{\alpha+\beta}$, and multiplying the two members of eq. (1.3) by V yields

$$(\alpha + \beta)\, V V_{\alpha+\beta} = (\alpha + \beta)\, V_{\alpha+\beta} V = V - V_{\alpha+\beta},$$

which is eq. (1.1) for $\alpha + \beta$. We have thus defined operators V_α satisfying eq. (1.1) for $\alpha \in [0, 2\|V\|^{-1}[$. The same pattern permits us then to define V_α for $\alpha \in [0, 4\|V\|^{-1}[$, and by iteration for all $\alpha > 0$. We notice that the map $\alpha \to V_\alpha$ is continuous for the uniform topology of operators.

We now show that the operators V_α satisfy the resolvent equation on $b\mathscr{E}$. Since $I + \alpha V$ is one-to-one, it suffices to prove that for $f \in b\mathscr{E}$

$$(I + \alpha V)\,(V_\alpha - V_\beta)\, f = (I + \alpha V)\,(\beta - \alpha)\, V_\alpha V_\beta f,$$

but this is an obvious consequence of eq. (1.1).

Finally if $\{f_n\}_{n \geqslant 0}$ is a sequence of functions in $b\mathscr{E}_+$ decreasing to zero, for every $\alpha > 0$,

$$\lim_n V_\alpha f_n \leqslant \lim_n V f_n = 0,$$

which proves that the operators V_α are the restrictions to $b\mathscr{E}$ of kernels $V_\alpha(x, \cdot\,)$. It is now easy to see that these kernels form the desired resolvent.

Corollary 1.6. *Under the same hypotheses, if moreover E is locally compact and V maps C_K into C_0, the kernels V_α have the same property.*

Proof. The above proof may be performed by using the Banach space C_0 instead of $b\mathscr{E}$.

To proceed with the general problem, we shall have to use resolvents of the kind just described. We shall need the notion of reduced functions in the cone of supermedian functions relative to this resolvent.

Proposition 1.7. *Let $\{V_\alpha\}_{\alpha>0}$ be a proper resolvent and $A \in \mathscr{E}$. With each supermedian function f we may associate a supermedian function $R_A f$ called the reduced function of f on A, which satisfies the following properties:*

(i) *$R_A f$ is the smallest supermedian function which dominates f on A; $R_A f = f$ on A and if f, g are two supermedian functions such that $f \leqslant g$ on A, then $R_A f \leqslant R_A g$;*

(ii) *if $\{f_n\}$ is a sequence of supermedian functions increasing to f, then $R_A f = \lim_n R_A f_n$;*

(iii) *if $h \in \mathscr{E}_+$ vanishes outside A, then $R_A V_0 h = V_0 h$;*

(iv) *$R_A(f + g) = R_A f + R_A g$;*

(v) *if $A \subset B$, then $R_A f \leqslant R_B f$.*

Proof. We use the notation of ch. 2 §6. We first prove that if $\alpha < \beta$, then the functions which are superharmonic for X^β are superharmonic for X^α. It suffices to consider the bounded potentials $f = (I + \beta V_0)\, h$, and then

$$\begin{aligned}\alpha V_\alpha f = \alpha V_\alpha (I + \beta V_0)\, h &= \alpha V_\alpha h + \beta (V_0 h - V_\alpha h) \\ &= (\alpha - \beta)\, V_\alpha h + \beta V_0 h \leqslant \beta V_0 h \leqslant f.\end{aligned}$$

Consequently, if f is supermedian $P^\alpha_A f \leqslant P^\beta_A f$, so that the limit

$$R_A f = \lim_{\alpha \to \infty} P^\alpha_A f$$

is well defined. We proceed to show that the operator R_A thus defined has the desired properties (i)–(v).

We first prove that $R_A f$ is a supermedian function. Pick a $\beta > 0$; for every $\alpha > \beta$, $P^\alpha_A f$ is P^β-superharmonic, so $R_A f$ is P^β-superharmonic. Since this is true for all β, the function $R_A f$ is supermedian. Next we note that if $f \leqslant g$ on A, then $P^\alpha_A f \leqslant P^\alpha_A g$, and consequently $R_A f \leqslant R_A g$; since clearly $R_A g \leqslant g$, it follows that $R_A f \leqslant g$, and the proof of (i) is complete.

The property (ii) follows at once from interchanging increasing limits. Now let h be finite and vanish outside A; by the known properties of balayage operators,

$$P_A^\alpha(I + \alpha V_0)\, h = (I + \alpha V_0)\, h,$$

whence, dividing by α and letting α tend to infinity, we get $R_A V_0 h = V_0 h$. The proof of (iii) is then completed by using approximation by potentials.

Properties (iv) and (v) are obvious.

We will also need the notion of *excessive* function which was introduced in Exercise 6.14 of ch. 2, the results of which will be used freely in the sequel.

Exercise 1.8. (1) Let V be the potential kernel of a proper Markovian resolvent $\{V_\alpha\}$. Define a new space $\tilde{E}$ by adjoining to E a point Δ, and let $\tilde{\mathscr{E}}$ be the σ-algebra $\sigma(\mathscr{E}, \{\Delta\})$. Define a kernel $\tilde{V}$ on $(\tilde{E}, \tilde{\mathscr{E}})$ be setting

$$\tilde{V}(x, \cdot\,) = V(x, \cdot\,) + \varepsilon_\Delta(\,\cdot\,), \qquad x \in E,$$

$$\tilde{V}(\Delta, \cdot\,) = \varepsilon_\Delta(\,\cdot\,).$$

Prove that $\tilde{V} \in$ (C.M.P.).

(2) Assume that there is a resolvent $\{\tilde{V}_\alpha\}$ on E such that $\tilde{V} = \tilde{V}_0$. Then $\alpha \tilde{V}_\alpha 1_E = 1_E$ for every $\alpha > 0$, and this is not consistent with the definition of $\tilde{V}$.

Exercise 1.9. Prove that a positive kernel V satisfying the C.M.P. for the functions in $\mathscr{B}_+$ satisfies the C.M.P. for the functions in $\mathscr{E}_+$.

Exercise 1.10. One could think of generalizing the problem dealt with in this section in the following manner. Let $V \in$ (C.M.P.) and $h \in b\mathscr{E}_+$. Does there exist a kernel V_h such that

(i) $V = V_h + V I_h V_h$, (ii) $V = V_h + V_h I_h V$, (iii) $V = \sum_n (V_h I_h)^n V_h$?

Clearly, if V_h satisfies (iii) it satisfies (i) and (ii).

(1) Prove that there exists at most one kernel V_h satisfying (i) and hence (iii). If h, k are in $b\mathscr{E}_+$ and $h \leqslant k$, and if there exist two operators V_h and V_k satisfying (i), then

$$V_h = V_k + V_h I_{k-h} V_k.$$

(2) Let $(E, \mathscr{E})$ be $\{-\infty\} \cup Z \cup \{+\infty\}$ with the discrete σ-algebra and define the kernel V by $V(x, y) = 0$ if $x < y$ and $V(x, y) = 1$ otherwise. Prove that $V \in$ C.M.P. and that there exists a kernel V_h satisfying (i) (resp. (ii)) if and

only if $\sum_{-\infty}^{0} h(z) < \infty$ (resp. $\sum_{0}^{+\infty} h(z) < \infty$). As a result a kernel V_h may satisfy (i) without satisfying (ii).

(3) If $h \in \mathscr{B}_+$, there does exist a kernel V_h satisfying (iii) and such that $V_h(h) = 1$. If f is V-supermedian (see next section) then $V_h(hf) \leqslant f$.

2. The reinforced principle. Construction of transient Markov chains

We now study a proper kernel G satisfying the R.C.M.P. for the functions of $\mathscr{B}_+$, and we shall give a necessary and sufficient condition under which it is the potential kernel of a transient Markov chain. A slight variation to Exercise 1.8 provides an example of a kernel which satisfies the R.C.M.P. and is nevertheless not the potential kernel of a transient Markov chain.

Recalling the results of ch. 2, we lay down the following

Definition 2.1. A function $f \in \mathscr{E}_+$ is called *G-superharmonic* if the inequality $Gg \leqslant f$, $g \in \mathscr{B}$, holds everywhere provided it holds on $\{g > 0\}$.

The following result provides important examples of G-superharmonic functions.

Proposition 2.2. *Let $a \in \mathbf{R}_+$ and $f \in \mathscr{B}_+$; the functions $a + Gf$ and $a + Gf - f$ are G-superharmonic.*

Proof. For the first one it follows at once from the fact that G satisfies the C.M.P. (cf. Proposition 1.1).

Assume next that, for $g \in \mathscr{B}$, we have

$$a + Gf - f \geqslant Gg \quad \text{on } \{g > 0\};$$

this implies

$$a + G(f + g^-) - (f + g^-) \geqslant Gg^+ \quad \text{on } \{g^+ > 0\},$$

hence everywhere, since $G \in$ (R.C.M.P.). It follows that

$$a + Gf - f \geqslant Gg$$

everywhere, which completes the proof.

By ch. 2 §1.14 and Proposition 1.2 there exists a bounded strictly positive function f such that the kernel $V = GI_f$ is a bounded kernel satisfying the C.M.P. for the functions in $b\mathscr{E}_+$ and such that $V(b\mathscr{E}_+) \subset G(\mathscr{B}_+)$. By the preceding section there exists one and only one resolvent $\{V_\alpha\}$, $\alpha > 0$, such that $V_0 = V$ and we have the following

Proposition 2.3. *The following three families of functions are equal:*

(i) *the supermedian functions with respect to* $\{V_\alpha\}$;

(ii) *the excessive functions with respect to* $\{V_\alpha\}$;

(iii) *the G-superharmonic functions.*

Proof. To prove the equality of (i) and (ii) it suffices by ch. 2, Exercise 6.14 to prove that the empty set is the only set of potential zero. But $V1_A = G(f1_A) = 0$ implies, from Proposition 1.3, that $f1_A = 0$, and since f is strictly positive, A is empty.

As f is strictly positive, the equality of (iii) and (i) follows from ch. 2, Proposition 6.5.

As a result, neither the cone of supermedian functions relative to $\{V_\alpha\}$ nor the operator R_A in this cone depend on the choice of f. Making use of Proposition 1.7, we may therefore speak of the reduced function R_Ah of the G-superharmonic function h on the set A, and the operator R_A enjoys the properties of Proposition 1.7. The following definition is basic.

Definition 2.4. A G-superharmonic function h is said to *vanish at the boundary* if there exists a decreasing sequence $\{A_n\}_{n\geq 0}$ of sets in $\mathscr{E}$ such that:

(i) $\bigcap_n A_n = \emptyset$ and $G(\cdot, A_n^c)$ is bounded for all n;

(ii) $\inf_n R_{A_n}h = 0$.

Let us further recall that if $h = f + g$, where f and g are G-superharmonic, then h is said to be *specifically greater* than f or g. This is the intrinsic or proper order in the cone of superharmonic functions.

Lemma 2.5. *Let* $\{h_p\}$ *be a specifically increasing sequence of G-superharmonic functions vanishing at the boundary relative to one and the same sequence* $\{A_n\}$. *If* $h = \sup h_p$ *is finite, it vanishes at the boundary relative to* $\{A_n\}$.

Proof. For every p we may write $h = h_p + g_p$, where g_p is G-superharmonic. Then by Proposition 1.7, $R_{A_n}h = R_{A_n}h_p + R_{A_n}g_p \leq R_{A_n}h_p + g_p$. Given $x \in E$ and $\varepsilon > 0$, we may find an integer p such that $g_p(x) < \varepsilon$, and then an integer n such that $R_{A_n}h_p(x) < \varepsilon$, which yields the desired result.

Proposition 2.6. *The following three conditions are equivalent:*

(i) *there exists a strictly positive function* $g \in \mathscr{B}_+$ *such that* Gg *vanishes at the boundary*;

(ii) *there exists a sequence* $\{K_m\}$ *of sets in* $\mathscr{E}$ *with* $\bigcup K_m = E$ *and such that* $G1_{K_m}$ *is bounded and vanishes at the boundary for all* m, *relative to one and the same sequence* $\{A_n\}$;

(iii) *for all* $f \in \mathscr{B}_+$, Gf *vanishes at the boundary relative to one and the same sequence* $\{A_n\}$.

Proof. (ii) implies (i). Indeed, there exists a sequence of numbers $b_m \in]0, 1]$ such that the series $\sum_m b_m 1_{K_m}$ and $\sum_m b_m G1_{K_m}$ converge uniformly. Setting $g = \sum_m b_m 1_{K_m}$, the result follows from Lemma 2.5.

Next, (i) implies (iii). Let $f \in \mathscr{B}_+$, and set $f_n = ng \wedge f$; then Gf_n increases specifically to Gf and each Gf_n vanishes at the boundary relative to any sequence $\{A_n\}$ for which Gg vanishes at the boundary. We conclude the result once more from Lemma 2.5.

Finally, that (iii) implies (ii) is obvious.

As has been proved in ch. 2, if G is the potential kernel of a Markov chain, then all potentials Gg vanish at the boundary. We are going to show that, conversely, this condition is sufficient to assert that G is the potential kernel of a Markov chain. We must first make some preparation.

Lemma 2.7. *Let* h *be a bounded superharmonic function, and set*

$$D_n h = nf(h - nV_n h);$$

then $GD_n h$ *increases to* h *as* n *tends to infinity*.

Proof. The lemma follows from the fact that h is excessive for $\{V_\alpha\}$, and the equalities

$$GD_n h = V(D_n h/f) = V(n(h - nV_n h) = nV_n h.$$

Lemma 2.8. *Let* h *be a bounded superharmonic function, and* $\{u_n\}$ *a sequence of functions in* $\mathscr{B}_+$ *such that* Gu_n *increases to* h. *The sequence* $\{u_n\}$ *converges boundedly to a function* u *and if* $h = Gg$ $(g \in \mathscr{B}_+)$, *then* $u = g$ *and* $Gu = h$.

Proof. From the inequalities $h \geqslant Gu_n \geqslant u_n$, we get that the sequence $\{u_n\}$ is uniformly bounded. Furthermore, since for $n \geqslant m$ we have $Gu_n \geqslant Gu_m$, it follows from Proposition 1.3 that $G(u_n - u_m) \geqslant u_n - u_m$, which can be written as

$$u_n \leqslant u_m + G(u_n - u_m).$$

Pick a point x in E, and two sequences $\{n_k\}$ and $\{m_k\}$ such that $n_k \geqslant m_k$,

$$\lim_k u_{n_k}(x) = \overline{\lim}\, u_n(x)$$

and

$$\lim_k u_{m_k}(x) = \underline{\lim}\, u_n(x);$$

letting k tend to infinity in the above inequality yields $\overline{\lim}\, u_n(x) \leqslant \underline{\lim}\, u_n(x)$, which proves the existence of the limit u. The inequality $Gu \leqslant h$ follows from Fatou's lemma.

If $h = Gg$, the inequality $G(g - u_n) \geqslant 0$ implies, as above, that $g \leqslant u_n + G(g - u_n)$ and, passing to the limit, that $g \leqslant u$. As a result, $h = Gg \leqslant Gu$; hence $Gg = Gu$. The operator G being one-to-one, it follows that $u = g$.

Lemma 2.9. *Let $\{u_n\}$ be a sequence in $\mathscr{B}_+$ converging to a function u and such that the potentials Gu_n are smaller than a G-superharmonic bounded function h vanishing at the boundary. Then $Gu = \lim_n Gu_n$.*

Proof. Let $\{A_n\}$ be a sequence relative to which h vanishes at the boundary, and choose a decreasing sequence of numbers a_n such that the series $\sum_n a_n G(\cdot, A_n^c)$ converges uniformly. The function

$$w = \sum_n a_n 1_{A_{n-1} \setminus A_n}$$

is in $\mathscr{B}_+$ and bounded from below on each set A_n^c.

We assume that for every x in E,

$$\lim_{a \to \infty} \int_{\{u_n > aw\}} G(x, \mathrm{d}y)\, u_n(y) = 0 \tag{2.1}$$

uniformly in n. By a classical argument we have

$$|Gu(x) - Gu_n(x)| \leqslant \int_{\{u_n > aw\}} G(x, \mathrm{d}y)\, (u_n(y) + u(y)) + \int_{\{u_n \leqslant aw\}} G(x, \mathrm{d}y)\, |u_n(y) - u(y)|;$$

the hypothesis entails that $Gu \leqslant h$, so we may choose a in order that the first integral be small, then letting n tend to infinity, we get the desired result.

It thus remains to prove the uniform integrability property of eq. (2.1). The integral therein is majorized by

$$G(u_n 1_{C_a})(x) = \int_{C_a} G(x, \mathrm{d}y)\, u_n(y),$$

where $C_a = \{h > aw\}$ contains $\{Gu_n > aw\}$, hence also $\{u_n > aw\}$. But on C_a we have

$$R_{C_a} Gu_p = Gu_p \geqslant G(u_p 1_{C_a}),$$

and since $R_{C_a} Gu_p$ is G-superharmonic, we get $R_{C_a} Gu_p \geqslant G(u_p 1_{C_a})$ everywhere. A fortiori $G(u_p 1_{C_a}) \leqslant R_{C_a} h$, and since h vanishes at the boundary, it suffices to show that $C_a \subset A_n$ for sufficiently large a. Since w is bounded from below by a_n on A_n^c we have $C_a \subset A_n$ as soon as $\|h\| < aa_n$, and the proof is complete.

Theorem 2.10. *Let G be a proper kernel satisfying the R.C.M.P.; then the two following conditions are equivalent:*

(i) *G is the potential kernel of a transition probability P;*

(ii) *every bounded potential vanishes at the boundary.*

Proof. We know from ch. 2 that (i) implies (ii). We proceed to prove the converse.

If (i) is true and if $g \in \mathscr{B}_+$, then $GPg = Gg - g$. We therefore set $Pg = \lim_n D_n(Gg - g)$, which is meaningful by Lemmas 2.7 and 2.8. Let us study the properties of the operator P thus defined on $\mathscr{B}_+$.

The G-superharmonic function $Gg - g$ vanishes at the boundary, since it is majorized by the bounded potential Gg. By Lemma 2.9, we have therefore $GPg = Gg - g$, so that Pg is also in $\mathscr{B}_+$, and we can define the iterates $P_n g$. From the equality $GP_n g = GP_{n-1} g - P_{n-1} g$, we derive

$$Gg = g + Pg + \cdots + P_n g + GP_{n+1} g.$$

This implies that $P_n g \to 0$ as $n \to \infty$, and since Gg vanishes at the boundary it follows from Lemma 2.9 that $GP_{n+1} g \to 0$, and hence that

$$Gg = g + Pg + \cdots + P_n g + \cdots.$$

Now let $\{g_n\}$ be a sequence of functions in $\mathscr{B}_+$ decreasing to zero. By Lebesgue's theorem the sequence $\{Gg_n\}$ also decreases to zero; since $Pg_n \geqslant 0$, it follows from Proposition 1.3 that $Pg_n \leqslant GPg_n = Gg_n - g_n$, and the sequence $\{Pg_n\}$ also decreases to zero.

Finally, let g be a function in $\mathscr{B}_+$ with $g \leqslant 1$; from the inequality $1 \geqslant g = G(g - Pg)$, we get

$$G(g - Pg)^+ \leqslant 1 + G(g - Pg)^- - (g - Pg)^-$$

on $\{(g - Pg)^+ > 0\}$, hence everywhere. This implies that

$$Pg = G(g - Pg) - (g - Pg) \leqslant 1.$$

The space $\mathscr{B}$ being clearly a lattice, the last two properties prove, by means of Daniell's theorem, that P is given by a T.P. still denoted P defined on the σ-algebra generated by $\mathscr{B}$. Since G is proper, the σ-algebra generated by $\mathscr{B}$ is equal to $\mathscr{E}$, and the proof is easily completed.

The above condition (ii) is not easy to check, but it yields some useful corollaries if some additional assumptions are made. One of these is given in Exercise 2.19; another is obtained in a topological setting.

Corollary 2.11. *If E is locally compact and countable at infinity, and if G is a kernel on E mapping C_K into C_0, then G is the potential of a Markov chain. If, moreover, $G(C_K)$ is dense in C_0, then the Markov chain is a Feller chain.*

Proof. The condition of Proposition 2.6(ii) is satisfied by taking for $\{K_m\}$ as well as for $\{A_n\}$ any increasing sequence of compact sets whose interiors cover E; consequently the preceding theorem applies, and G is the potential kernel of a transition probability P.

If $g \in C_K$, from $PGg = g + Gg$ it is easily deduced that PGg is in C_0, and the second part of the corollary follows.

We now apply the same methods to the problem of kernels satisfying the C.M.P.

Definition 2.12. Let V be a kernel satisfying the C.M.P. A function $h \in \mathscr{E}_+$ is called *supermedian* if for every $g \in \mathscr{B}$, the inequality $Vg \leqslant h$ holds everywhere whenever it holds on $\{g > 0\}$.

The potentials $Vf (f \in \mathscr{B}_+)$ are supermedian functions.

Since V is proper, there is a strictly positive function f in $\mathscr{B}_+$. The kernel $\tilde{V} = VI_f$ is bounded and satisfies the C.M.P., so that there is a resolvent $\{\tilde{V}_\alpha\}$ such that $\tilde{V}_0 = \tilde{V}$. The supermedian functions are the same for V and $\tilde{V}$, and therefore by following the pattern of §1 we can define the reduced functions in the cone of supermedian functions. We define the functions vanishing at the boundary in the same way as above and clearly Proposition 2.6 is still available.

For every $\alpha > 0$, the kernel $G^\alpha = I + \alpha V$ satisfies the R.C.M.P., and the space $\mathscr{B}$ is the same for V and G^α. Furthermore, the supermedian functions are G^α-superharmonic, as is easily checked. As a result, if h is supermedian, its reduced function on A in the cone of G^α-superharmonic functions is smaller than its reduced function on A in the cone of supermedian functions. We may thus state

Lemma 2.13. *Let $\{u_n\}$ be a sequence of functions converging to u and such that $\{G^\alpha u_n\}$ is majorized by a potential Vg vanishing at the boundary; then $G^\alpha u = \lim_n G^\alpha u_n$.*

Proof. The same as the proof of Lemma 2.9.

This enables us to state

Theorem 2.14. *For a proper kernel V satisfying the C.M.P., the following two statements are equivalent:*

(i) *there exists a submarkovian resolvent $\{V_\alpha\}$ such that $V_0 = V$;*

(ii) *every bounded potential vanishes at the boundary.*

Proof. We first prove that (ii) implies (i). Let $g \in \mathscr{B}_+$. Since Vg is G^α-superharmonic, by Lemma 2.7 there is a sequence $\{l_n\}$ such that $G^\alpha l_n$ increases to Vg. Since the sequence $\{G^\alpha l_n - l_n\}$ increases and is majorized by Vg, it converges, and consequently the sequence

$$\{l_n\} = \{G^\alpha l_n - (G^\alpha l_n - l_n)\}$$

converges to a limit which we call $V_\alpha g$. By Lemma 2.13, we have $G^\alpha V_\alpha g = Vg$, and therefore $V_\alpha g$ does not depend on the particular sequence $\{l_n\}$, since G^α is one-to-one. Finally $V_\alpha g$ lies in $\mathscr{B}_+$.

As in the proof of Theorem 2.10, it is verified that if $g \leqslant 1$, then $\alpha V_\alpha g \leqslant 1$. That the operators V_α satisfy the resolvent equation is seen as in the proof of Theorem 1.5. Finally Daniell's theorem enables us to show that V_α is given by a kernel. It remains to show that $V_0 = V$.

We already know that $V_\alpha g \leqslant Vg$ for every α, hence that $V_0 g \leqslant Vg$. Now the identity $(I + \alpha V)\, V_\alpha g = Vg$ implies that, for every n,

$$(I + \alpha V)\,(\alpha V_\alpha)^n g = \alpha V(\alpha V_\alpha)^{n-1} g,$$

and summing up these equations yields

$$\sum_{1}^{n} (\alpha V_\alpha)^k g + \alpha V(\alpha V_\alpha)^n g = Vg.$$

This proves that $V(\alpha V_\alpha)^n g \leqslant Vg$ and $\lim_n (\alpha V_\alpha)^n g = 0$, which implies $\lim_n \alpha V(\alpha V_\alpha)^n g = 0$. Finally, it follows that

$$\sum_{1}^{\infty} (\alpha V_\alpha)^k g = \alpha Vg,$$

which proves, on account of ch. 2, Proposition 6.4, that $V_0 = V$.

We now turn to proving that (i) implies (ii). Let $\{\alpha_n\}$ be a sequence of real numbers decreasing to zero. Let f be strictly positive and in $\mathscr{B}_+$. The sets $A_n' = \{V_{\alpha_n} f < \alpha_n\}$ decrease to $N = \{Vf = 0\}$. We have $V(x, \cdot\,) = 0$ for every x in N because f is strictly positive, hence $V(\cdot\,, N) = 0$ on N and $V(\cdot\,, N) = 0$ everywhere by the C.M.P. If we set $A_n = A_n' \backslash N$, we have therefore

$$V(\cdot\,, A_n^c) = V(\cdot\,, A_n'^c) \leqslant \alpha_n^{-1} VV_{\alpha_n} f \leqslant \alpha_n^{-2} \|Vf\| < \infty,$$

and the sequence $\{A_n\}$ decreases to the empty set.

We will complete the proof by showing that $R_{A_n} Vf$ goes to 0 as n tends to infinity. Since $A_n \subset A_n'$ we have $R_{A_n} Vf \leqslant R_{A_n'} Vf$; moreover, on A_n', we have, by the resolvent equation,

$$Vf \leqslant \alpha_n + \alpha_n VV_{\alpha_n} f,$$

and since the function on the right side is supermedian,

$$R_{A_n'} Vf \leqslant \alpha_n + \alpha_n VV_{\alpha_n} f = \alpha_n + (Vf - V_{\alpha_n} f).$$

Letting n tend to infinity yields the desired result.

As before we have the corollary

Corollary 2.15. *If E is locally compact and countable at infinity, and if V maps C_K into C_0, there is a sub-markovian resolvent $\{V_\alpha\}$ such that $V_0 = V$. If, in addition, $V(C_K)$ is dense in C_0, then V_α maps C_0 into C_0 for each $\alpha > 0$.*

Proof. The same as for Corollary 2.11.

Remark. This corollary is not the best available, as is pointed out in the exercises and in the notes and comments.

Exercise 2.16. Prove that $h - Gu$ in Lemma 2.8 is superharmonic.

Exercise 2.17. Show that every bounded G-superharmonic function vanishing at the boundary is a potential.

Exercise 2.18. The kernel $I + 1 \otimes \nu$, where ν is a non-zero σ-finite measure, satisfies the R.C.M.P. and is not the potential kernel of a Markov chain.

Exercise 2.19. Prove that if $V \in$ (C.M.P.) and $V1$ is finite, then there is a submarkovian resolvent $\{V_\alpha\}$ such that $V_0 = V$. Solve the same problem for a kernel $G \in$ (R.C.M.P.).

Exercise 2.20. Resume the hypotheses of Theorem 2.14. Let $\{K_n\}$ be an increasing sequence of sets such that $\bigcup K_n = E$ and $V1_{K_n}$ is bounded for all n. Let $\{b_n\}$ be a sequence of numbers in $]0, 1[$ such that if we set $a = \sum_n b_n 1_{K_n}$, then a and Va are bounded. Set $a_n = na \wedge 1$ and $V^n = VI_{a_n}$; define $\{V^n_\alpha\}$ to be the resolvent with $V^n_0 = V$. Finally define $\mathscr{E}_K$ to be the set of functions g such that $\{g \neq 0\} \subset K_m$ for some m.

(1) Prove that if $g \in \mathscr{E}^+_K$, then for fixed α, $V^n_\alpha(g/a_n)$ is a decreasing function of n.

(2) Give a proof of the first part of Theorem 2.14 by setting $V_\alpha g = \lim_n V^n_\alpha(g/a_n)$.

(3) Prove that Corollary 2.15 holds without the density hypothesis.

Exercise 2.21. With the hypothesis of Corollary 2.11 (resp. 2.15), if in addition E is a group and G (resp. V) is translation-invariant, the corresponding kernel P (resp. αV_α) is the T.P. of a random walk.

3. The semi-complete maximum principle

The goal of this section is essentially to raise the following problem. As we have seen in ch. 8, the recurrent counterpart of the C.M.P. is the so-called semi-complete maximum principle; does there exist a characterization of recurrent potential kernels by means of this principle similar to the characterization found in the foregoing sections for the transient kernels? To give a full answer to this question would take too much space to be attempted here, and we shall content ourselves with a few remarks and a very incomplete answer.

Once for all, a measure m is given on $(E, \mathscr{E})$, and any word such as "negligible" will refer to this measure.

Definition 3.1. A kernel $W(\Gamma)$ is said to satisfy the (reinforced) *semi-complete maximum principle* if, for every bounded integrable function f such that $m(|f|) > 0$, $m(f) = 0$, and $Wf(\Gamma f)$ is bounded, the inequality $Wf(\Gamma f) \leqslant a$, $a \in \mathbf{R}$, on $\{f > 0\}$ implies $Wf \leqslant a$ everywhere ($\Gamma f \leqslant a - f^-$ everywhere, or equivalently $\Gamma f - f \leqslant a$ everywhere).

As in the preceding sections, we shall use the simplifying notations R.S.C.M. and S.C.M. We call $\mathcal{N}$ the space of functions f having the properties in the definition; of course $\mathcal{N}$ depends on m and on the kernel studied. It may happen that some kernels satisfy the principles only for subclasses of $\mathcal{N}$. We may now state our problem in a proper way: is the kernel Γ in Definition 3.1 the potential kernel of a Harris Markov chain?

We start with some easy remarks. In what follows we suppose $\Gamma \in$ (R.S.C.M.).

Proposition 3.2. *If $W \in$ (S.C.M.), then $I + \alpha W \in$ (R.S.C.M.) for every $\alpha > 0$.*

Proof. Easy and left to the reader.

Proposition 3.3. *If $f \in \mathcal{N}$ then Γf cannot be constant on $\{f \neq 0\}$. In particular, Γ is one-to-one on $\mathcal{N}$.*

Proof. If $\Gamma f = a$ on $\{f \neq 0\}$, in particular $\Gamma f = a$ on $\{f > 0\}$; hence $\Gamma f \leqslant a - f^-$ everywhere, and in particular on $\{f^- > 0\}$. It follows that $f^- = 0$, and consequently that f^+ is negligible.

A clue to the difficulty of the problem is given by the following theorem, which states that Harris chains are not the only chains which give rise to potential kernels satisfying R.S.C.M. We do not strive for the greatest generality.

Theorem 3.4. *Let E be L.C.C.B. and G be a proper kernel on E satisfying the R.C.M.P. and mapping C_K into C_b. If the set of measures $\{G(x, \cdot\,),\, x \in E\}$ has a non-zero cluster point m as $x \to \Delta$, then G satisfies the R.S.C.M. with respect to m for functions in $\mathcal{N} \cap C_K$.*

Proof. There exists a function $u \in C_0$ which is strictly positive everywhere and such that the kernel $V = GI_u$ is bounded and satisfies the C.M.P. There exists therefore a resolvent $\{V_\alpha\}$ such that $V_0 = V$, and we denote by P_A^α the balayage operator relative to A and the chain X^α of T.P. αV_α. If h vanishes outside A, then

$$P_A^\alpha(I + \alpha V)\, h = (I + \alpha V)\, h.$$

Let $f \in C_K$ be non-zero and such that $m(f) = 0$, $m(f^+) > 0$, and $Gf \leqslant a$ on $\{f > 0\}$. If we let $\tilde{f} = fu^{-1}$ then $\langle um, \tilde{f}\rangle = 0$ and $V\tilde{f} \leqslant a$ on $A = \{\tilde{f} > 0\}$. We have

$$(I + \alpha V)\, \tilde{f}(x) \leqslant (\|\tilde{f}\| + \alpha a)\, P_A^\alpha 1(x).$$

Let $\{x_n\}$ be a sequence converging to Δ such that $V(x_n, \cdot\,)$ converges to um. Passing to the limit along $\{x_n\}$ in the above inequality yields

$$(\|\tilde{f}\| + \alpha a)\, \overline{\lim_n}\, P_A^\alpha 1(x_n) \geqslant 0.$$

Now, passing to the limit in the same way in the relation

$$(I + \alpha V)\, f^+ = P_A^\alpha(I + \alpha V)\, f^+,$$

we get

$$0 < \alpha\langle um, f^+\rangle \leqslant (\|f^+\| + \alpha\|Vf^+\|)\, \overline{\lim_n}\, P_A^\alpha 1(x_n).$$

It follows that $\overline{\lim}_n\, P_A 1(x_n) > 0$; hence $\|\tilde{f}\| + \alpha a \geqslant 0$ for all α, and consequently $a \geqslant 0$. As a result we have

$$Gf^+ \leqslant a + Gf^- - f^- \quad \text{on } \{f^+ > 0\} \quad \text{with } a \in \mathbf{R}_+$$

and we get the desired result by applying the R.C.M.P. to G.

This theorem holds as well with the C.M.P. and S.C.M. in place of the reinforced principles, and if G is the potential of a transient chain or resolvent, then m is an excessive measure. We remark that there are indeed many kernels satisfying the hypotheses of this theorem: for instance, the potential kernels of transient random walks of type II. Furthermore, if G satisfies the R.S.C.M. with respect to m, then the kernel $G + 1 \otimes \nu + \varphi \otimes m$, where φ is a bounded function and ν a measure such that $\nu(|f|) < \infty$ for every $f \in \mathcal{N}$, also satisfies the R.S.C.M. This shows that there are plenty of kernels satisfying the R.S.C.M. that are nevertheless not the potential kernels of Harris chains. This will be made plain by the following result.

Finally, it is worth recording that the kernels dealt with in the above result are precisely those for which Corollaries 2.11 and 2.15 do not apply.

We now turn to a result parallel to Corollary 1.4. The notion of modification refers to m.

Proposition 3.5 (uniqueness theorem). *If $(E, \mathscr{E})$ is separable, and if G is a proper kernel satisfying the R.S.C.M., there exists at most one chain up to modification such that G is its potential kernel either in the sense of ch. 2 or in the sense of Harris chains.*

Proof. If P is a T.P. such that $G = \sum_0^\infty P_n$, then for every $f \in b\mathscr{E}_+$ and such that $Gf \in b\mathscr{E}_+$, $G(I - P)f = 0$. If P is a Harris chain such that G is one of its potential kernels in the sense of ch. 6 §5, then there is a measure ν such that $G(I - P)f = \nu(f)$.

Now let P and $\tilde{P}$ be two T.P.'s for which G is a potential kernel in any of the two meanings which we deal with. It follows that if $g \in b\mathscr{E}_+ \cap \mathscr{L}^1(m)$ is such that Gf is bounded, then $G(Pf - \tilde{P}f)$ is constant. Proposition 3.3 implies therefore that $Pf = \tilde{P}f$ m-a.e. The usual arguments prove that P is a modification of $\tilde{P}$.

Remark. It suffices of course that G satisfies the R.S.C.M. for a sufficiently rich subclass of $\mathscr{N}$.

We will now give a partial answer to our problem. We will assume that E is L.C.C.B. and that m is a Radon measure such that the only open set with zero measure is the empty set. (See ch. 8 §4.) For any compact set K, we denote by $\mathscr{B}(K)$ the space of Borel functions vanishing outside K and by $\mathscr{C}(K)$ the space of functions continuous on K and vanishing outside K. We study a kernel G satisfying the R.S.C.M. for the functions with compact support; but we assume further that G may be written $G = I + U$, where U has the following property: for every compact K, the map $f \to (Uf)\, 1_K$ is compact from $b\mathscr{E}$ into $\mathscr{C}(K)$, the spaces being endowed with the usual supremum norm. These conditions are satisfied by the potential kernels of Harris chains whose T.P. is strong Feller in the strict sense, hence in particular if E is countable.

The following proposition should be compared with ch. 8, Theorem 2.1.

Proposition 3.6. *For every compact K such that $m(K) > 0$, there exists a unique Radon measure v_K on K such that*:

(i) $\langle v_K, 1\rangle = 1$;

(ii) $\langle v_K, Gf\rangle = 0$ *for every* $f \in \mathscr{N} \cap \mathscr{B}(K)$.

Proof. In view of our hypotheses, the map $f \to (Uf)\, 1_K$ is compact from $\mathscr{C}(K)$ into $\mathscr{C}(K)$, hence from $\mathscr{N} \cap \mathscr{C}(K)$ into $\mathscr{C}(K)$. The subspace $\mathscr{N} \cap \mathscr{C}(K)$ has co-dimension 1 in $\mathscr{C}(K)$. The operator I is thus of index 1 from $\mathscr{N} \cap \mathscr{C}(K)$ into $\mathscr{C}(K)$; hence by the index theorem, the operator

$$f \to f + (Uf)\, 1_K = (Gf)\, 1_K$$

is also of index 1. By Proposition 3.3, this operator is one-to-one on $\mathscr{N}$; its range is thus of co-dimension 1, and therefore is closed by an application of Banach's theorem. On the other hand, Proposition 3.3 asserts also that 1_K is not in this range. There exists, therefore, a unique probability measure v_K on K such that $\langle v_K, Gf \rangle = 0$ for every $f \in \mathscr{N} \cap \mathscr{C}(K)$.

Now let f be any function in $\mathscr{C}(K)$; the function $m(K)\, f - m(f)\, 1_K$ is in $\mathscr{N} \cap \mathscr{C}(K)$, and therefore

$$0 = v_K\, G(m(K)\, f - m(f)\, 1_K) = m(K)\, v_K Gf - m(f)\, \langle v_K, G1_K \rangle.$$

The measure $v_K G$ is thus a multiple of m on K, and the proof is complete.

For $g \in \mathrm{b}\mathscr{E}$ define $h^K = (g - v_K(g))\, 1_K$; the function h^K is in $\mathscr{B}(K)$ and $v_K(h^K) = 0$. By the above result there is a function f^K in $\mathscr{N} \cap \mathscr{B}(K)$ such that $h^K = (Gf^K)\, 1_K$. Keeping in mind the formulae of ch. 8, Theorem 2.1 we define

$$P_K g = Gf^K + v_K(g),$$

$$\Pi_K g = Gf^K + v_K(g) - f^K = P_K g - f^K = Uf^K + v_K(g).$$

We notice that if $g \in \mathscr{B}(K)$, then $P_K g = g$, and that $\Pi_K g$ is a continuous function. On the other hand, $P_K g = \Pi_K g$ on K^c.

Finally, we have:

Proposition 3.7. *The operators P_K and Π_K are given by transition probabilities. The measures $P_K(x, \cdot)$ and $\Pi_K(x, \cdot)$ vanish outside K and furthermore*

(i) $mI_K \Pi_K = mI_K$;

(ii) $(I - \Pi_K)\, Gf = f$ *for every* $f \in \mathscr{N} \cap \mathscr{B}(K)$.

Proof. If g vanishes on K, then $h^K = 0$; hence $f^K = 0$ and consequently $P_K g = \Pi_K g = 0$. If $g \in \mathscr{B}(K)_+$, since $P_K g = g$ on K, we have $Gf^K + v_K(g) \geq 0$ on K; hence

$$G(-f^K) \leq v_K(g) \quad \text{on } \{f^K \neq 0\},$$

and by R.S.C.M., $G(-f^K) \leq v_K(g) - (-f^K)^-$ everywhere. As a result, $P_K g \geq 0$, and

$$\Pi_K g = Gf^K + v_K(g) - f^K \geq 0.$$

That P_K and Π_K are kernels is clear from the form of Π_K as a function of U. Finally if $g = 1$, then $h^K = f^K = 0$, and therefore $P_K 1 = \Pi_K 1 = v_K(1) = 1$.

Next, for any $g \in b\mathscr{E}_+$, we have

$$\langle m1_K, \Pi_K g\rangle = \langle m1_K, P_K g - f^K\rangle = \langle m1_K, P_K g\rangle = \langle m, 1_K g\rangle,$$

since f^K is in $\mathscr{N} \cap \mathscr{B}(K)$ and $P_K g = g$ on K. This proves (i).

To prove (ii), pick f in $\mathscr{N} \cap \mathscr{B}(K)$; then Gf is in $b\mathscr{E}$ and $v_K(Gf) = 0$. If we put $g = Gf$, we have $h^K = g1_K$; hence $f^K = f$, and consequently

$$\Pi_K Gf = Gf - f.$$

The T.P.'s Π_K and P_K must be the balayage kernels of the chain we are looking for. The following result goes further in this direction.

Lemma 3.8. *If H and K are two compact sets with $H \subset K$, then*

(i) $P_K P_H = P_H$;

(ii) $\Pi_K P_H = \Pi_H$;

(iii) *for any* $g \in \mathscr{B}(H)_+$, $\Pi_K(g) \leqslant \Pi_H(g)$.

Proof. Set $h = P_H g = Gf^H + v_H(g)$, where $f^H \in \mathscr{N} \cap \mathscr{B}(H)$. There is a function $f^K \in \mathscr{N} \cap \mathscr{B}(K)$ such that $P_K h = Gf^K + v_K(h)$. Since $P_K h = h$ on K, we have

$$G(f^H - f^K) = v_K(h) - v_H(g) \quad \text{on } K;$$

hence, by Proposition 3.3, $f^H = f^K$ and $v_K(h) = v_H(g)$, so that $P_K P_H g = P_H g$.

Next, from $f^H = f^K$ we derive

$$\Pi_K P_H g = P_K P_H g - f^K = P_H g - f^H = \Pi_H g.$$

Finally, if $g \in \mathscr{B}(H)_+$, then $P_H g \geqslant g$, and therefore $\Pi_K g \leqslant \Pi_H g$. We are now ready to state

Theorem 3.9. *There exists a strong Feller transition probability P with excessive measure m and such that, for any $f \in \mathscr{N}$ with compact support,*

$$(I - P)\, Gf = f.$$

Proof. Let $\{K_n\}$ be a sequence of compact sets whose interiors cover E and are increasing to E. If $g \in b\mathscr{E}_+$ has compact support, the preceding lemma implies that the sequence $\Pi_{K_n}g$ decreases everywhere (at least for n sufficiently large) to a limit that we call Pg. We have

$$Pg(x) = \lim_n \Pi_{K_n}g(x) = \lim_n (Uf^{K_n}(x) + v_{K_n}(f^{K_n})).$$

As soon as K_n contains the support of g, we have $\Pi_{K_n}g = g - f^{K_n}$; hence $\|f^{K_n}\| \leqslant 2\|g\|$. It follows that the family $\{Uf^{K_n}\}$ is equi-continuous on each compact set; one may therefore find a sub-sequence converging uniformly on every compact set, and as a result the function Pg is continuous. Clearly P is a transition probability.

Furthermore $1_{K_n}\Pi_{K_n}g$ converges to Pg, and since

$$\langle m, 1_{K_n}\Pi_{K_n}g\rangle = \langle m, 1_{K_n}g\rangle,$$

by Fatou's lemma, passing to the limit yields $\langle m, Pg\rangle \leqslant m(g)$, and thus m is excessive for P.

Now let $f \in \mathcal{N}$ have compact support, and set

$$g = Gf + \|Gf\| = f + Uf + \|Gf\|.$$

By Proposition 3.7, as soon as f vanishes outside K_n, we have $\Pi_{K_n}g = Gf - f + \|Gf\|$; as a result, the lower semi-continuity of g and the fact that $\Pi_{K_n}(x, \cdot)$ converges to $P(x, \cdot)$ vaguely entail

$$Pg \leqslant \varliminf \Pi_{K_n}g = Gf - f + \|Gf\|;$$

hence $PGf \leqslant Gf - f$. Applying this to $(-f)$ yields

$$PGf = Gf - f,$$

and the proof is complete.

Unfortunately we do not know of a satisfactory criterion to decide whether the T.P. of the preceding theorem is recurrent or transient. (See, however, Exercise 3.12 and the notes and comments.) But we have the following partial answer.

Theorem 3.10. *If m is bounded, the T.P. of Theorem 3.9 is Harris, and m is its invariant measure.*

Proof. If m is bounded we may, in the preceding proof, apply Lebesgue's theorem instead of Fatou's lemma, and we get the equality $m(Pg) = m(g)$;

thus m is invariant by P.

Since m is bounded and invariant, it is easily seen that for $f \in C_K^+$ the series $\sum_0^\infty P_n f$ is either $+\infty$ or 0, or equivalently that P is a conservative contraction of $L^1(m)$. We prove that it is ergodic. If P were not ergodic, we could find two functions f, g in $L^1(m)$ such that $h = f - g$ is in $\mathcal{N}$, and a set A with $m(A) > 0$ such that for $x \in A$, $\sum P_n f(x) = +\infty$ and $\sum P_n g(x) = 0$. From the equality $(I - P)\, Gh = h$ we derive

$$\sum_0^k P_n(f - g)\,(x) \leqslant Gh(x) - P_{n+1}Gh(x).$$

Since Gh is bounded, we get a contradiction on letting k tend to infinity.

Thus, for any $A \in \mathscr{E}$ with $m(A) > 0$, we have $U_A(1_A) = 1$ m-a.s.; but since P is strong Feller, it follows by the usual continuity argument that $U_A(1_A) = 1$, and hence that P is Harris. This proof could also be completed by calling upon ch. 4 §4.

If E is compact, then m is bounded and the preceding theorem applies.

Exercise 3.11. Assume that the operator U used in the proof of Proposition 3.6 satisfies the S.C.M., and then prove that it is the potential operator of a strong Feller resolvent. Assume further that m is bounded, and prove that the resolvent is Harris.

Exercise 3.12. Let E be countable and m a measure on E; G a kernel on E satisfying the R.S.C.M. for the functions of $\mathcal{N}$ (relative to m) with compact supports.

(1) Fix an arbitrary state e in E and define

$$G^e(x, y) = G(x, y) - G(e, y) - (G(x, e) - G(e, e))\, m(y)/m(e).$$

Prove that the kernel G^e satisfies the R.C.M.P. on $E \backslash \{e\}$. The reader is invited to look at ch. 8, Exercise 2.11 to see the significance of G^e.

(2) Prove that if every bounded potential $G^e f$ vanishes at the boundary in the sense of § 2, then there is a Harris chain on E for which G is a potential operator. Prove that if m is bounded, then the above condition on G^e-potentials is satisfied, which thus gives another proof of Theorem 3.10 for countable state space.

NOTES AND COMMENTS

Before we proceed to comment on the text we feel that the title of the book deserves an explanation. There has been over the years some confusion on what is meant by "chain". According to the author, this word refers sometimes to the discreteness of the set of time parameters, and sometimes to the discreteness of the state space. In our opinion the most important feature is the time: a denumerable state space is a special case of a general state space, whereas discrete time is not a particular case of continuous time. In particular, the set of integers has only one accumulation point, namely $+\infty$, whereas the set of positive reals has plenty. As a result, most of the problems in discrete-time processes concern the asymptotic behaviour at $+\infty$. For continuous-time processes, the analogous problems are often solved by straightforward applications of the results known for discrete time. But a host of new problems, with no discrete counterpart, arises in connection with the other accumulation points: for instance, the study of local properties of sample paths. A good terminology must thus stress the nature of the set of time parameters. We therefore stick to using "chain" when the time is discrete and "process" when it is continuous. If one wants to specify the kind of state space or analytical properties, one can add an adjective or a noun. For instance a chain can be Harris, discrete, ..., a process can be a diffusion process, a Feller process, a Hunt process.

Although this book is devoted to chains, (continuous-time) processes will be alluded to by means of resolvents.

Chapter 1

The first three sections are of standard character. Some proofs as well as some exercises are taken from Blumenthal and Getoor [1]. The proof of Theorem 2.8 is from Ionescu–Tulcea [1]. The notion of admissible σ-algebra is taken from Doob [2].

When the space is discrete, transition probabilities may be written and handled as matrices as is done extensively in the book by Kemeny et al. [1].

Section 5. The important Lemma 5.3 is taken from Doob [2] and Theorem 5.7 from Mokobodzki [1]. Some results of Mokobodzki on what may be said for σ-finite kernels in the line of Lemma 5.3 are described in Meyer and Yor [1].

An ergodic theory without measure has been developed for Feller kernels. We refer the reader to the expository paper by Foguel [2].

Chapter 2

Most of the results of the first two sections are originally due to Deny [1] and Doob [3]. We have also used Meyer [2] and Neveu [1]. The introduction and systematic use of the kernels U_h is due to Neveu [5]. It is interesting to describe their continuous-time counterpart. If X is a standard process with resolvent $\{V_\alpha\}$, and we study the chain with T.P. $P = V_1$, then for $h \in \mathscr{U}_+$ and $f \in \mathscr{E}_+$,

$$U_h f(x) = E_x \int_0^\infty M_t f(X_t)\, dt$$

where $M_t = \exp[-\int_0^t h(X_s)\, ds]$.

The proof of Theorem 2.3 has been borrowed from Meyer [6]. The proof of Theorem 2.2 is related to the work of Rost [1, 2] and Mokobodzki [2]. For a more thorough study of reduced functions see the third part of Dellacherie and Meyer [1] where one will also find a general method of conducting computations about kernels.

Let us mention that the identity between harmonic functions for random walks on symmetric spaces and harmonic functions in the classical sense is shown by Furstenberg [1]. We also refer to Baxter [2]. Exercise 1.24 and especially the subsequent ones dealing with Birth and Death chains are taken from Karlin and McGregor [1]. This was generalized by Kemeny [3].

Section 3. The fruitful idea of relating harmonic functions and invariant events is due to Blackwell [3].

Section 5. The filling scheme procedure was introduced as a tool in ergodic theory (see Chacon and Ornstein [1]) and was used in many contexts, for instance random walks (Ornstein [1]). The material in this section as well as the exercises are mainly from Rost. Our presentation borrows to Meyer [7] and to Baxter and Chacon [1]. Exercise 5.15 is from Baxter [1].

Section 6. The goal of this section is to prepare for the discussion in Chapter 10 as well as to show in numerous exercises of the following chapters how some of the results obtained for chains may be translated to he continuous time situation.

Exercise **6.13** is taken from Meyer [3] and Exercises **6.14–6.15** from Watanabe [2].

Chapter 3

Section 1 is standard, at least for the results, and is intended as an introduction to the sequel as well as to provide examples for the more general situations. The existence of the invariant measure for irreducible recurrent discrete chains was first proved in Derman [1].

Section 2. The main results on irreducible chains, namely Theorems **2.3** and 2.5 are due to Jain and Jamison [1], but the proofs in the text, as well as most of the subsequent ones, are from Neveu [5]. The whole subject was started by Harris who proved the implication (iii) $\Rightarrow$ (ii) in Proposition 2.7. The proof of Proposition 2.2 uses a pattern similar to Harris' original proof. The useful remark in Exercise **2.14** is from Doob [1] and Exercise **2.21** owes to Blackwell [1].

The theory of R-recurrence roughly sketched in Exercise **2.23** has been developed in the papers of Tweedie [1]. It is important in as much as the number R thus isolated is the inverse of the largest possible value for an eigenvalue of the operator P acting on functions or on measures. In some contexts, for instance random walks (see Guivarc'h [3]) it is better to assume a topological irreducibility, but the proofs go through the same way.

Section 3. The first results are from Chung and Fuchs [1] in the case of the real line. The extension to arbitrary groups is straightforward. Theorem **3.9** seems to have been widely known, at least for abelian groups; the present proof is probably new. The proof in Exercise **3.21** was given by Bretagnolle and Dacunha–Castelle [1]. Exercise **3.13** was taken from the basic book by Spitzer [1] and Exercise **3.14** from Chung and Ornstein [1]. Spread-out (in French *étalées*) random walks are generally referred to in the literature as "non-singular".

Section 4. For the few results of harmonic analysis which are used in this section we refer to Rudin [1] and to Hewitt and Ross [1].

The main result of this section, namely Criterion 5.2, is due to Chung and Fuchs [1] in the case of the real line. It was extended to arbitrary abelian groups by Loynes [1]. Theorem **5.11** was proved by Dudley [1]. The structure theorem used in its proof may be found in Kurosh [1] and in Hewitt and Ross [1].

For the characterization of non-abelian recurrent groups the state of the subject may be found in Guivarc'h et al. [1] and in Baldi [1].

In this book, we aimed at keeping the use of Fourier transform to a minimum, as is seen from chs. 5 and 9 on random walks. The reader who wishes to study random walks by means of Fourier transforms is referred to the book by Spitzer [1] and to the papers by Port and Stone, especially their final paper [1].

Chapter 4

Section 1. This owes much to the fifth chapter of the book by Neveu [2]. Exercise 1.10 is from Chacon and Krengel [1]. For extensions of Proposition 1.8 see Getoor [1].

Section 2. The results of this section are due to Hopf [1]. The simple proof of the basic Maximal ergodic lemma is from Garsia [1].

Exercise 2.12 is from Krengel [1], Exercise 2.13 from Ackoglu and Brunel [1] and Exercise 2.18 from Brunel [unpublished]. The proofs of the ergodic Lemma in Exercises 2.19 and 2.20 are taken respectively from Neveu [8] and Meyer [7].

Section 3. The proof given here of the Chacon–Ornstein Theorem is due to Neveu [3], Lemma 3.1 being an original lemma of Chacon and Ornstein. Exercise 3.14 is a trivial generalization of a result of Oxtoby [1].

Section 4. The characterization of Harris chains given in Theorem 4.6 is taken as their definition when one wants to perform their study by purely analytical means. We refer the reader to Foguel [1, 4]. This characterization is useful to implement the following general "principle": if a conservative chain has some strong property of discrete irreducible recurrent chains then it is Harris.

Theorem 4.5 is from Neveu [6] (see also Jain [2]). The use of the measures Π_N to prove the duality is suggested in Foguel [1]. The result in Exercise 4.10 was proved by Doeblin (at least the existence of the limit) and was the first result of this kind upon which the following ones were patterned. The dichotomy property of Exercise 4.12 was first proved in Hennion and Roynette [1].

Let us finally mention that there exist many central-limit theorems for functionals of positive Harris chains which are too numerous to be listed here. One can see for example Orey [5] and Maigret [1].

Section 5. The form of Brunel's lemma given here is due to Ackoglu [1] and its proof is from Garsia [2]. The other proofs of this section are borrowed from Meyer [1].

Exercise 5.8 was taken in Meyer [7] and the proof of the Chacon–Ornstein theorem hinted at in Exercise 5.9 is from Neveu [8].

Section 6. The bulk of this section is due to Kingman [1] whom we follow rather closely in our exposition. The proof of the decomposition theorem however is borrowed from Ackoglu and Sucheston [1] who deal with the more general case of markovian operators and prove a ratio limit theorem for superadditive processes. The subadditive convergence theorem turned out to be very useful in widespread situations as is illustrated in the expository paper of Kingman [2] from which we have borrowed some of our exercises (see also Derrienic [5]). The result in Exercise 6.15 was originally due to Spitzer (see Spitzer [1]). Another proof of the convergence theorem which avoids the decomposition theorem may be found in Derrienic [3].

Finally, we must record that the hypothesis of invariance of the probability measure may be weakened by the use of ideas of Krengel [2] (see Abid [1]).

This chapter might have included a section on invariant measures. We refer to the books by Foguel [1] and Friedman [1] and also to the paper by Brunel [2].

Chapter 5

Section 1. The Choquet–Deny theorem was announced in Choquet and Deny [2] and its proof given in Deny [2]. The proof presented here is from Guivarc'h, Lemma 1.1 being from Brunel [2]. The proof in Exercise 1.7 is from Doob et al. [1]. Example 1.10 is taken from the fundamental paper by Furstenberg [1]. Further work in this direction was accomplished by Azencott [1], Guivarc'h [1], and especially Raugi [1].

Section 2. For the history of the renewal theorem, we refer the reader to the book by Feller [4]. To write this section, we used the book by Feller already quoted and the paper by Herz [1], from which the proof of Theorem 2.5 is taken. Exercise 2.8 is from Bougerol and Elie [1] and Exercise 2.9 from Spitzer [1].

Section 3. The most general form of the Renewal Theorem for the real line was given by Feller and Orey [1]. We follow here the presentation of Feller [4].

Section 4. The Renewal Theorem 4.4 was shown by Port and Stone [1] after Kesten and Spitzer [2] had opened the way. For non-abelian groups the results are now almost complete thanks to the work of Elie [1] and Sunyach [2].

Section 5. Theorem 5.1 was first shown in Stone [1] for the real line and was extended to general abelian groups in Port and Stone [1]. The present proof is from Bellaïche–Fremont and Sueur–Pontier [1]. Proposition 5.6 is taken from Spitzer [1] in the case of integers and from Bretagnolle and Dacunha–Castelle [1] for the real line. Further results on recurrent sets of transient random walks may be found in Spitzer [1] and Jain and Orey [1]; Exercise 5.13 is borrowed from Spitzer [1].

Let us also mention that there is an enormous literature on the Renewal theory for Semi-Markov chains for the definition of which we refer to Çinlar [1] and Jacod [1].

Chapter 6

Section 1. "Zero-two" laws have been discovered by Ornstein and Sucheston [1] and were extended to continuous time by Winkler. In our presentation we follow closely Derrienic [4] which stresses the probabilistic point of view. We also refer to Foguel [3].

Section 2. The important advances for the results of this section were made by Orey. He proved the existence of Cyclic classes in [1], then proved the limit theorem for aperiodic chains (Property 2.2(i)) in the case of irreducible recurrent discrete chains. Another method of proof of this result was given by Blackwell and Freedman [1] and finally carried over to the general Harris case by Jamison and Orey [1]. The presentation adopted here, which rests on the zero-two laws is that of Revuz [5].

The original method of Orey described in Orey [5] and in the first edition of this book, is by means of a differentiation lemma, to prove the existence of a set C such that $\inf_{x,y\in C} p_k(x, y) > 0$ for some k. This set can then be used as points were used in the discrete case (see ch. 3, Exercise 1.17) to exhibit the cyclic classes. Athreya and Ney [1] and also Nummelin [1] have taken advantage of the existence of such so-called "C-sets" to give a proof of the existence of the invariant measure and of the limit theorem relying on Renewal theory ideas along the line described in Exercise 3.14 of ch. 5. This is the "splitting" technique which can be used to give proofs of many results on Harris chains among which we would like to mention the result on the speed of convergence in the limit theorem given in Nummelin and Tweedie [1]. In connection with these results let us also draw attention to the papers of Griffeath [1, 2]. Finally, another approach to cyclic classes will be found in Foguel [1, 4].

Theorem 2.6 and its proof are from Jain [1]. An analogous limit theorem for L^p functions is proved by Horowitz [2]. Exercise 2.10 is from Cogburn [1] and Exercise 2.19 from Duflo [1].

The question of what can be said in place of Exercise 2.7(2) when X is null has been investigated by Krickeberg [1, 2]. This is related to the study of the asymptotic behaviour of the ratios $P_{n+m}(x, A)/P_n(y, B)$, which has been tackled by several authors. We refer the reader interested in this topic to Orey [5]. For the study of point transformations and the notions of mixing, see the book by Friedman [1].

Section 3. This section is intended mainly as a first step in the study of special functions and it is Theorem 3.5 which is important in this respect. Condition (ii) therein appears in Brunel [3] together with the fact that $I - P$ is an isomorphism of $b^\circ\mathscr{E}$. The proof of (i) $\Rightarrow$ (ii) uses ideas of Foguel [4] to which we refer the reader for another more general approach to many results on Markov chains. Many results, especially the conditions in Theorem 3.7 are from Horowitz [1]. Finally the general study of quasi-compact operators is due to Yosida and Kakutani [1] and our presentation borrows much to Neveu [2]. For the equivalence between quasi-compactness and Doeblin's condition see Doob [2] and Fortet [1]. Finally some further results and an exposition of the spectral theory for quasi-compact operators may be found in Brunel and Revuz [2].

Section 4. Special functions which generalize the former "bounded sets" (Orey [5], Metivier [1], Brunel [3]) appear in Neveu [5]. Together with the potential kernels of the next section, they constitute a major progress in the theory of Harris chains, permitting one to simplify and sharpen all the previous results. Neveu proved all the results of this section up to Proposition 4.8. Proposition 4.9 and 4.10 are from Brunel and Revuz [1]. The idea of relating "boundedness" of sets to quasi-compactness appears in Brunel [3]. Horowitz [1] has shown that if a chain is conservative and ergodic in the sense of ch. 4 and possesses a "bounded" set, then it is Harris, a result which was extended to special functions by Lin [2]. (See Exercise 4.15.) The results on Harris resolvents of this and the following section are mostly from Brancovan [1]. For Exercise 4.16 see Nummelin [1].

Section 5. Almost all this section is taken from Neveu [5] where, however, Theorem 5.2 is proved only for strictly positive functions; Proposition 5.4 was first stated by Orey [4] in the case of discrete chains, and enables one to simplify the proof of Corollary 5.5. Proposition 5.7 and its sharpening in Exercise 5.16 had been obtained before for "bounded" sets by Ornstein [1], Metivier [1] and Duflo [1].

Section 6. The first ratio limit theorem for recurrent discrete chains was given by Doeblin. The proofs of this section are due to Neveu [7]; they are patterned after Levitan [1]. The reader will find another interesting proof in Metivier [2]. For earlier work, see Jain [1], Isaac [2] and Metivier [1]. A counterexample which shows that the most general ratio limit theorem fails to be true was designed by Krengel [1]. The important Exercise 6.8 is from Brunel et al. [1]; the result is still true for general recurrent random walks.

Chapter 7

Section 1. The results on Martingales needed in this section may be found in Neveu [2] or Dellacherie and Meyer (in particular on page 21 in Volume II).

The ideas in the first part of the section are from Blackwell [2]. The ideas involved in the proof of Theorem 1.11 are originally due to Hunt [2] in the case of discrete chains and were used by Abrahamsee [1] for general state space. The observation contained in Theorem 1.13 is from Roudier [1]. For the important Exercise 1.18 see Furstenberg [1] and Azencott [1].

Section 2. The Martin boundary for Markov chains was introduced independently by Doob [3] and Watanabe [1]. The metric we use here to complete the space is the metric of Hunt [2]. To extend the theory to a case more general than discrete chains we have used the paper by Derriennic [1].

Section 3. The boundary M of §2 is not easily computed. The third section gives a way of computing its interesting part and at the same time gives the desired integral representation of harmonic functions. What lies beneath the present discussion is the theory of convex cones of Choquet. We refer the reader to Neveu [1] for an exposition of the Martin boundary relying on the theory of Choquet. In this line, the results of Mokobodzky (see Meyer [5]) would perhaps lead to some generalizations in more general state spaces.

The Martin boundary has now be worked out for several important classes of discrete chains. The reader should look in Doob et al. [1], Dynkin and Malyutov [1], Lamperti and Snell [1], Blackwell and Kendall [1] and Ney and Spitzer [1]. For random walks on groups and symmetric spaces see also Furstenberg [2] and Azencott and Cartier [1]. Their results use the theory of Poisson spaces, sketched in Exercise 1.18. The most general non-bounded harmonic functions are not known for all groups. See, however, Furstenberg [3] and Deny [2].

Chapter 8

Section 1. This is taken from Brunel and Revuz [1].

Section 2. This is taken mostly from Neveu [5], but these kinds of results were already known for discrete chains (see, for instance, Neveu [1] or Kemeny et al. [1]). The observation in Exercise 2.11 is due to Orey [4].

Exercise 2.13 is due to Chung in the case of discrete chains; the proof hinted at is from Lin [2].

Section 3. The theory of normal chains is due to Kemeny and Snell [1] in the case of discrete chains. Their results are also contained in the book by Kemeny et al. [1]. It was shown by Orey [2] that there exist non-normal chains (see Exercise 3.11). Exercise 3.9 for the case in which X is discrete was taken from Kemeny et al. [1].

Section 4. The theory of boundaries is only sketched in this section. More complete results for discrete chains will be found in the last chapter of the book of Kemeny et al. [1] where in particular Theorem 4.10 is proved by another method. We also refer to the paper of Orey [4] and the exposition in Neveu [1] where integral representations are given for potential kernels. Theorem 4.13 is in some sense a version of these results which have been carried over to a more general setting by Bronner [3].

A general theory for Feller–Harris chains generalizing all the results known either for discrete chains or for random walks has yet to be written. To this end the methods of Brunel and Revuz [3, V] may possibly prove useful. In the case of random walks it is shown that the conditions of Theorem 4.10 hold, but furthermore that the convergence of the powers P_n hold for all special and co-special functions, whether continuous or not. This is due in particular to the fact that for Harris random walks, P_n is "asymptotically" strong Feller. Are similar results true for all chains? More precisely, if a chain satisfies the equivalent conditions of Theorem 4.9 what can be said of the class of functions on which P_n converges?

Chapter 9

The study of potential theory for recurrent random walks was initiated by Spitzer, who treated the case of $\mathbf{Z}^p$ and discovered the main features such as the classification into types I and II. His results are collected in his basic book [1].

The theory was enlarged to abelian discrete groups by Kesten and Spitzer [2], to discrete groups by Kesten [4], to the real line by Ornstein [1], and to general abelian locally compact groups by Port and Stone [1]. They deal also with non-Harris recurrent random walks. Finally, in Brunel and Revuz [2] the case of general non-abelian groups is dealt with in the more general setting of special functions.

Chapter 9 follows mostly the paper by Brunel and Revuz [2] with the exception of the end of §4, which comes from Port and Stone [1], but the original ideas are due to Kesten and Spitzer [2]. Many exercises are borrowed from Spitzer's book. The Fourier criterion (2.20) is taken from Kesten and Spitzer [2]. It is also true for general groups and non-Harris recurrent random walks, as was shown by Ornstein [1] and Stone [3] for the real line and by Port and Stone [1] in full generality.

Many results which lie beyond the scope of this book, especially asymptotic theorems of probabilistic nature, have been shown for random walks. We refer the reader to the papers by Kesten, Spitzer, Ornstein, Port and Stone and Bougerol.

Chapter 10

The subject was started by Hunt in his fundamental paper [1]. He proves the basic Theorem 1.5 and Corollary 2.15 by means of probabilistic methods. In fact he proves not only the existence of a resolvent but also of a semi-group.

Section 1. Aside from 1.5, the results in this section are taken mostly from Meyer [3]. Exercise 1.13 is from Bronner [1] to which we refer for further results in this line.

Section 2. The main result, namely Theorem 2.10, as well as its proof, are from Meyer [3]. The proof of 2.14 follows Kondo [1] for the sufficiency and Hirsch [2] and Taylor [2] for the necessity. Another proof, outlined in Exercise 2.20, is due to Taylor [1]. It follows in part the proof given by Lion [1], who showed that the density hypothesis in 2.15 may be dropped, and gave a purely analytical proof. Actually it is known from the work of Hirsch [1] that the countability assumption on E may also be dropped. The reader is also referred to Hansen [1] and Mokobodzki and Sibony [1].

The notion of "vanishing at the boundary" was emphasized in Meyer [3]. Proposition 2.6, which clarifies the matter, is due to Taylor [1]. In view of Proposition 1.3, one could think of deriving Theorem 2.10 from Theorem 2.14 or Theorem 2.14 from Theorem 2.10 without having to repeat the proof. Unfortunately this does not seem to work.

Theorem 2.14 may be generalized to the problem raised in Exercise 1.13. We refer the reader to Hirsch [2] and Taylor [2].

Section 3. This section is only intended as an introduction to a subject where the best results are those of Sunyach [1]. Theorems 3.9 and 3.10 were given by Kondo [2] for countable E and the proofs in the text are merely an extension of Kondo's proofs to strong Feller kernels. We also refer to Oshima [1] and Kondo and Oshima [1].

REFERENCES

A.I.F. = *Ann. Inst. Fourier*; *A.I.H.P.* = *Ann. Inst. Henri Poincaré*; *A.M.S.* = *Ann. Math. Statist.*; *A.P.* = *Ann. Probability*; *C.R.A.S.* = *Comptes Rendus Acad. Sci. Paris*; *I.J.M.* = *Illinois J. Math.*; *T.A.M.S.* = *Trans. Amer. Math. Soc.*; *T.P.* = *Theory Probability Appl.* (Translation of *Teoriëë Primeneniya*); *Z.W.* = *Z. Wahrscheinlichkeitstheorie Verw. Geb.*

M. Abid

[1] Un théorème ergodique pour des processus sous-additifs et sur stationnaires, *C.R.A.S.* **287** (1978) 149–152.

A. F. Abrahamsee

[1] The Martin Potential kernel for improperly essential chains, *Z.W.* **18** (1971) 2.18–2.26.

M. A. Ackoglu

[1] An ergodic lemma, *Proc. Am. Math. Soc.* **16** (1965) 388–392.

[2] Pointwise ergodic theorems, *T.A.M.S.* **125** (1966) 296–309.

M. A. Ackoglu and A. Brunel

[1] Contractions on L^1-spaces, *T.A.M.S.* **155** (1971) 315–325.

M. A. Ackoglu and R. W. Sharpe

[1] Ergodic theories and boundaries, *T.A.M.S.* **132** (1968) 447–460.

M. A. Ackoglu and L. Sucheston

[1] A ratio ergodic theorem for superadditive processes, *Z.W.* **44** (1978) 269–278.

K. B. Athreya and P. Ney

[1] A new approach to the limit theory of recurrent Markov chains, *T.A.M.S.* **245** (1978) 493–501.

R. Azencott

[1] *Espaces de Poisson des Groupes Localement Compacts*, Lecture Notes in Math. **148** (Springer, Berlin, 1970).

R. Azencott and P. Cartier

[1] Martin boundaries of random walks on locally compact groups, in: *Proc. 6th Berkeley Symp. on Mathematical Statistics and Probability*, Vol. III (Univ. of California Press, Berkeley, Calif., 1970) 87–129.

P. BALDI

[1] Caractérisation des groupes de Lie connexes récurrents, *A.I.H.P.* **17** (1981) 281–308.

J. R. BAXTER

[1] Balayage in least time, unpublished.

[2] Restricted mean values and harmonic functions, *T.A.M.S.* **167** (1972) 451–463.

J. R. BAXTER and R. V. CHACON

[1] Stopping times for recurrent Markov processes, *I.J.M.* **20** (1976) 467–475.

C. BELLAÏCHE-FREMONT and M. SUEUR-PONTIER

[1] Théorème de renouvellement pour groupes abéliens localement compacts, in: *Astérisque* **4** (Soc. Math. de France, Paris, 1973).

D. BLACKWELL

[1] The existence of anormal chains, *Bull. Amer. Math. Soc.* **51** (1945) 465–468.

[2] A renewal theorem, *Duke Math. J.* **15** (1948) 145–150.

[3] On transient Markov processes with a countable number of states and stationary transition probabilities, *A.M.S.* **26** (1955) 654–658.

D. BLACKWELL and D. FREEDMAN

[1] The tail σ-field of a Markov chain and a theorem of Orey, *A.M.S.* **35** (1964) 1291–1295.

D. BLACKWELL and D. KENDALL

[1] The Martin boundary for Polya's urn scheme and an application to stochastic population growth, *J. Appl. Probab.* **1** (1964) 284–296.

A. BLANC-LAPIERRE and R. FORTET

[1] *Théories des Fonctions Aléatoires* (Masson, Paris, 1953).

R. M. BLUMENTHAL and R. K. GETOOR

[1] *Markov Processes and Potential Theory* (Academic Press, New York, 1968).

P. BOUGEROL

[1] Fonctions de concentration sur certains groupes localement compacts, *Z.W.* **45** (1978) 135–157.

[2] Comportement asymptotique des puissances de convolution d'une probabilité sur un espace symétrique, in: *Astérisque* **74** (Soc. Math. France, Paris, 1980) 29–45.

[3] Théorème central limite local sur certains groupes de Lie, *Ann. Sci. Ecole Norm. Sup.* **14** (1981) 403–432.

P. BOUGEROL and L. ELIE

[1] Sur une propriété de compacité du noyau potentiel associé à une probabilité sur un groupe, *Z.W.* **52** (1980) 59–68.

M. BRANCOVAN

[1] Quelques propriétés des resolvantes récurrentes au sens de Harris, *A.I.H.P.* **9** (1973) 1–18.

J. BRETAGNOLLE and D. DACUNHA-CASTELLE
[1] Sur une classe de marches aléatoires, *A.I.H.P.* **3** (1967) 403–431.

F. BRONNER
[1] Principe du maximum et résolvantes sous-markoviennes, *C.R.A.S.* **277** (1973) 221–223.
[2] Méthodes probabilistes pour la détermination des résolvantes sous-markoviennes, *A.I.H.P.* **11** (3) (1975) 253–264.
[3] Frontière de Martin d'un processus récurrent au sens de Harris, *Z.W.* **44** (1978) 227–251.

A. BRUNEL (see Ackoglu)
[1] Sur un lemme voisin du lemme de E. Hopf et sur une de ses applications, *C.R.A.S.* **256** (1963) 5481–5484.
[2] New conditions for existence of invariant measures, in: *Contributions to Ergodic and Probability*, Lecture Notes in Math. **160** (Springer, Berlin, 1970).
[3] Chaines abstraites de Markov vérifiant une condition de Orey. Extention à ce cas d'un théorème ergodique de M. Métivier, *Z.W.* **19** (1971) 323–329.
[4] Sur les sommes d'itérés d'un opérateur positif, in: *Théorie Ergodique*, Lecture Notes in Math. **532** (Springer, Berlin, 1976) 19–34.

A. BRUNEL, P. CRÉPEL, Y. GUIVARC'H and M. KEANE
[1] Marches aléatoires récurrentes sur les groupes localement compacts, *C.R.A.S.* **275** (1972).

A. BRUNEL and D. REVUZ
[1] Un critère probabiliste de compacité des groupes, *A.P.* **2** (4) (1974) 475–476.
[2] Quelques applications probabilistes de la quasi-compacité, *A.I.H.P.* **10** (1974) 301–337.
[3] Marches de Harris sur les groupes localement compacts I, *Ann. Sci. Ecole Norm. Sup.* **7** (1974) 273–310; II, *Bull. Soc. Math. France* **104** (1976) 3–31; III, *Symposia Math.* **XXI** (1977) 55–63; IV, *Ann. Math.* **105** (1977) 361–396; V, *A.I.H.P.* **15** (1979) 205–234.
[4] Sur le théoreme du renouvellement pour les groupes non abéliens, *Israel J. Math.* **20** (1975) 46–56.

R. V. CHACON (see Baxter)
[1] Identification of the limit of operator averages, *J. Math. Mech.* **11** (1962) 961–968.
[2] Convergence of operator averages, in: *Ergodic Theory* (Academic Press, New York, 1963) 82–120.

R. V. CHACON and U. KRENGEL
[1] Linear modulus of a linear operator, *Proc. Amer. Math. Soc.* **15** (1964) 553–559.

R. V. CHACON and D. S. ORNSTEIN
[1] A general ergodic theorem, *I.J.M.* **4** (1960) 153–160.

E. ÇINLAR
[1] *Introduction to Stochastic Processes* (Prentice-Hall, Englewood Cliffs, NJ, 1975).

G. Choquet

[1] *Lectures in Analysis* (Benjamin, New York, 1969).

G. Choquet and J. Deny

[1] Modèles finis en théorie du potentiel, *J. Anal. Math.* **5** (1956/1957) 77–135.

[2] Sur l'équation de convolution $\mu = \mu * \sigma$, *C.R.A.S.* **250** (1960) 799–801.

K. L. Chung

[1] The general theory of Markov processes according to Doeblin, *Z.W.* **2** (1964) 230–250.

[2] *Markov Chains with Stationary Transition Probabilities*, 2nd ed. (Springer, Berlin, 1967).

K. L. Chung and W. H. Fuchs

[1] On the distribution of values of sums of random variables, *Mem. Am. Math. Soc.* **6** (1951).

K. L. Chung and D. S. Ornstein

[1] On the recurrence of sums of random variables, *Bull. Am. Math. Soc.* **68** (1962) 30–32.

R. Cogburn

[1] A uniform theory for sums of Markov chains transition probabilities, *A.P.* **3** (1975) 191–214.

P. Crepel

[1] Fonctions spéciales pour des contractions de L^1, in: *Astérisque* **4** (Soc. Math. France, Paris, 1973).

C. Dellacherie and P. A. Meyer

[1] *Probabilités et Potentiels* (Hermann, Paris, I, 1975, II, 1980, III, to appear).

J. Deny (see Choquet)

[1] Familles fondamentales, noyaux associés, *A.I.F.* **3** (1951) 73–101.

[2] Sur l'équation de convolution $\mu = \mu * \sigma$, in: *Sem. de Théorie du Potentiel* (Inst. Henri Poincaré, Paris, 1959/60) 4me année.

C. Derman

[1] A solution to a set of fundamental equations in Markov chains, *Proc. Amer. Math. Soc.* **5** (1954) 332–334.

Y. Derriennic

[1] Sur la frontière de Martin des processus a temps discret, *A.I.H.P.* **9** (3) (1973) 233–258.

[2] Marche aléatoire sur le groupe libre et frontière de Martin, *Z.W.* **32** (1975) 261–276.

[3] Sur le théorème ergodique sous-additif, *C.R.A.S.* **281** (1975) 985–988.

[4] Lois "zero ou deux" pour les processus de Markov. Applications aux marches aléatoires, *A.I.H.P.* **12** (2) (1976) 111–129.

[5] Quelques applications du théorème ergodique sous-additif, in: *Astérisque* **74** (Soc. Math. France, Paris, 1980) 183–201.

Y. Derriennic and Y. Guivarc'h
[1] Théorème de renouvellement pour les groupes non moyennables, *C.R.A.S.* **277** (1973) 613–615.

Y. Derriennic and M. Lin
[1] Sur la tribu asymptotique des marches aléatoires sur les groupes, *C.R.A.S.*

M. D. Donsker and S. R. S. Varadhan
[1] Asymptotic evaluation of certain Markov process expectations for large time, *Comm. Pure Appl. Math.* I, **27** (1975) 1–47; III, **29** (1976) 389–461.

J. L. Doob
[1] Asymptotic properties of Markoff transition probabilities, *T.A.M.S.* **63** (1948) 393–421.
[2] *Stochastic Processes* (Wiley, New York, 1953).
[3] Discrete potential theory and boundaries, *J. Math. Mech.* **8** (1959) 433–458.

J. L. Doob, J. L. Snell and R. E. Williamson
[1] Application of boundary theory to sums of independent random variables, in: *Contributions to Probability and Statistics* (Stanford Univ. Press, Stanford, Calif., 1960).

R. M. Dudley
[1] Random walk on abelian groups, *Proc. Am. Math. Soc.* **13** (1962) 447–450.

M. Duflo
[1] Opérateurs potentiels des chaines et des processus de Markov irréductibles, *Bull. Soc. Math. France* **98** (1970) 127–163.

N. Dunford and J. T. Schwarz
[1] *Linear Operators, Part I: General Theory* (Interscience, New York, 1953).

E. B. Dynkin and M. B. Maljutov
[1] Random walks on groups with a finite number of generators, *Soviet Math. Dokl.* **2** (1961) 399–402 (English Transl.).

D. A. Edwards
[1] On potentials and general ergodic theorems for resolvents, *Z.W.* **20** (1971) 1–8.

L. Elie (see Bougerol)
[1] Théorie du renouvellement sur les groupes, *Ann. Sci. Ecole Norm. Sup.* **15** (1982) 257–364.

J. Feldman
[1] Subinvariant measures for Markov operators, *Duke Math. J.* **29** (1962) 71–98.
[2] Integral kernels and invariant measures for Markov transition functions, *Ann. Math. Statist.* **36** (1965) 517–523.

W. Feller
[1] Boundaries induced by non-negative matrices, *T.A.M.S.* **83** (1956) 19–54.
[2] Non-Markovian processes with the semi-group property, *A.M.S.* **30** (1959) 1252–1253.

[3] A simple proof for renewal theorems, *Commun. Pure Appl. Math.* **14** (1961) 285–293.

[4] *An Introduction to Probability Theory and its Applications*, Vol. 2 (Wiley, New York, **1966**).

W. Feller and S. Orey

[1] A renewal theorem, *J. Math. Mech.* **10** (1961) 619–624.

S. R. Foguel

[1] *The Ergodic Theory of Markov Processes* (Van Nostrand, New York, **1969**).

[2] The ergodic theory of positive operators on continuous functions, *Ann. Scuola Norm. Sup. Pisa* **27** (1973) 19–51.

[3] More on the zero-two law, *Proc. Amer. Math. Soc.* **61** (1976) 262–264.

[4] Harris operators, *Israel J. Math.* **33** (1979) 281–309.

S. R. Foguel and N. Ghoussoub

[1] Ornstein–Métivier–Brunel Theorem revisited, *A.I.H.P.* **15** (1979) 293–301.

S. R. Foguel and M. Lin

[1] Some ratio limit theorems for Markov operators, *Z.W.* **23** (1972) 55–66.

R. Fortet (see Blanc–Lapierre)

[1] Condition de Doeblin et quasi-compacité, *A.I.H.P.* **14** (1978) 379–390.

D. Freedman (see Blackwell)

[1] *Markov Chains* (Holden–Day, San Francisco, **1971**).

N. A. Friedman

[1] *Introduction to Ergodic Theory* (Van Nostrand, New York, **1970**).

H. Furstenberg

[1] A Poisson formula for semi-simple Lie groups, *Ann. Math* (2) **77** (1963) 335–386.

[2] Non-commuting random products, *T.A.M.S.* **108** (1963) 377–428.

[3] Translation-invariant cones of functions on semi-simple Lie groups, *Bull. Am. Math. Soc.* **71** (2) 271–326 (1965).

L. Gallardo and Schott

[1] Marches aléatoires sur les espaces homogènes de certains groupes de Lie de type rigide, in: *Astérisque* **74** (Soc. Math. France, Paris, 1980) 149–170.

A. Garsia

[1] A simple proof of E. Hopf's maximal ergodic theorem, *J. Math. Mech.* **14** (1965) 381–382.

[2] More about the maximal ergodic lemma of Brunel, *Proc. Nat. Acad. Sci.* **67** (1967) 21–24.

[3] *Topics in Almost Everywhere Convergence*, Lectures in Adv. Math. (Markham, Chicago, Ill., **1970**).

R. K. Getoor (see Blumenthal)

[1] On the construction of kernels, in: *Séminaire de Probabilités IX*, Lecture Notes in Math. **465** (Springer, Berlin, 1975).

I. V. Girsanov

[1] Strong Feller processes I. General properties, *T.P.* **5** (1960) 7–28.

N. Ghoussoub (see Foguel)

[1] Processus de Harris abstraits, *A.I.H.P.* **11** (1975) 381–395.

D. Griffeath

[1] A maximal coupling for Markov chains, *Z.W.* **31** (1975) 95–106.

[2] Partial coupling and loss of memory for Markov chains, *A.P.* **4** (1976) 850–858.

Y. Guivarc'h (see Brunel, Derriennic)

[1] Croissance polynomiale et périodes des fonctions harmoniques, *Bull. Soc. Math. France* **101** (1973) 333–379.

[2] Sur la loi des grands nombres et le rayon spectral d'une marche aléatoire, in: *Astérisque* **74** (Soc. Math. France, Paris, 1980) 47–98.

[3] Théorèmes quotient pour les marches aléatoires, in: *Astérisque* **74** (Soc. Math. France, Paris, 1980) 15–28.

Y. Guivarc'h, M. Keane and B. Roynette

[1] *Marches Aléatoires sur les Groupes de Lie*, Lecture Notes in Math. **624** (Springer, Berlin, 1977).

W. Hansen

[1] Konstruktion von Halbgruppen und Markoffschen Prozessen, *Invent. Math.* **3** (1967) 179–214.

T. E. Harris

[1] Recurrent Markov processes II, *A.M.S.* **26** (1955) 152–153 (abstract).

[2] The existence of stationary measures for certain Markov processes, in: *Proc. 3rd Berkeley Symp. on Mathematical Statistics and Probability* Vol. 2 (Univ. of California Press, Berkeley, Calif., 1956) 113–124.

P. L. Hennequin

[1] Les processus de Markov en cascades, *A.I.H.P.* (12) **18** (1963) 109–196.

H. Hennion and B. Roynette

[1] Un théorème de dichotomie pour une marche aléatoire sur un espace homogène, in: *Astérisque* **74** (Soc. Math. France, Paris, 1980) 99–122.

C. S. Herz

[1] Les théorèmes de renouvellement, *A.I.F* **15** (1) (1965) 169–188.

K. Hewitt and K. A. Ross

[1] *Abstract Harmonic Analysis I* (Springer, Berlin, 1963).

F. Hirsch

[1] Familles résolvantes, générateurs, cogénérateurs, potentiels, *A.I.F.* **22** (1972) 89–210.

[2] Conditions nécessaires et suffisantes d'existence de résolvantes, *Z.W.* **29** (1974) 73–85.

F. Hirsch and J. C. Taylor

[1] Renouvellement et existence de résolvantes, in: *Séminaire de Théorie du Potentiel*, Lecture Notes in Math. **624** (Springer, Berlin, 1977).

E. Hopf

[1] The general temporally discrete Markov process, *J. Rational. Math. Mech. Anal.* **3** (1954) 13–45.

[2] On the ergodic theorem for positive linear operators, *J. Reine Angew. Math.* **205** (1960/61) 101–106.

S. Horowitz

[1] Transition probabilities and contractions of L^∞, *Z.W.* **24** (1972) 263–274.

[2] Pointwise convergence of the iterates of a Harris-recurrent Markov operator, *Israel J. Math.* **33** (1979) 177–180.

G. A. Hunt

[1] Markov processes and potentials I, II, III, *I.J.M.* **1** (1957) 44–93, 316–389, **2** (1958) 151–213.

[2] Markov chains and Martin boundaries, *I.J.M.* **4** (1960) 119–132.

[3] La théorie du potentiel et les processus récurrents, *A.I.F.* **15** (1) (1965) 3–12.

C. Ionescu-Tulcea

[1] Méasures dans les espaces produits, *Atti Accad. Naz. Lincei. Rend.* **7** (1949) 208–211.

R. Isaac

[1] Non-singular recurrent Markov processes have stationary measures, *A.M.S.* **35** (1964) 869–871.

[2] On the ratio-limit theorem for Markov processes recurrent in the sense of Harris, *I.J.M.* **11** (1967) 608–615.

J. Jacod

[1] Théorème de renouvellement et classification pour les chaines semi-markoviennes, *A.I.H.P.* **7** (1971) 83–129.

N. C. Jain

[1] Some limit theorems for a general Markov process, *Z.W.* **6** (1966) 206–223.

[2] A note on invariant measures, *A.M.S.* **37** (1966) 729–732.

N. C. Jain and B. Jamison

[1] Contributions to Doeblin's Theory of Markov processes, *Z.W.* **8** (1967) 19–40.

N. C. Jain and S. Orey

[1] Some properties of random walk paths, *J. Math. Anal. Appl.* **43** (1973) 795–815.

B. Jamison and S. Orey

[1] Tail σ-field of Markov processes recurrent in the sense of Harris, *Z.W.* **8** (1967) 41–48.

A. Del Junco
[1] On the decomposition of a subadditive stochastic process, *A.P.* **5** (1977) 298–302.

S. Karlin and J. McGregor
[1] Random walks, *I.J.M.* **3** (1959) 66–81.

J. G. Kemeny
[1] Representation theory for denumerable Markov chains, *T.A.M.S.* **125** (1966) 47–62.

J. G. Kemeny and J. L. Snell
[1] Potentials for denumerable Markov chains, *J. Math. Anal. Appl.* **3** (1961) 196–260.
[2] Notes on discrete potential theory, *J. Math. Anal. Appl.* **3** (1961) 117–121.
[3] Boundary theory for recurrent Markov chains, *T.A.M.S.* **106** (1963) 495–520.
[4] A new potential operator for recurrent Markov chains, *J. London Math. Soc.* **38** (1963) 359–371.

J. G. Kemeny, J. L. Snell and A. W. Knapp
[1] *Denumerable Markov Chains* (Van Nostrand, New York, 1966).

J. H. B. Kemperman
[1] *The Passage Problem for a Stationary Markov Chain* (Chicago Univ. Press, Chicago, Ill., 1961).

H. Kesten
[1] Symmetric random walks on groups, *T.A.M.S.* **92** (1959) 336–354.
[2] Full Banach mean values on countable groups, *Math. Scand.* **7** (1959) 146–156.
[3] Ratio theorems for random walks II, *J. Anal. Math.* **11** (1963) 323–379.
[4] The Martin boundary of recurrent random walks on countable groups, in: *Proc. 5th Berkeley Symp. on Mathematical Statistics and Probability*, Vol. II (Univ. of California Press, Berkeley, Calif., 1967) 51–75.

H. Kesten and F. Spitzer
[1] Ratio theorems for random walks I, *J. Anal. Math.* **11** (1963) 285–322.
[2] Random walk on countably infinite Abelian groups, *Acta. Math.* **114** (1965) 237–265.

J. F. C. Kingman
[1] The ergodic theory of subadditive stochastic processes, *J. Roy. Statist. Soc.* Ser. B **30** (1968) 499–510.
[2] Subadditive ergodic theory, *A.P.* **1** (1973) 883–909.

R. Kondo
[1] On potential kernels satisfying the complete maximum principle, *Proc. Japan. Acad.* **44** (1968) 193–197.
[2] A construction of recurrent Markov chains, *Osaka. J. Math.* **6** (1969) 13–28.
[3] A construction of recurrent Markov chains II, *T.P.* **15** (1970) 499–507.

R. KONDO and Y. OSHIMA

[1] A characterization of weak potential kernels for strong Feller recurrent Markov chains, *Proc. Soc. Japan–USSR Symp. Prob. Theory*, Lecture Notes in Math. **330** (Springer, Berlin, 1973).

U. KRENGEL (see Chacon)

[1] On the global limit behaviour of Markov chains and of general nonsingular Markov processes, *Z.W.* **6** (4) (1966) 302–316.

[2] Un théorème ergodique pour les processus surstationnaires, *C.R.A.S.* **282** (1976) 1019–1021.

K. KRICKEBERG

[1] Strong mixing properties of Markov chains with infinite invariant measures, in: *Proc. 5th Berkeley Symp. on Mathematical Statistics and Probability* (California Univ. Press, Berkeley, Calif., 1966).

[2] Mischende Transformationen auf Mannigfaltigkeiten unendlichen Maßes, *Z.W.* **7** (1967) 235–247.

A. G. KUROSH

[1] *The Theory of Groups* (Chelsea, New York, 1955).

J. LAMPERTI and J. L. SNELL

[1] Martin boundaries for certain Markov chains, *J. Math. Soc. Japan* **15** (1963) 113–128.

M. L. LEVITAN

[1] Some ratio limit theorems for a general state space Markov process, *Z.W.* **15** (1970) 29–50.

[2] A generalized Doeblin ratio limit theorem, *A.M.S.* **42** (1971) 904–911.

M. LIN (see Derriennic, Foguel)

[1] Mixed ratio limit theorems for Markov processes, *Israel J. Math.* **8** (1970) 357–366.

[2] Mixing for Markov operators, *Z.W.* **19** (1971) 231–242.

[3] On quasi-compact operators, *A.P.* **2** (1974) 464–475.

[4] On the "zero-two" law for conservative Markov processes, *Z. W.* **61** (1982) 513–525.

G. LION

[1] Familles d'opérateurs et frontière en théorie du potentiel, *A.I.F.* **16** (1966) 389–453.

M. LOÈVE

[1] Probability Theory, 2nd edition (Van Nostrand, New York, 1960).

R. M. LOYNES

[1] Products of independent random elements in a topological group, *Z.W.* **1** (1963) 446–455.

N. MAIGRET
[1] Théorème de limite centrale fonctionnel pour une chaine de Markov récurrente au sens de Harris et positive, *A.I.H.P.* **14** (1978) 425–440.

J. F. MERTENS, E. SAMUEL–CAHN and S. ZAMIR
[1] Necessary and sufficient conditions for recurrence and transience of Markov chains, *J. Appl. Probability* **15** (1978) 848–851.

M. METIVIER
[1] Existence of an invariant measure and an Ornstein's ergodic theorem, *A.M.S.* **40** (1969) 74–96.
[2] Théorème limite quotient pour les chaines de Markov récurrentes au sens de Harris, *A.I.H.P.* **8** (2) (1972) 93–105.

P. A. MEYER
[1] Théorie ergodique et potentiels, *A.I.F.* **15** (1) (1965) 89–102.
[2] *Probability and Potentials* (Blaisdell, Waltham, Mass., 1966).
[3] Caractérisation des noyaux potentiels des semi-groupes discrets, *A.I.F.* **16** (2) (1966) 225–240.
[4] Les resolvantes fortement Felleriennes d'après Mokobodzki, in: *Sém. de Probabilités II*, Lecture Notes in Math. **51** (Springer, Berlin, 1968).
[5] Representation intégrale des fonctions excessives, résultats de Mokobodzki, in: *Sém. de Probabilités V*, Lecture Notes in Math. **191** (Springer, Berlin, 1971).
[6] Deux petits résultats du théorie du potentiel, in: *Sém. de Probabilités V*, Lecture Notes in Math. **191** (Springer, Berlin, 1971).
[7] Travaux de H. Rost et théorie du balayage, in: *Sém. de Probabilités V*, Lecture Notes in Math. **191** (Springer, Berlin, 1971).
[8] Solutions de l'équation de Poisson dans le cas récurrent, in: *Sém. de Probabilités V*, Lecture Notes in Math. **191** (Springer, Berlin, 1971).

P. A. MEYER and M. YOR
[1] Sur l'extension d'un théorème de Doob à un noyau σ-fini, in: *Séminaire de Probabilités XII*, Lecture Notes in Math. **649** (Springer, Berlin, 1978).

G. MOKOBODZKI
[1] Noyaux absolument mesurables et opérateurs nucléaires, *C.R.A.S.* **270** (1970) 1673–1675.
[2] Densité relative de deux potentiels comparables, in: *Séminaire de Probabilités IV*, Lecture Notes in Math. **124** (Springer, Berlin, 1970).

S. T. C. MOY
[1] Period of an irreducible positive operator, *I.J.M.* **11** (1967) 24–39.
[2] λ-continous Markov chains, *T.A.M.S.* **117** (1965) 68–99.

J. NEVEU
[1] Potentiels markoviens discrets, *Ann. Univ. Clermont* **24** (1964) 37–89.
[2] *Mathematical Foundations of the Calculus of Probability* (Holden Day, San Francisco, Calif., 1965).

[3] Relations entre la théorie des martingales et la théorie ergodique, *A.I.F.* **15** (1965) 31–42.

[4] Existence of bounded invariant measures in ergodic theory, in: *Symp. 5th Berkeley Symp. on Mathematical Statistics and Probability* (Univ. of California Press, Berkeley, Calif., 1966).

[5] Potentiel markovien recurrent des chaines de Harris, *A.I.F.* **22** (2) (1972) 85–130.

[6] Sur l'irréductibilité des chaines de Markov, *A.I.H.P.* **8** (3) (1972) 249–254.

[7] Généralisation d'un théorème limite-quotient, in: *Trans. 6th Prague Conf. on Information Theory, Statistical Decision Functions, Random Processes* (Czech. Acad. Sci., Prague, 1973).

[8] The filling scheme and the Chacon–Ornstein theorem, *Israel J. Math.* **33** (1979) 368–377.

P. Ney and F. Spitzer

[1] The Martin boundary for random walk, *T.A.M.S.* **121** (1966) 116–132.

E. Nummelin

[1] A splitting technique for Harris recurrent Markov chains, *Z.W.* **43** (1978) 309–318.

[2] Strong ratio limit theorems for φ-recurrent Markov chains, *A.P.* **7** (1979) 639–650.

[3] On the Poisson equation for φ-recurrent Markov chains, to appear.

E. Nummelin and R. L. Tweedie

[1] Geometric ergodicity and R-positivity for general Markov chains, *A.P.* **6** (1978) 404–420.

S. Orey (see Feller, Jain and Jamison)

[1] Recurrent Markov chains, *Pacific J. Math.* **9** (1959) 805–827.

[2] Sums arising in the theory of Markov chains, *Proc. Am. Math. Soc.* **12** (1961) 847–856.

[3] An ergodic theorem for recurrent Markov chains, *Z.W.* **1** (1962) 174–176.

[4] Potential kernels for recurrent Markov chains, *J. Math. Anal. Appl.* **8** (1964) 104–132.

[5] *Limit Theorems for Markov Chain Transition Probabilities* (Van Nostrand, New York, 1971).

D. S. Ornstein (see Chacon and Chung)

[1] Random walks I, II, *T.A.M.S.* **138** (1969) 1–43, 45–60.

[2] The sums of the iterates of a positive operator, *Advances in Probability and Related Topics* **2** (1970) 85–115.

D. S. Ornstein and L. Sucheston

[1] An operator theorem on L^1-convergence to zero with applications to Markov kernels, *A.M.S.* **41** (1970) 1631–1639.

Y. Oshima (see Kondo)

[1] A necessary and sufficient condition for a kernel to be a weak potential kernel of a recurrent Markov chain, *Osaka J. Math.* **6** (1969) 29–37.

J. C. Oxtoby

[1] Ergodic sets, *Bull. Amer. Math. Soc.* **58** (1952) 116–136.

F. Papangelou

[1] A martingale approach to the convergence of iterates of a transition function, *Z.W.* **37** (1977) 211–226.

J. Pitman

[1] Uniform rates of convergence for Markov chain transition probabilities, *Z.W.* **29** (1974) 199–227.

S. C. Port

[1] Limit theorems involving capacities, *J. Math. Mech.* **15** (1966) 805–832.

S. C. Port and C. J. Stone

[1] Potential theory of random walks on abelian groups, *Acta Math.* **122** (1969) 19–114.

A. Raugi

[1] Fonctions harmoniques sur les groupes localement compacts à base dénombrable, *Mem. Soc. Math. France* **54** (1977) 5–118.

D. Revuz (see Brunel)

[1] Théorèmes limites pour les résolvantes récurrentes, *Rend. Circ. Mat. Palermo* **19** (1970) 294–300.

[2] Sur la théorie du potentiel pour les processus de Markov récurrents, *A.I.F.* **21** (1971) 245–262.

[3] Le principe semi-complet du maximum, in: *Sém. de Probabilités VI*, Lecture Notes in Math. **258** (Springer, Berlin, 1972).

[4] On the filling scheme for recurrent Markov chains, *Duke Math. J.* **45** (1978) 681–689.

[5] Sur la définition des classes cycliques des chaines de Harris, *Israel J. Math.* **33** (1979) 378–383.

[6] Sur le théorème de dichotomie d'Hennion–Roynette, to appear.

J. Roudier

[1] Chaine de Markov μ-continue à l'infini, *A.I.H.* **8** (3) (1972) 241–248.

H. Rost

[1] Darstellung einer Ordnung von Massen durch Stoppzeiten, *Z.W.* **15** (1970) 19–28.

[2] Markoff-Ketten bei sich füllenden Löchern im Zustandsraum, *A.I.F.* **21** (1) (1971) 253–270.

B. ROYNETTE (see Guivarc'h)

[1] Marches aléatoires sur le groupe des déplacements de $\mathbf{R}^d$, *Z.W.* **31** (1974) 25–34.

[2] Théorème central-limite pour le groupe des déplacements de $\mathbf{R}^d$, *A.I.H.P.* **10** (1974) 391–398.

W. RUDIN

[1] *Fourier Analysis on Groups* (Interscience, New York, 1967).

F. SPITZER (see Kesten, Ney)

[1] *Principle of Random Walks* (Van Nostrand, New York, 1964).

A. J. STAM

[1] On shifting iterated convolutions I, *Compositio Math.* **17** (1966) 268–280.

C. J. STONE (see Port)

[1] On absolutely continuous components and renewal theory, *A.M.S.* **37** (1966) 271–275.

[2] Ratio limit theorem for random walks on groups, *T.A.M.S.* **125** (1966) 86–100.

[3] On the potential operator for one-dimensional recurrent random walks, *T.A.M.S.* **136** (1969) 413–426.

C. SUNYACH

[1] Principes du maximum récurrent et construction de résolvantes récurrentes, *C.R.A.S.* **282** (1976) 747–750.

[2] Capacités et théorie du renouvellement I, *Bull. Soc. Math. France* **109** (1981) 283–296.

J. C. TAYLOR (see Hirsch)

[1] On the existence of sub-Markovian resolvents, *Invent. Math.* **17** (1972) 85–93.

[2] A characterization of the kernels $\lim_{\lambda \downarrow 0} V_\lambda$ for submarkovian resolvents (V_λ), *A.P.* **3** (1975) 355–357.

P. TUOMINEN and R. L. TWEEDIE

[1] Markov chains with continuous components. *Proc. London Math. Soc.* **38** (1979) 89–114.

R. L. TWEEDIE (see Nummelin, Tuominen)

[1] *R*-theory for Markov chains on a general state space I and II, *A.P.* **2** (1974) 840–864 and 865–878.

[2] Criteria for classifying general Markov chains, *Adv. Appl. Probability* **8** (1976) 737–771.

[3] Topological aspects of Doeblin decompositions for Markov chains, *Z.W.* **46** (1979) 299–305.

T. UENO

[1] Some limit theorems for temporally discrete Markov processes, *J. Fac. Sci. Univ. Tokyo* (I) **7** (1957).

T. WATANABE

[1] On the theory of Martin boundaries induced by countable Markov processes, *Mem. Coll. Sci. Univ. Kyoto* (A) **33** (1960) 39–108.

[2] On balayees of excessive measures and functions with respect to resolvents, in: *Sém. de Probabilités V*, Lecture Notes in Math. **191** (Springer, Berlin, 1971).

W. WINKLER

[1] A note on a continuous parameter zero-two law, *A.P.* **1** (1973) 341–344.

K. YOSIDA and S. KAKUTANI

[1] Operator-theoretical treatment of Markov's process and mean ergodic theorem, *Ann. Math.* **42** (1941) 188–228.

INDEX OF NOTATION

INDEX OF TERMS

Printed and bound by CPI Group (UK) Ltd, Croydon, CR0 4YY
14/10/2024
01773727-0001